CAMBRIDGE LIBRARY COLLECTION

Books of enduring scholarly value

Earth Sciences

In the nineteenth century, geology emerged as a distinct academic discipline. It pointed the way towards the theory of evolution, as scientists including Gideon Mantell, Adam Sedgwick, Charles Lyell and Roderick Murchison began to use the evidence of minerals, rock formations and fossils to demonstrate that the earth was older by millions of years than the conventional, Bible-based wisdom had supposed. They argued convincingly that the climate, flora and fauna of the distant past could be deduced from geological evidence. Volcanic activity, the formation of mountains, and the action of glaciers and rivers, tides and ocean currents also became better understood. This series includes landmark publications by pioneers of the modern earth sciences, who advanced the scientific understanding of our planet and the processes by which it is constantly re-shaped.

Fossil Plants

A.C. Seward (1863–1941) was an eminent English geologist and botanist who pioneered the study of palaeobotany. After graduating from St John's College, Cambridge, in 1886 Seward was appointed a University Lecturer in Botany in 1890. In 1898 he was elected a Fellow of the Royal Society, and was appointed Professor of Botany in 1906. These volumes, published to great acclaim between 1898 and 1919, provide a detailed discussion and study of an emerging science. In the early nineteenth century, research and critical literature concerning palaeobotany was scattered across disciplines. In these volumes Seward synthesised and revised this research and also included a substantial amount of new material. Furnished with concise descriptions of fossil plants, detailed figures and extensive bibliographies these volumes became the standard reference for palaeobotany well into the twentieth century. Volume 1, published in 1898, contains an overview of palaeobotany with systematic descriptions of fossil plants.

Cambridge University Press has long been a pioneer in the reissuing of out-of-print titles from its own backlist, producing digital reprints of books that are still sought after by scholars and students but could not be reprinted economically using traditional technology. The Cambridge Library Collection extends this activity to a wider range of books which are still of importance to researchers and professionals, either for the source material they contain, or as landmarks in the history of their academic discipline.

Drawing from the world-renowned collections in the Cambridge University Library, and guided by the advice of experts in each subject area, Cambridge University Press is using state-of-the-art scanning machines in its own Printing House to capture the content of each book selected for inclusion. The files are processed to give a consistently clear, crisp image, and the books finished to the high quality standard for which the Press is recognised around the world. The latest print-on-demand technology ensures that the books will remain available indefinitely, and that orders for single or multiple copies can quickly be supplied.

The Cambridge Library Collection will bring back to life books of enduring scholarly value (including out-of-copyright works originally issued by other publishers) across a wide range of disciplines in the humanities and social sciences and in science and technology.

Fossil Plants

*A Text-Book for Students of
Botany and Geology*

VOLUME 1

A.C. SEWARD

CAMBRIDGE
UNIVERSITY PRESS

CAMBRIDGE UNIVERSITY PRESS

Cambridge, New York, Melbourne, Madrid, Cape Town, Singapore,
São Paolo, Delhi, Dubai, Tokyo, Mexico City

Published in the United States of America by Cambridge University Press, New York

www.cambridge.org
Information on this title: www.cambridge.org/9781108015950

This edition first published 1898
This digitally printed version 2010

ISBN 978-1-108-01595-0 Paperback

𝕮𝖆𝖒𝖇𝖗𝖎𝖉𝖌𝖊 𝕹𝖆𝖙𝖚𝖗𝖆𝖑 𝕾𝖈𝖎𝖊𝖓𝖈𝖊 𝕸𝖆𝖓𝖚𝖆𝖑𝖘.

BIOLOGICAL SERIES.

GENERAL EDITOR :—ARTHUR E. SHIPLEY, M.A.

FELLOW AND TUTOR OF CHRIST'S COLLEGE, CAMBRIDGE.

FOSSIL PLANTS.

London: C. J. CLAY AND SONS,

CAMBRIDGE UNIVERSITY PRESS WAREHOUSE,

AVE MARIA LANE,

AND

H. K. LEWIS,

136, GOWER STREET, W.C.

Glasgow: 263, ARGYLE STREET.

Leipzig: F. A. BROCKHAUS.

New York: THE MACMILLAN COMPANY.

Bombay: E. SEYMOUR HALE.

TREE STUMPS IN A CARBONIFEROUS FOREST. VICTORIA PARK, GLASGOW.

Frontispiece to face Title.

FOSSIL PLANTS

FOR STUDENTS OF BOTANY AND GEOLOGY

BY

A. C. SEWARD, M.A., F.G.S.

ST JOHN'S COLLEGE, CAMBRIDGE,
LECTURER IN BOTANY IN THE UNIVERSITY OF CAMBRIDGE.

WITH ILLUSTRATIONS.

VOL. I.

CAMBRIDGE:
AT THE UNIVERSITY PRESS.

1898

Cambridge:

PRINTED BY J. AND C. F. CLAY,

AT THE UNIVERSITY PRESS.

PREFACE.

IN acceding to Mr Shipley's request to write a book on Fossil Plants for the Cambridge Natural History Series, I am well aware that I have undertaken a work which was considered too serious a task by one who has been called a "founder of modern Palaeobotany." I owe more than I am able to express to the friendship and guidance of the late Professor Williamson; and that I have attempted a work to which he consistently refused to commit himself, requires a word of explanation. My excuse must be that I have endeavoured to write a book which may render more accessible to students some of the important facts of Palaeobotany, and suggest lines of investigation in a subject which Williamson had so thoroughly at heart.

The subject of Palaeobotany does not readily lend itself to adequate treatment in a work intended for both geological and botanical students. The Botanist and Geologist are not always acquainted with each other's subject in a sufficient degree to appreciate the significance of Palaeobotany in its several points of contact with Geology and recent Botany. I have endeavoured to bear in mind the possibility that the following pages may be read by both non-geological and non-botanical students. It needs but a slight acquaintance with Geology for a Botanist to estimate the value of the most important applications of Palaeobotany; on the other hand, the bearing of fossil plants on the problems of phylogeny and descent cannot be adequately understood without a fairly intimate knowledge of recent Botany.

The student of elementary geology is not as a rule required to concern himself with vegetable palaeontology, beyond a general acquaintance with such facts as are to be found in geological text-books. The advanced student will necessarily find in these pages much with which he is already familiar; but this is to some extent unavoidable in a book which is written with the dual object of appealing to Botanists and Geologists. While considering those who may wish to extend their botanical or geological knowledge by an acquaintance with Palaeobotany, my aim has been to keep in view the requirements of the student who may be induced to approach the subject from the standpoint of an original investigator. As a possible assistance to those undertaking research in this promising field of work, I have given more references than may seem appropriate to an introductory treatise, and there are certain questions dealt with in greater detail than an elementary treatment of the subject requires. In several instances references are given in the text or in footnotes to specimens of Coal-Measure plants in the Williamson cabinet of microscopic sections. Now that this invaluable collection of slides has been acquired by the Trustees of the British Museum, the student of Palaeobotany has the opportunity of investigating for himself the histology of Palaeozoic plants.

My plan has been to deal in some detail with certain selected types, and to refer briefly to such others as should be studied by anyone desirous of pursuing the subject more thoroughly, rather than to cover a wide range or to attempt to make the list of types complete. Of late years there has been a much wider interest evinced by Botanists in the study of fossil plants, and this is in great measure due to the valuable and able work of Graf zu Solms-Laubach. His *Einleitung in die Palaeophytologie* must long remain a constant book of reference for those engaged in palaeobotanical work. While referring to

authors who have advanced the study of petrified plants of the
Coal period, one should not forget the valuable services that
have been rendered by such men as Butterworth, Binns, Wilde,
Earnshaw, Spencer, Nield, Lomax and Hemingway, by whose
skill the specimens described by Williamson and others were
first obtained and prepared for microscopical examination.

I am indebted to many friends, both British and Continental,
for help of various kinds. I would in the first place express my
thanks to Professor T. McKenny Hughes for having originally
persuaded me to begin the study of recent and fossil plants.
I am indebted to Prof. Nathorst of Stockholm, Dr Hartz of
Copenhagen, Prof. Zeiller, Dr Renault and Prof. Munier-Chalmas
of Paris, Prof. Bertrand of Lille, Prof. Stenzel and the late
Prof. Roemer of Breslau, Dr Sterzel of Chemnitz, the late
Prof. Weiss of Berlin, the late Dr Stur of Vienna, and other
continental workers, as well as to Mr Knowlton of Washington,
for facilities afforded me in the examination of fossil plant
collections. My thanks are due to the members of the Geological and Botanical departments of the British Museum;
also to Mr E. T. Newton of the Geological Survey, and to
those in charge of various provincial museums, for their
never-failing kindness in offering me every assistance in the
investigation of fossil plants under their charge. Prof. Marshall
Ward has given me the benefit of his criticism on the section
dealing with Fungi; and my friend Mr Alfred Harker has
rendered me a similar service as regards the chapter on
Geological History. I am especially grateful to my colleague,
Mr Francis Darwin, for having read through the whole of the
proofs of this volume. To Mr Shipley, as Editor, I am under a
debt of obligation for suggestions and help in various forms. I
would also express my sense of the unfailing courtesy and skill
of the staff of the University Press.

My friend Mr Kidston of Stirling has always generously
responded to my requests for the loan of specimens from
his private collection. Prof. Bayley Balfour of Edinburgh,

Mr Wethered of Cheltenham and others have assisted me in a similar manner. I would also express my gratitude to Dr Hoyle of Manchester, Mr Platnauer of York, and Mr Rowntree of Scarborough for the loan of specimens.

To Dr Henry Woodward of the British Museum I am indebted for the loan of the woodblocks made use of in figs. 10, 47, 60, 66, and 101, and to Messrs Macmillan for the process-block of fig. 25.

For the photographs reproduced in figs. 15, 34, 68, 102 and 103 I owe an acknowledgment to Mr Edwin Wilson of Cambridge, and to my friend Mr C. A. Barber for the micro-photograph made use of in fig. 40.

In conclusion I wish more particularly to thank my wife, who has drawn by far the greater number of the illustrations, and has in many other ways assisted me in the preparation of this Volume.

In Volume II the Systematic treatment of Plants will be concluded, and the last chapters will be devoted to such subjects as geological floras, plants as rock-builders, fossil plants and evolution, and other general questions connected with Palaeobotany.

A. C. SEWARD.

BOTANICAL LABORATORY, CAMBRIDGE.
March, 1898.

TABLE OF CONTENTS.

CHAPTER IV

THE PRESERVATION OF PLANTS AS FOSSILS.
Pp. 54—92.

Old surface-soils. Fossil wood. Conditions of fossilisation. Drifting of trees. Meaning of the term 'Fossil.' Incrustations. Casts of trees. Fossil casts. Plants and coal. Fossils in half-relief. Petrified trees. Petrified wood. Preservation of tissues. Coal-balls. Fossil nuclei. Fossil plants in volcanic ash. Conditions of preservation.

CHAPTER V.

DIFFICULTIES AND SOURCES OF ERROR IN THE DETERMINATION OF FOSSIL PLANTS. Pp. 93—109.

External resemblance. Venation characters. Decorticated stems. Imperfect casts. Mineral deposits simulating plants. Traces of wood-borers in petrified tissue. Photography and illustration.

CHAPTER VI.

NOMENCLATURE. Pp. 110—115.

Rules for nomenclature. The rule of priority. Terminology and convenience.

PART II. SYSTEMATIC.

CHAPTER VII.

THALLOPHYTA. Pp. 116—228.

CHAPTER VIII.

BRYOPHYTA. Pp. 229—241.

CHAPTER IX.

PTERIDOPHYTA (VASCULAR CRYPTOGRAMS).
Pp. 242—294.

CHAPTER XI.

SPHENOPHYLLALES. Pp. 389—414.

LIST OF ILLUSTRATIONS.

FRONTISPIECE. TREE STUMPS IN A CARBONIFEROUS FOREST. Drawn
from a photograph. (M. Seward.) PAGE 57.

Note. The references in the footnotes require a word of
explanation. The titles of the works referred to will be found in
the Bibliography at the end of the volume. In this list the authors'
names are arranged alphabetically and the papers of each author are
in chronological order. The numbers in brackets after the author's
name in the footnotes, and before his name in the bibliographical
list, refer to the year of publication. Except in cases where the
works were published prior to 1800, the first two figures are
omitted: thus Ward (84) refers to a paper published by L. F. Ward
in 1884. This system was suggested by Dr H. H. Field in the
Biologisches Centralblatt, vol. XIII. 1893, p. 753. (*Ueber die Art der
Abfassung naturwissenschaftlicher Litteraturverzeichnisse.*)

PART I. GENERAL.

CHAPTER I.

HISTORICAL SKETCH.

"But particular care ought to be had not to consult or take relations from any but those who appear to have been both long conversant in these affairs, and likewise persons of Sobriety, Faithfulness and Discretion, to avoid the being misled and imposed upon either by falsehood, or the ignorance, credulity, and fancifulness, that some of these people are but too obnoxious unto." JOHN WOODWARD, 1728.

THE scientific study of fossil plants dates from a comparatively recent period, and palaeobotany has only attained a real importance in the eyes of botanists and geologists during the last few decades of the present century. It would be out of place, in a short treatise like the present, to attempt a detailed historical sketch, or to give an adequate account of the gradual rise and development of this modern science. An excellent *Sketch of Palaeobotany* has recently been drawn up by Prof. Lester Ward[1], of the United States Geological Survey, and an earlier historical retrospect may be found in the introduction to an important work by an eminent German palaeobotanist, the late Prof. Göppert[2]. In the well-known work by Parkinson on *The Organic Remains of a Former World*[3] there is much interesting information as to the early history of our knowledge

---

[1] Ward (84). [2] Göppert (36). [3] Parkinson (11), vol. I.

of fossil plants, as well as a good exposition of the views held
at the beginning of this century.

As a means of bringing into relief the modern development
of the science of fossil plants, we may briefly pass in review
some of the earlier writers, who have concerned themselves in a
greater or less degree with a descriptive or speculative treatment
of the records of a past vegetation. In the early part of the
present century, and still more in the eighteenth century, the
occurrence of fossil plants and animals in the earth's crust
formed the subject of animated, not to say acrimonious,
discussion. The result was that many striking and ingenious
theories were formulated as to the exact manner of formation of fossil remains, and the part played by the waters
of the deluge in depositing fossiliferous strata. The earlier
views on fossil vegetables are naturally bound up with the
gradual evolution of geological science. It is from Italy that
we seem to have the first glimmering of scientific views;
but we are led to forget this early development of more
than three hundred years ago, when we turn to the writings
of English and other authors of the eighteenth century.
"Under these white banks by the roadside," as a writer on
Verona has expressed it, "was born, like a poor Italian gipsy,
the modern science of geology." Early in the sixteenth century
the genius of Leonardo da Vinci[1] compelled him to adopt
a reasonable explanation of the occurrence of fossil shells in
rocks far above the present sea-level. Another Italian writer,
Fracastaro, whose attention was directed to this matter by the
discovery of numerous shells brought to light by excavations
at Verona, expressed his belief in the organic nature of the
remains, and went so far as to call in question the Mosaic
deluge as a satisfactory explanation of the deposition of fossil-
bearing strata.

The partial recognition by some observers of the true
nature of fossils marks the starting point of more rational
views. The admission that fossils were not mere sports
of nature, or the result of some wonderful vis lapidifica,'

---

[1] For an account of the early views on fossils, v. Lyell (67), Vol. i. Vide
also Leonardo da Vinci (83).

was naturally followed by numerous speculations as to the
manner in which the remains of animals and plants came to be
embedded in rocks above the sea-level. For a long time, the
'universal flood' was held responsible by nearly all writers
for the existence of fossils in ancient sediments. Dr John
Woodward, in his *Essay toward a Natural History of the
Earth*, propounded the somewhat revolutionary theory, that
"the whole terrestrial globe was taken all to pieces and dissolved at the Deluge, the particles of stone, marble, and all
solid fossils dissevered, taken up into the water, and there
sustained together with sea-shells and other animal and vegetable bodies: and that the present earth consists, and was
formed out of that promiscuous mass of sand, earth, shells,
and the rest falling down again, and subsiding from the
water[1]." In common with other writers, he endeavoured to
fix the exact date of the flood by means of fossil plants.
Speaking of some hazel-nuts, which were found in a Cheshire
moss pit, he draws attention to their unripened condition, and
adds: "The deluge came forth at the end of May, when nuts
are not ripe." As additional evidence, he cites the occurrence of
"Pine cones in their vernal state," and of some Coal-Measure
fossils which he compares with Virginian Maize, "tender,
young, vernal, and not ripened[2]." Woodward (1665—1728)
was Professor of Physic in Gresham College; he bequeathed
his geological collections to the University of Cambridge, and
founded the Chair which bears his name.

Another writer, Mendes da Costa, in a paper in the *Philosophical Transactions* for 1758, speaks of the impressions of
"ferns and reed-like plants" in the coal-beds, and describes
some fossils (*Sigillaria* and *Stigmaria*) as probably unknown
forms of plant life[3].

Here we have the suggestion that in former ages there
were plants which differed from those of the present age.
Discussing the nature of some cones (*Lepidostrobi*) from the ironstone of Coalbrookdale in Shropshire, he concludes: "I firmly
believe these bodies to be of vegetable origin, buried in the

---

[1] Woodward, J. (1695), Preface.   [2] Woodward, J. (1728), p. 59.
[3] Mendes da Costa (1758), p. 232.

strata of the Earth at the time of the universal deluge recorded by Moses." Scheuchzer of Zurich, the author of one of the earliest works on fossil plants and a "great apostle of the Flood Theory," figures and describes a specimen as an ear of corn, and refers to its size and general appearance as pointing to the month of May as the time of the deluge[1]. Another English writer, Dr Parsons, in giving an account of the well-known 'fossil fruits and other bodies found in the island of Sheppey,' is disposed to dissent from Woodward's views as to the time of the flood. He suggests that the fact of the Sheppey fruits being found in a perfectly ripe condition, points to the autumn as the more probable time for the occurrence of the deluge[2].

In looking through the works of the older writers, and occasionally in the pages of latter-day contributors, we frequently find curiously shaped stones, mineral markings on rock surfaces, or certain fossil animals, described as fossil plants. In Plot's *Natural History of Oxfordshire*, published in 1705, a peculiarly shaped stone, probably a flint, is spoken of as one of the 'Fungi lethales non esculenti[3]'; and again a piece of coral[4] is compared with a 'Bryony root broken off transversely.' On the other hand, that we may not undervalue the painstaking and laborious efforts of those who helped to lay the foundations of modern science, we may note that such authors as Scheuchzer and Woodward were not misled by the mosslike or dendritic markings of oxide of manganese on the surface of rocks, which are not infrequently seen to-day in the cabinets of amateurs as specimens of fossil plants.

The oldest figures of fossil plants from English rocks which are drawn with any degree of accuracy are those of Coal-Measure ferns and other plants in an important work by Edward Lhwyd published at Oxford in 1760[5].

Passing beyond these prescientific speculations, brief reference may be made to some of the more eminent pioneers of palaeobotany. The Englishman Artis[6] deserves mention for

---

[1] Scheuchzer (1723), p. 7, Pl. I. fig. 1.   [2] Parsons (1757), p. 402.
[3] Plot (1705), p. 125, Pl. VI. fig. 2.   [4] *Ibid.* Pl. VI. fig. 2.
[5] Lhywd (1760).   [6] Artis (25).

the quality rather than the quantity of his contributions to Palaeozoic botany; and among American authors Steinhauer's[1] name must hold a prominent place in the list of those who helped to found this branch of palaeontology. Among German writers, Schlotheim stands out prominently as one who first published a work on fossil plants which still remains an important book of reference. Writing in 1804, he draws attention to the neglect of fossils from a scientific standpoint; they are simply looked upon, he says, as "unimpeachable documents of the flood[2]." His book contains excellent figures of many Coal-Measure plants, and we find in its pages occasional comparisons of fossil species with recent plants of tropical latitudes. Among the earlier authors whose writings soon become familiar to the student of fossil plants, reference must be made to Graf Sternberg, who was born three years before Schlotheim, but whose work came out some years later than that of the latter. His great contribution to Fossil Botany entitled *Versuch einer geognostisch-botanischen Darstellung der Flora der Vorwelt*, was published in several parts between the years 1820 and 1838; it was drawn up with the help of the botanist Presl, and included a valuable contribution by Corda[3]. In addition to descriptions and numerous figures of plants from several geological horizons, this important work includes discussions on the formation of coal, with observations on the climates of past ages.

Sternberg endeavoured to apply to fossil plants the same methods of treatment as those made use of in the case of recent species. About the same time as Sternberg's earlier parts were published, Adolphe Brongniart[4] of Paris began to enrich palaeobotanical science by those splendid researches which have won for him the title of the "Father of palaeobotany." In Brongniart's *Prodrome*, and *Histoire des végétaux fossiles*, and later in his *Tableau des genres de végétaux fossiles*, we have not merely careful descriptions and a systematic arrangement of the known species of fossil plants, but a masterly scientific

---

[1] Steinhauer (18).   [2] Schlotheim (04).
[3] Sternberg (20).   [4] Brongniart (28) (28²) (49).

treatise on palaeobotany in its various aspects, which has to a
large extent formed the model for the best subsequent works
on similar lines. From the same author, at a later date, there
is at least one contribution to fossil plant literature which
must receive a passing notice even in this short sketch. In
1839 he published an exhaustive account of the minute
structure of one of the well-known Palaeozoic genera, *Sigillaria;*
this is not only one of the best of the earliest monographs on
the histology of fossil species, but it is one of the few existing
accounts of the internal structure of this common type[1]. The
fragment of a Sigillarian stem which formed the subject of
Brongniart's memoir is in the Natural History Museum in the
Jardin des Plantes, Paris. It affords a striking example of
the perfection of preservation as well as of the great beauty
of the silicified specimens from Autun, in Central France.
Brongniart was not only a remarkably gifted investigator,
whose labours extend over a period connecting the older and
more crude methods of descriptive treatment with the modern
development of microscopic analysis, but he possessed the
power of inspiring a younger generation with a determination
to keep up the high standard of the palaeobotanical achieve-
ments of the French School. In some cases, indeed, his disciples
have allowed a natural reverence for the Master to warp
their scientific judgement, where our more complete knowledge
has naturally led to the correction of some of Brongniart's conclusions. Without attempting to follow the history of the science
to more recent times, the names of Heer, Lesquereux, Zigno,
Massalongo, Saporta and Ettingshausen should be included
among those who rendered signal service to the science of fossil
plants. The two Swiss writers, Heer[2] and Lesquereux[3], contributed numerous books and papers on palaeobotanical subjects,
the former being especially well known in connection with the
fossil floras of Switzerland and of Arctic lands, and the latter
for his valuable writings on the fossil plants of his adopted
country, North America. Zigno[4] and Massalongo[5] performed
like services for Italy, and the Marquis of Saporta's name will

---

[1] Brongniart (39).   [2] Heer (55) (68) (76).
[3] Lesquereux (66) (70) (80) etc.   [4] Zigno (56).   [5] Massalongo (51).

always hold an honourable and prominent position in the list
of the pioneers of scientific palaeobotany; his work on the
Tertiary and Mesozoic floras of France being specially note-
worthy among the able investigations which we owe to his
ability and enthusiasm[1]. In Baron Ettingshausen[2] we have
another representative of those students of ancient vegetation
who have done so much towards establishing the science of
fossil plants on a philosophical basis.

As in other fields of Natural Science, so also in a marked
degree in fossil botany, a new stimulus was given to scientific
inquiry by the application of the microscope to palaeobotanical
investigation. In 1828 Sprengel published a work entitled
Commentatio de Psarolithis, ligni fossilis genere[3]; in which he
dealt in some detail with the well-known silicified fern-stems
of Palaeozoic age, from Saxony, basing his descriptions on the
characteristics of anatomical structure revealed by microscopic
examination.

In 1833 Henry Witham of Lartington brought out a
work on *The Internal Structure of Fossil Vegetables*[4]; this
book, following the much smaller and less important work
by Sprengel, at once established palaeobotany on a firmer
scientific basis, and formed the starting point for those
more accurate methods of research, which have yielded such
astonishing results in the hands of modern workers. In
the introduction Witham writes, "My principal object in
presenting this work to the public, is to impress upon geologists the advantage of attending more particularly to the
intimate organization of fossil plants; and should I succeed in
directing their efforts towards the elucidation of this obscure
subject, I shall feel a degree of satisfaction which will amply
repay my labour[5]."

On another page he writes as follows,—"From investigations made by the most active and experienced botanical
geologists, we find reason to conclude that the first appearance

---

[1] Saporta (72) (73).
[2] Ettingshausen (79). Also numerous papers on fossil plants from Austria
and other countries.
[3] Sprengel (28).   [4] Witham (33).   [5] *ibid.*, p. 3.

of an extensive vegetation occurred in the Carboniferous series;
and from a recent examination of the mountain-limestone
groups and coal-fields of Scotland, and the north of England,
we learn that these early vegetable productions, so far from
being simple in their structure, as had been supposed, are as
complicated as the phanerogamic plants of the present day.
This discovery necessarily tends to destroy the once favourite
idea, that, from the oldest to the most recent strata, there has
been a progressive development of vegetable and animal forms,
from the simplest to the most complex[1]." Since Witham's
day we have learnt much as to the morphology of Palaeozoic
plants, and can well understand the opinions to which he thus
gives expression.

It would be difficult to overrate the immense importance of
this publication from the point of view of modern palaeobotany.
The art of making transparent sections of the tissues of
fossil plants seems to have been first employed by Sanderson,
a lapidary, and it was afterwards considerably improved by
Nicol[2]. This most important advance in methods of examination gave a new impetus to the subject, but it is somewhat
remarkable that the possibilities of the microscopical investigation of fossil plants have been but very imperfectly realised
by botanical workers until quite recent years. As regards such
a flora as that of the Coal-Measures, we can endorse the
opinion expressed at the beginning of the century in reference
to the study of recent mosses—" Ohne das Göttergeschenk des
zusammengesetzten Mikroskops ist auf diesem Felde durchaus
keine Ernte[3]." A useful summary of the history of the study
of internal structure is given by Knowlton in a memoir
published in 1889[4]. Not long after Witham's book was issued
there appeared a work of exceptional merit by Corda[5], in which
numerous Palaeozoic plants are figured and fully described,
mainly from the standpoint of internal structure. This author

---

[1] Witham (33), p. 5.

[2] Nicol (34). See note by Prof. Jameson on p. 157 of the paper quoted, to the
effect that he has long known of this method of preparing sections.

[3] Limpricht (90) in Rabenhorst, vol. IV. p. 73.

[4] Knowlton (89).　　　　　　　　　　　[5] Corda (45).

lays special stress on the importance of studying the microscopical structure of fossil plants.

Without pausing to enumerate the contributions of such
well-known continental authors as Göppert, Cotta, Schimper,
Stenzel, Schenk and a host of others, we may glance for a
moment at the services rendered by English investigators to
the study of palaeobotanical histology. Unfortunately we
cannot always extend our examination of fossil plants beyond
the characters of external form and surface markings; but in
a few districts there are preserved remnants of ancient floras in
which fragments of stems, roots, leaves and other structures
have been petrified in such a manner as to retain with wonderful completeness the minute structure of their internal tissues.
During the deposition of the coal seams in parts of Yorkshire
and Lancashire the conditions of fossilisation were exceptionally favourable, and thus English investigators have been
fortunately placed for conducting researches on the minute
anatomy of the Coal-Measure plants. The late Mr Binney of
Manchester did excellent service by his work on the internal
structure of some of the trees of the Coal Period forests. In
his introductory remarks to a monograph on the genus
Calamites, after speaking of the desirability of describing
our English specimens, he goes on to say, "When this is
done, we are likely to possess a literature on our Carboniferous
fossils worthy of the first coal-producing country[1]." The con-
tinuation and extension of Binney's work in the hands of
Carruthers, Williamson, and others, whose botanical qualifica-
tions enabled them to produce work of greater scientific value,
has gone far towards the fulfilment of Binney's prophecy.

In dealing with the structure of Palaeozoic plants, we shall
be under constant obligation to the splendid series of memoirs
from the pen of Prof. Williamson[2]. As the writer of a
sympathetic obituary notice has well said: "In his fifty-fifth
year he began the great series of memoirs which mark the
culminating point of his scientific activity, and which will
assure to him, for all time, in conjunction with Brongniart, the

---

[1] Binney (68), Introductory remarks.   [2] Williamson (71), etc.

honourable title of a founder of modern Palaeobotany[1]." If we
look back through a few decades, and peruse the pages of Lindley
and Hutton's classic work[2] on the Fossil flora of Great Britain,
a book which is indispensable to fossil botanists, and read the
description of such a genus as *Sigillaria* or *Stigmaria*; or if we
extend our retrospect to an earlier period and read Woodward's
description of an unusually good specimen of a *Lepidodendron*,
and finally take stock of our present knowledge of such plants,
we realise what enormous progress has been made in palaeobotanical studies. Lindley and Hutton, in the preface to the
first volume of the Flora, claim to have demonstrated that both
Sigillaria and *Stigmaria* were plants with "the highest degree
of organization, such as *Cactaeae*, or *Euphorbiaceae*, or even
Asclepiadeae"; Woodward describes his *Lepidodendron* (Fig. 1) as
"an ironstone, black and flat, and wrought over one surface very
finely, with a strange cancellated work[3]." Thanks largely to

FIG. 1.   Four leaf-cushions of a *Lepidodendron*.   Drawn from a specimen in
the Woodward Collection, Cambridge.   (Nat. size.)

the work of Binney, Carruthers, Hooker, Williamson, and to the
labours of continental botanists, we are at present almost as
familiar with *Lepidodendron* and several other Coal-Measure

[1] Solms-Laubach (95), p. 442.   [2] Lindley and Hutton (31).
[3] Woodward (1729), Pt. ii. p. 106.

genera as with the structure of a recent forest tree. While emphasizing the value of the microscopic methods of investigation, we are not disposed to take such a hopeless view of the possibilities of the determination of fossil forms, in which no internal structure is preserved, as some writers have expressed. The preservation of minute structure is to be greatly desired from the point of view of the modern palaeobotanist, but he must recognise the necessity of making such use as he can of the numberless examples of plants of all ages, which occur only in the form of structureless casts or impressions.

In looking through the writings of the earlier authors we cannot help noticing their anxiety to match all fossil plants with living species; but by degrees it was discovered that fossils are frequently the fragmentary samples of extinct types, which can be studied only under very unfavourable conditions. In the absence of those characters on which the student of living plants relies as guides to classification, it is usually impossible to arrive at any trustworthy conclusions as to precise botanical affinity. Brongniart and other authors recognised this fact, and instituted several convenient generic terms of a purely artificial and provisional nature, which are still in general use. The dangers and risks of error which necessarily attend our attempts to determine small and imperfect fragments of extinct species of plants, will be briefly touched on in another place.

CHAPTER II.

RELATION OF PALAEOBOTANY TO BOTANY AND GEOLOGY.

"La recherche du plan de la création, voilà le but vers lequel nos efforts peuvent tendre aujourd'hui." GAUDRY, 1883.

SINCE the greater refinements and thoroughness of scientific methods and the enormous and ever-increasing mass of literature have inevitably led to extreme specialisation, it is more than ever important to look beyond the immediate limits of one's own subject, and to note its points of contact with other lines of research. A palaeobotanist is primarily concerned with the determination and description of fossil plants, but he must at the same time constantly keep in view the bearing of his work on wider questions of botanical or geological importance. From the nature of the case, we have in due measure to adapt the methods of work to the particular conditions before us. It is impossible to follow in the case of all fossil species precisely the same treatment as with the more complete and perfect recent plants; but it is of the utmost importance for a student of palaeobotany, by adhering to the methods of recent botany, to preserve as far as he is able the continuity of the past and present floras. Palaeontological work has often been undertaken by men who are pure geologists, and whose knowledge of zoology or botany is of the most superficial character, with the result that biologists have not been able to avail themselves, to any considerable extent, of the records of extinct forms of life

They find the literature is often characterised by a special palaeontological phraseology, and by particular methods of treatment, which are unknown to the student of living plants and animals. From this and other causes a purely artificial division has been made between the science of the organic world of to-day and that of the past.

Fossils are naturally regarded by a stratigraphical geologist as records which enable him to determine the relative age of fossil-bearing rocks. For such a purpose it is superfluous to inquire into the questions of biological interest which centre round the relics of ancient floras. Primarily concerned, therefore, with fixing the age of strata, it is easy to understand how geologists have been content with a special kind of palaeontology which is out of touch with the methods of systematic zoology or botany. On the other hand, the botanist whose observations and researches have not extended beyond the limits of existing plants, sees in the vast majority of fossil forms merely imperfect specimens, which it is impossible to determine with any degree of scientific accuracy. He prefers to wait for perfect material ; or in other words, he decides that fossils must be regarded as outside the range of taxonomic botany. It would seem, then, that the unsatisfactory treatment or comparative neglect of fossil plants, has been in a large measure due to the narrowness of view which too often characterises palaeobotanical literature. This has at once repelled those who have made a slight effort to recognise the subject, and has resulted in a one-sided and, from a biological standpoint, unscientific treatment of this branch of science. It must be admitted that palaeobotanists have frequently brought the subject into disrepute by their over-anxiety to institute specific names for fragments which it is quite impossible to identify. This over-eagerness to determine imperfect specimens, and the practice of drawing conclusions as to botanical affinity without any trustworthy evidence, have naturally given rise to considerable scepticism as to the value of palaeobotanical records. Another point, which will be dealt with at greater length in a later chapter, is that geologists have usually shown a distinct prejudice against fossil plants as indices of geological age; this again, is no doubt to a large extent the result of imperfect and inaccurate methods of description, and of the neglect of and consequent imperfect acquaintance with fossil plants as compared with fossil animals.

The student of fossil plants should endeavour to keep before him the fact that the chief object of his work is to deal with the available material in the most natural and scientific manner; and by adopting the methods of modern botany, he should always aim to follow such lines as may best preserve the continuity of past and present types of plants. Descriptions of floras of past ages and lists of fossil species, should be so compiled that they may serve the same purpose to a stratigraphical geologist, who is practically a geographer of former periods of the Earth's history, as the accounts of existing floras to students of present day physiography. The effect of carrying out researches on some such lines as these, should be to render available to both botanists and geologists the results of the specialist's work.

In some cases, palaeobotanical investigations may be of the utmost service to botanical science, and of little or no value to geology. The discovery of a completely preserved gametophyte of *Lepidodendron* or *Calamites*, or of a petrified Moss plant in Palaeozoic rocks would appeal to most botanists as a matter of primary importance, but for the stratigraphical geologist such discoveries would possess but little value. On the other hand the discovery of some characteristic species of Coal-Measure plants from a deep boring through Mesozoic or Tertiary strata might be a matter of special geological importance, but to the botanist it would be of no scientific value. In very many instances, however, if the palaeobotanist follows such lines as have been briefly suggested, the results of his labours should be at once useful and readily accessible to botanists and geologists. As Humboldt has said in speaking of Palaeontology, "the analytical study of primitive animal and vegetable life has taken a double direction; the one is purely morphological, and embraces especially the natural history and physiology of organisms, filling up the chasms in the series of still living species by the fossil structures of the primitive world. The second is more specially geognostic, considering fossil remains in their relations to the superposition and relative age of the sedimentary formations[1]."

To turn for a moment to some of the most obvious connections between palaeobotany and the wider sciences of botany and geology. The records of fossil species must occupy a prominent position in the data by which we may hope to solve some at least of the problems of plant evolution. From the point of view of distribution, palaeobotany is of considerable value, not only to the student of geographical botany, but to the geologist, who endeavours to map out the positions of ancient continents with the help of palaeontological evidence. The present distribution of plants and animals represents but one chapter in the history of life on the Earth ; and to understand or appreciate the facts which it records, we have to look back through such pages as have been deciphered in the earlier chapters of the volume. The distribution of fossil plants lies at the foundation of the principles of the present grouping of floras on the Earth's surface. Those who have confined their study of distribution to the plant geography of the present age, must supplement their investigations by reference to the work of palaeobotanical writers. If the lists of plant species drawn up by specialists in fossil botany, have been prepared with a due sense of the important conclusions which botanists may draw from them from the standpoint of distribution, they will be readily accepted as sound links in the chain of evidence. Unfortunately, however, if many of the lists of ancient floras were made use of in such investigations, the conclusions arrived at would frequently be of little value on account of the untrustworthy determinations of many of the species. In the case of particular genera the study of the distribution of the former species both in time and space, that is geologically and geographically, points to rational explanations of, or gives added significance to, the facts of present day distribution. That isolated conifer, *Ginkgo biloba* L. now restricted to Japan and China, was in former times abundant in Europe and in other parts of the world. It is clearly an exceedingly ancient type, isolated not only in geographical distribution but in

[1] Humboldt (48), vol. I. p. 274.

botanical affinities, which has reached the last stage in its natural life. The Mammoth trees of California (*Sequoia sempervirens* Endl., and *S. gigantea* Lindl. and Gord.) afford other examples of a parallel case. The North American Tulip tree and other allied forms are fairly common in the Tertiary plant beds of Europe, but the living representatives are now exclusively North American. Such differences in distribution as are illustrated by these dicotyledonous forest trees in Tertiary times and at the present day, have been clearly explained with the help of the geological record. Forbes, Darwin, Asa Gray [1] and others have been able to explain many apparent anomalies in the distribution of existing plants, and to reconcile the differences between the past and present distribution of many genera by taking account of the effect on plant life of the glacial period. As the ice gradually crept down from the polar regions and spread over the northern parts of Europe, many plants were driven further south in search of the necessary warmth. In the American continent such migration was rendered possible by the southern land extension; in Europe on the other hand the southerly retreat was cut off by impassable barriers, and the extinction of several genera was the natural result.

The comparatively abundant information which we possess as to the past vegetation of polar regions and the value of such knowledge to geologists and botanists alike is in striking contrast to the absence of similar data as regards Antarctic fossils. Darwin in an exceedingly interesting letter to Hooker *à propos* of a forthcoming British Association address, referring to this subject writes as follows:—

"The extreme importance of the Arctic fossil plants is self-evident. Take the opportunity of groaning over our ignorance of the Lignite plants of Kerguelen Land, or any Antarctic land. It might do good [2]."

In working out any collection of fossil plants, it would be well, therefore, to bear in mind that our aim should be rather to reproduce an accurate fragment of botanical history, than to

[1] *Vide* Hooker, J. D. (81), for references to other writers on this subject; also Darwin (82), ch. XII.

[2] Darwin (87), vol. III. p. 247.

perform feats of determination with hopelessly inadequate specimens. Had this principle been generally followed, the number of fossil plant species would be enormously reduced, but the value of the records would be considerably raised.

Our knowledge of plant anatomy, and of those laws of growth which govern certain classes of plants to-day and in past time, has been very materially widened and extended by the facts revealed to us by the detailed study of Coal-Measure species. The modern science of Plant Biology, refounded by Charles Darwin, has thrown considerable light on the laws of plant life, and it enables us to correlate structural characteristics with physiological conditions of growth. Applying the knowledge gained from living plants to the study of such extinct types as permit of close microscopic examination, we may obtain a glimpse into the secrets of the botanical binomics of Palaeozoic times. The wider questions of climatic conditions depend very largely upon the evidence of fossil botany for a rational solution. As an instance of the best authenticated and most striking alternation in climatic conditions in comparatively recent times, we may cite the glacial period or Ice-Age. The existence of Arctic conditions has been proved by purely geological evidence, but it receives additional confirmation, and derives a wider importance from the testimony of fossil plants. In rocks deposited before the spread of ice from high northern latitudes, we find indubitable proofs of a widely distributed subtropical flora in Central and Northern Europe. Passing from these rocks to more recent beds there are found indications of a fall in temperature, and such northern plants as the dwarf Birch, the Arctic Willow and others reveal the southern extension of Arctic cold to our own latitudes.

The distribution of plants in time, that is the range of classes, families, genera and species of plants through the series of strata which make up the crust of the earth, is a matter of primary importance from a botanical as well as from a geological point of view.

Among the earlier writers, Brongniart recognised the marked differences between the earlier and later floras, and attempted to correlate the periods of maximum development of certain classes of plants with definite epochs of geological history. He gives the following classification in which are represented the general outlines of plant development from Palaeozoic to Tertiary times[1].

I.   Reign of Acrogens $\left\{\begin{array}{l} \text{1. Carboniferous epoch.} \\ \text{2. Permian epoch.} \end{array}\right.$

II.   Reign of Gymnosperms $\left\{\begin{array}{l} \text{3. Triassic epoch.} \\ \text{4. Jurassic epoch (including} \\ \quad \text{the Wealden).} \end{array}\right.$

III.   Reign of Angiosperms $\left\{\begin{array}{l} \text{5. Cretaceous epoch.} \\ \text{6. Tertiary epoch.} \end{array}\right.$

Since Brongniart's time this method of classification has been extended to many of the smaller subdivisions of the geological epochs, and species of fossil plants are often of the greatest value in questions of correlation. In recent years the systematic treatment of Coal-Measure and other plants in the hands of various Continental and English writers has clearly demonstrated their capabilities for the purpose of subdividing a series of strata into stages and zones[2]. The more complete becomes our knowledge of any flora, the greater possibility there is of making use of the plants as indices of geological age[3].

Not only is it possible to derive valuable aid in the correlation of strata from the facts of plant distribution, but we may often follow the various stages in the history of a particular genus as we trace the records of its occurrence through the geologic series. In studying the march of plant life through past ages, the botanist may sometimes follow the progress of a genus from its first appearance, through the time of maximum development, to its decline or extinction. In the Palaeozoic forests there was perhaps no more conspicuous or common tree than the genus long known under the name of *Calamites*.

---

[1] Brongniart (49), p. 94.

[2] Grand'Eury (77), Potonié (96), Kidston (94), &c.

[3] Ward (92), Knowlton (94), Grand'Eury (90), p. 155.

This plant attained a height of fifty or a hundred feet, with a proportionate girth, and increased in thickness in a manner precisely similar to that in which our forest trees grow in diameter. The exceptionally favourable conditions under which specimens of calamitean plants have been preserved, have enabled us to become almost as familiar with the minute structure of their stems and roots, as well as with their spore-producing organs, as with those of a living species. In short, it is thoroughly established that *Calamites* agrees in most essential respects with our well known *Equisetum*, and must be included in the same order, or at least sub-class, as the recent genus of *Equisetaceae*. As we ascend the geologic series from the Coal-Measures, a marked numerical decline of *Calamites* is obvious in the Permian period, and in the red sandstones of the Vosges, which belong to the same series of rocks as the Triassic strata of the Cheshire plain, the true *Calamites* is replaced by a large *Equisetum* apparently identical in external appearance and habit of growth with the species living to-day. In the more recent strata the Horsetails are still represented, but the size of the Tertiary species agrees more closely with the comparatively small forms which have such a wide geographical distribution at the present time. Thus we are able to trace out the history of a recent genus of Vascular Cryptogams, and to follow a particular type of organisation from the time of its maximum development, through its gradual transition to those structural characters which are represented in the living descendants of the arborescent Calamites of the coal-period forests. The pages of such a history are frequently imperfect and occasionally missing, but others, again, are written in characters as clear as those which we decipher by a microscopical examination of the tissues of a recent plant.

As one of the most striking instances in which the microscopic study of fossil plants has shown the way to a satisfactory solution of the problems of development, we may mention such extinct genera as *Lyginodendron*, *Myeloxylon* and others. Each of these genera will be dealt with at some length in the systematic part of the book, and we shall afterwards discuss the importance of such types, from the point of view of plant evolution.

The botanist who would trace out the phylogeny of any existing class or family, makes it his chief aim to discover points of contact between the particular type of structure which he is investigating, and that of other more or less closely related classes or families.

Confining himself to recent forms, he may discover, here and there, certain anatomical or embryological facts, which suggest promising lines of inquiry in the quest after such affinities as point to a common descent. Without recourse to the evidence afforded by the plants of past ages, we must always admit that our existing classification of the vegetable kingdom is an expression of real gaps which separate the several classes of plants from one another. On the other hand our recently acquired and more accurate knowledge of such genera as have been alluded to, has made us acquainted with types of plant structure which enable us to fill in some of the lacunae in our existing classification.   In certain instances we find merged in a single species morphological characteristics which, in the case of recent plants, are regarded as distinctive features of different sub-divisions.   It has been clearly demonstrated that in *Lyginodendron*, we have anatomical peculiarities typical of recent cycads, combined with structural characteristics always associated with existing ferns.   In rare cases, it happens that the remarkably perfect fossilisation of the tissues of fossil plants, enables us not only to give a complete description of the histology of extinct forms, but also to speak with confidence as to some of those physiological processes which governed their life.

So far, palaeobotany has been considered in its bearings on the study of recent plants. From a geological point of view the records of ancient floras have scarcely less importance. In recent years, facts have been brought to light, which show that plants have played a more conspicuous part than has usually been supposed as agents of rock-building. As tests of geologic age, there are good grounds for believing that the inferiority of plants to animals is more apparent than real

This question, however, must be discussed at greater length in a later chapter.

Enough has been said to show the many-sided nature of the science of Fossil Plants, and the wide range of the problems which the geologist or botanist may reasonably expect to solve, by means of trustworthy data afforded by scientific palaeobotanical methods.

CHAPTER III.

GEOLOGICAL HISTORY.

"But how can we question dumb rocks whose speech is not clear[1]?"

IN attempting to sketch in briefest outline the geological history of the Earth, the most important object to keep in view is that of reproducing as far as possible the broad features of the successive stages in the building of the Earth's crust. It is obviously impossible to go into any details of description, or to closely follow the evolution of the present continents; at most, we can only refer to such facts as may serve as an introduction of the elements of stratigraphical geology to non-geological readers. For a fuller treatment of the subject reference must be made to special treatises on geology.

For the sake of convenience, it is customary in strati-graphical geology as also in biology, to make use of our imperfect knowledge as an aid to classification. If we possessed complete records of the Earth's history, we should have an unbroken sequence, not merely of the various forms of life that ever existed, but of the different kinds of rocks formed in the successive ages of past time. As gaps exist in the chain of life, so also do we find considerable breaks in the sequence of strata which have been formed since the beginning of geologic time. The danger as well as the convenience of artificial classification must be kept in view. This has been

---

[1] Old Persian writer, quoted by E. G. Browne in *A Year among the Persians*, p. 220, London, 1893.

well expressed by Freeman, in speaking of architectural
styles,—"Our minds," he says, "are more used to definite
periods; they neglect or forget transitions which do indeed
exist[1]." The idea of definite classification is liable to narrow
our view of uniformity and the natural sequence of events.

Composing that part of the earth which is accessible to
us,—or as it is generally called the earth's crust,—there are
rocks of various kinds, of which some have been formed by
igneous agency, either as lavas or beds of ashes, or in the
form of molten magmas which gradually cooled and became
crystalline below a mass of superincumbent strata. With
these rocks we need not concern ourselves.

A large portion of the earth's crust consists of such
materials as sandstones, limestones, shales, and similar strata
which have been formed in precisely the same manner as
deposits are being accumulated at the present day. The
whole surface of the earth is continually exposed to the
action of destructive agencies, and suffers perpetual decay; it
is the products of this ceaseless wear and tear that form the
building materials of new deposits.

The operation of water in its various forms, of wind,
changes of temperature, and other agents of destruction
cannot be fully dealt with in this short summary.

A river flowing to the sea or emptying itself into an inland
lake, carries its burden of gravel, sand, and mud, and sooner or
later, as the rate of flow slackens, it deposits the materials
in the river-bed or on the floor of the sea or lake.

Fragments of rock, chipped off by wedges of ice, or
detached in other ways from the parent mass, find their way
to the mountain streams, and if not too heavy are conveyed
to the main river, where the larger pieces come to rest as more
or less rounded pebbles. Such water-worn rocks accumulate
in the quieter reaches of a swiftly flowing river, or are thrown
down at the head of the river's delta. If such a deposit of
loose water-worn material became cemented together either
by the consolidating action of some solution percolating
through the general mass, or by the pressure of overlying

---

[1] W. R. W. Stephens, *Life of Freeman*, p. 132, London, 1895.

deposits, there would be formed a hard rock made up of rounded fragments of various kinds of strata derived from different sources. Such a rock is known as a CONGLOMERATE. The same kind of rock may be formed equally well by the action of the sea; an old sea-beach with the pebbles embedded in a cementing matrix affords a typical example of a coarse conglomerate. Plant remains are occasionally met with in conglomerates, but usually in a fragmentary condition.

From a conglomerate composed of large water-worn pebbles, to a fine homogeneous sandstone there are numerous intermediate stages. A body of water, with a velocity too small to carry along pebbles of rock in suspension or to roll them along the bed of the channel, is still able to transport the finer fragments or grains of sand, but as a further decrease in the velocity occurs, these are eventually deposited as beds of coarse or fine sand. The stretches of sand on a gradually shelving sea shore, or the deposits of the same material in a river's delta, have been formed by the gradual wearing away and disintegration of various rocks, the detritus of which has been spread out in more or less regular beds on the floor of a lake or sea. Such accumulations of fine detrital material, if compacted or cemented together, become typical SANDSTONES.

In tracing beds of sandstone across a tract of country, it is frequently found that the character of the strata gradually alters; mud or clay becomes associated with the sandy deposit, until finally the sandstone is replaced by beds of dark coloured shale. Similarly the sandy detritus on the ocean floor, or in an inland lake, when followed further and further from the source from which the materials were derived, passes by degrees into argillaceous sand, and finally into sheets of dark clay or mud. The hardened beds of clay or fine grained mud become transformed into SHALES. As a general rule, then, shales are rocks which have been laid down in places further from the land, or at a greater distance from the source of origin of the detrital material, than sandstones or conglomerates. The conglomerates, or old shingle beaches, usually occur in somewhat irregular patches, marking old shore-lines or the head of a river delta. Coarse sandstones, or grits, may occur

in the form of regularly bedded strata stretching over a wide
area; and shales or clays may be followed through a considerable
extent of country. The finer material composing the clays
and shales has been held longer in suspension and deposited
in deeper water in widespread and fairly horizontal layers.
In some districts sandstones occur in which the individual
grains show a well marked rounding of the angles, and in
which fossils are extremely rare or entirely absent. The close
resemblance of such deposits to modern desert sands suggests
a similar method of formation; and there can be no doubt that
in some instances there have been preserved the wind-worn
desert sands of former ages. Aeolian or wind-formed accumulations, although by no means common, are of sufficient importance to be mentioned as illustrating a certain type of rock.

The thick masses of limestone which form so prominent
a feature in parts of England and Ireland, have been formed
in a manner different from that to which sandstones and
shales owe their origin. On the floor of a clear sea, too far
from land to receive any water-borne sediment, there is
usually in process of formation a mass of calcareous material,
which in a later age may rise above the surface of the water
as chalk or LIMESTONE. Those organisms living in the sea,
which are enclosed either wholly or in part by calcareous
shells, are agents of limestone-building; their shells constantly
accumulating on the floor of the sea give rise in course of
time to a thick mass of sediment, composed in great part of
carbonate of lime. Some of the shells in such a deposit may
retain their original form, the calcareous body may on the other
hand be broken up into minute fragments which are still recognisable with the help of a microscope, or the shells and other
hard parts may be dissolved or disintegrated beyond recognition,
leaving nothing in the calcareous sediment to indicate its
method of formation.

Not a few limestones consist in part of fossil corals, and
owe their origin to colonies of coral polyps which built up
reefs or banks of coral in the ancient seas.

In the white cliffs of Dover, Flamborough Head and other
places, we have a somewhat different form of calcareous rock,

which in part consists of millions of minute shells of Foraminifera, in part of broken fragments of larger shells of extinct molluscs, and to some extent of the remains of siliceous sponges. As a general rule, limestones and chalk rocks are ancient sediments, formed in clear and comparatively deep water, composed in the main of carbonate of lime, in some cases with a certain amount of carbonate of magnesium, and occasionally with a considerable admixture of silica.

In such rocks land-plants must necessarily be rare. There are, however, limestones which wholly or in part owe their formation to masses of calcareous algae, which grew in the form of submarine banks or on coral reefs. Occasionally the remains of these algae are clearly preserved, but frequently all signs of plant structure have been completely obliterated. Again, there occur limestone rocks formed by chemical means, and in a manner similar to that in which beds of travertine are now being accumulated.

Granites, basalts, volcanic lavas, tuffs, and other igneous rocks need not claim our attention, except in such cases as permit of plant remains being found in association with these materials. Showers of ashes blown from a volcano, may fall on the surface of a lake or sea and become mixed with sand and mud of subaerial origin. Streams of lava occasionally flow into water, or they may be poured from submarine vents, and so spread out on the ocean bed with strata of sand or clay.

Passing from the nature and mode of origin of the sedimentary strata to the manner of their arrangement in the Earth's crust, we must endeavour to sketch in the merest outline the methods of stratigraphical geology. The surface of the Earth in some places stands out in the form of bare masses of rock, roughly hewn or finely carved by Nature's tools of frost, rain or running water; in other places we have gently undulating ground with beds of rock exposed to view here and there, but for the most part covered with loose material such as gravel, sands, boulder clay and surface soil.

In the flat lands of the fen districts, the peat beds and low-lying salt marshes form the surface features, and are the connecting links between the rock-building now in progress and

the deposits of an earlier age.   If we could remove all these
surface accumulations of sand, gravel, peat and surface soil,
and take a bird's eye view of the bare surface of the rocky
skeleton of the earth's crust, we should have spread before
us the outlines of a geological map.   In some places fairly
horizontal beds of rock stretching over a wide extent of
country, in another the upturned edges of almost vertical
strata form the surface features; or, again, irregular bosses of
crystalline igneous rock occur here and there as patches in
the midst of bedded sedimentary or volcanic strata.   A map
showing the boundaries and distribution of the rocks as seen
at the surface, tells us comparatively little as to the relative
positions of the different rocks below ground, or of the relative
ages of the several strata.   If we supplement this superficial
view by an inspection of the position of the strata as shown on
the walls of a deep trench cut across the country, we at once
gain very important information as to the relative position of the
beds below the earth's surface.   The face of a quarry, the side
of a river bed or a railway cutting, afford HORIZONTAL SECTIONS
or PROFILES which show whether certain strata lie above or
below others, whether a series of rocks consists of parallel and
regularly stratified beds, or whether the succession of the strata
is interfered with by a greater or less divergence from a parallel
arrangement.   If, for example, a section shows comparatively
horizontal strata lying across the worn down edges of a series
of vertical sedimentary rocks, we may fairly assume that
some such changes as the following have taken place in that
particular area.

The underlying beds were originally laid down as more or
less horizontal deposits; these were gradually hardened and
compacted, then elevated above sea-level by a folding of the
earth's crust; the crests of the folds were afterwards worn down
by denudation, and the eroded surface finally subsided below
sea-level and formed the floor on which newer deposits were
built up.   Such breaks in the continuity of stratified deposits
are known as UNCONFORMITIES; in the interval of time which
they represent great changes took place of which the records
are either entirely lost, or have to be sought elsewhere.

In certain more exceptional cases, it is possible to obtain what is technically known as a VERTICAL SECTION; for example if a deep boring is sunk through a series of rocks, and the core of the boring examined, we have as it were a sample of the earth's crust which may often teach us valuable lessons which cannot be learnt from maps or horizontal sections.

It is obvious, that in a given series of beds, which are either horizontal or more or less obliquely inclined, the underlying strata were the first formed, and the upper beds were laid down afterwards. If, however, we trusted solely to the order of superposition in estimating relative age, our conclusions would sometimes be very far from the truth. Recent geological investigations have brought to light facts well nigh incredible as to the magnitude and extent of rock-foldings. In regions of great earth-movements, the crust has been broken along certain lines, and great masses of strata have been thrust for miles along the tops of newer rocks. Thus it may be brought about that the natural sequence of a set of beds has been entirely altered, and older rocks have come to overlie sediments of a later geological age. Facts such as these clearly illustrate the difficulties of correct geological interpretation.

In the horizontal section (Fig. 2), from the summit of Büzi-stock on the left to Saasterg on the right, we have a striking case of intense rock-folding and dislocation[1]. Prof. Heim[2] of Geneva has given numerous illustrations of the almost incredible positions assumed in the Swiss Mountains by vast thicknesses of rocks, and in the accompanying section taken from a recent work by Rothpletz we have a compact example of the possibilities of earth-movements as an agent of rock-folding. The section illustrates very clearly an exception to the rule that the order of superposition of a set of beds indicates the relative age of the strata. The horizontal line at the base is drawn at a height of 1650 metres above sea-level, and the summit of Büzistock reaches a height of 2340 m. The youngest rocks seen in the diagram are the Eocene beds e, at the base and as small isolated patches on the right-hand end of the section;

---

[1] Rothpletz (94).   [2] Heim (78).

the main mass of material composing the higher ground has
been bodily thrust over the Eocene rocks, and in this process
some of the beds, *b* and *c*, have been folded repeatedly on them-

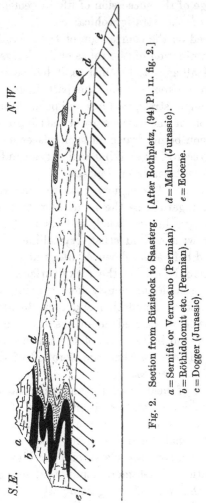

Fig. 2. Section from Büzistock to Saasterg. [After Rothpletz, (94) Pl. II. fig. 2.]

a = Sernift or Verrucano (Permian).
b = Röthidolomit etc. (Permian).
c = Dogger (Jurassic).
d = Malm (Jurassic).
e = Eocene.

selves. Similar instances of the overthrusting of a considerable
thickness of strata have been described in the North-west
Highlands of Scotland[1] and elsewhere in the British Isles. It
is important therefore to draw attention to cases of extreme

Geikie (93), p. 706.

folding, as such phenomena are by no means exceptional in many parts of the world.

The order of superposition of strata has afforded the key to our knowledge of the succession of life in geologic time, and the refinements of the stratigraphical correlation of sedimentary rocks are based on the comparison of their fossil contents. By a careful examination of the relics of fossil organisms obtained from rocks of all ages and countries, it has been found possible to restore in broken outline the past history of the Earth. By means, then, of stratigraphical and palaeontological evidence, a classification of the various rocks has been established, the lines of division being drawn in such places as represent gaps in the fossil records, or striking and widespread unconformities between different series of deposits.

It is only in a few regions that we find rocks which can reasonably be regarded as the foundation stones of the Earth. As the globe gradually cooled, and its molten mass became skinned over with a solid crust, crystalline rocks must have been produced before the dawn of life, and before water could remain in a liquid form on the rocky surface. As soon as the temperature became sufficiently low, running water and rain began the work of denudation and rock disintegration which has been ceaselessly carried on ever since. In this continual breaking down and building up of the Earth's surface, it would be no wonder if but few remnants were left of the first formed sediments of the earliest age.

The action of heat, pressure and chemical change accompanying rock-foldings and crust-wrinklings, often so far alters sedimentary deposits, that their original form is entirely lost, and sandstone, shales and limestones become metamorphosed into crystalline quartzites, slates and marbles.

The operation of metamorphism is therefore another serious difficulty in the way of recognising the oldest rocks. The earliest animals and plants which have been discovered are not such as we should expect to find as examples of the first products of organic life. Below the oldest known fossiliferous rocks, there must have been thousands of feet of sedimentary material, which has either been altered beyond recognition, or

from some cause or other does not form part of our present
geological record.

As a general introduction to geological chronology, a short
summary may be given of the different formations or groups of
strata, to which certain names have been assigned to serve as
convenient designations for succeeding epochs in the world's
evolution. The following table (Fig. 3, pp. 32, 33) represents
the geological series in a convenient form ; the most characteristic rocks of each period are indicated by the usual conventional
shading, and the most important breaks or lacunae in the records
are shown by gaps and uneven lines. The relative thickness
of the rocks of each period is approximately shown; but the
vertical extent of the oldest or Archaean rocks as shown in
Fig. 3 represents what is without doubt but a fraction of
their proportional thickness. This table is taken, with certain
alterations, from a paper by Prof. T. McKenny Hughes in the
Cambridge Philosophical Proceedings for 1879. Speaking
of the graphic method of showing the geological series,
the author of the paper says, " It is convenient to have a
table of the known strata, and although we cannot arrange all
the rocks of the world in parallel columns, and say that *ABC*
of one area are exactly synchronous with *A'B'C'* of another,
still if we take any one country and establish a grouping for it,
we find so many horizons at which equivalent formations can
be identified in distant places that we generally make an
approximation to HOMOTAXIS as Huxley called it. The most
convenient grouping is obviously to bracket together locally
continuous deposits, *i.e.* all the sediment which was formed
from the time when the land went down and accumulation
began, to the time when the sea bottom was raised and the
work of destruction began. In the accompanying table I have
given the rocks of Great Britain classified on this system, and
bearing in mind that waste in one place must be represented
by deposit elsewhere, I have represented the periods of degra-
dation by intervals estimated where possible by the amount of
denudation known to have taken place between the periods of
deposition in the same district[1]."

[1] Hughes (79), p. 248.

TABLE OF STRATA.

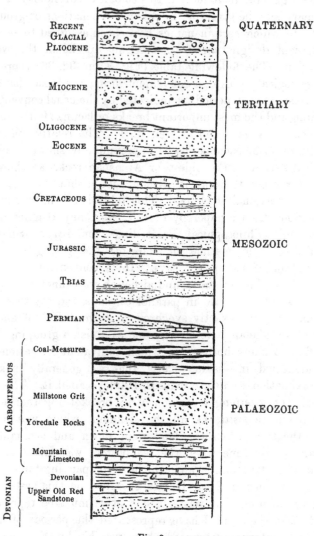

Fig. 3.

TABLE OF STRATA (*continued*).

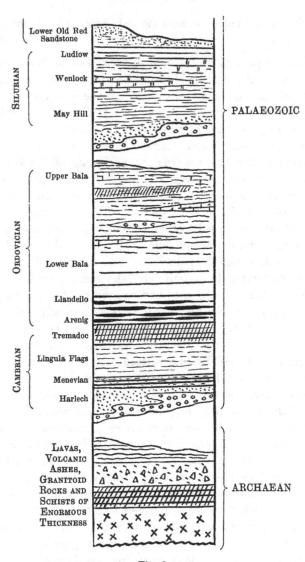

Fig. 3.

I. Archaean.

"Men can do nothing without the make-believe of a beginning."
GEORGE ELIOT.

There is perhaps no problem at once so difficult and so full
of interest to the student of the Earth's history, as the interpretation of the fragmentary records of the opening stages in
geological and organic evolution. In tracing the growth and
development of the human race, it becomes increasingly
difficult to discover and decipher written documents as we
penetrate farther back towards the beginning of the historical
period; the records are usually incomplete and fragmentary, or
rendered illegible by the superposed writings of a later date. So
in the records of the rocks, as we pass beyond the oldest strata in
which clearly preserved fossils are met with, we come to older
rocks which afford either no data as to the period in which they
were formed, or like the palimpsest, with its original characters
almost obliterated by a late MS. the older portions of the
Earth's crust have been used and re-used in the rock-building
of later ages. In the first place, it is exceedingly difficult to
determine with any certainty what rocks may be regarded as
trustworthy fragments of a primaeval land. Throughout the
geological eras the Earth's surface has been subjected to
foldings and wrinklings, volcanic activity has been almost
unceasing, and there is abundant evidence to show how the
original characters of both igneous and sedimentary rocks may
be entirely effaced by the operation of chemical and physical
forces. It was formerly held that coarsely crystalline rocks
such as granite are the oldest portions of the crust, but
modern geology has conclusively proved that many of the
so-called fundamental masses of rock are merely piles of ancient
sediments which have been subjected to the repeated operation
of powerful physical and chemical forces, and have undergone a
complete rearrangement of their substance. As the result of
more detailed investigations, many regions formerly supposed
to consist of the foundation stones of the Earth's crust, are
now known to have been centres of volcanic disturbance and

wide-spread metamorphism, and to be made up of post-archaean rocks.

The first formed rocks no doubt became at once the prey of denudation and disintegration, and on their surface would be accumulated the products of their own destruction; newer strata would entirely cover up portions of the original land, to be in their turn succeeded by still later deposits. There is reason to believe that in the remotest ages of the Earth's history, the forces of denudation and igneous activity were more potent than in later times, and thus the oldest rocks could hardly retain their original structure through the long ages of geologic time. The earliest representatives of organic life were doubtless of such a perishable nature that their remains could not be preserved in a fossil state even under the most favourable conditions. Such organisms, whether plants or animals, as possessed any resistant tissues or hard skeletons might be preserved in the oldest rocks, but as these strata became involved in earth-foldings or were penetrated by injections of igneous eruptions, the relics of life would be entirely destroyed. It is, in short, practically hopeless to look for any fragments of the primitive crust except such as have undergone very considerable metamorphism, and equally futile to search for any recognisable remains of primitive life.

In many parts of the world vast thicknesses of rock occur below the oldest known fossiliferous strata; these consist largely of laminated crystalline masses composed of quartz, felspar, and other minerals, having in fact the same composition as granite, but differing in the regular arrangement of the constituent parts. To such rocks the terms gneiss and schist have been applied. Rocks of this kind are by no means always of Archaean age, but many of the earliest known rocks consist of gneisses of various kinds, associated with altered lavas, metamorphosed ashes, breccias and other products of volcanic activity; with these there may be limestones, shales, sandstones, and other strata more or less closely resembling sedimentary deposits. Such a succession of gneissic rocks has been described as occupying a wide area in the basin of the St Lawrence river, and to these enormously thick and widespread masses a late

Director of the Canadian Geological Survey applied the term Laurentian. These Laurentian rocks, with similar strata in Scandinavia, the north-west Highlands of Scotland, in certain parts of such mountain ranges as the Alps, Pyrenees, Carpathians, Himalayas, Andes, Atlas, &c., have been classed together as members of the oldest geological period, and are usually referred to under the name of Archaean, or less frequently Azoic rocks. In some of the uppermost Archaean rocks there have been recently discovered a few undoubted traces of fossil animals, but with this exception no fossils are known throughout the great mass of Archaean strata. It is true that some authorities regard the beds of graphite and other rocks as a proof of the abundance of plant life, but this supposition is not supported by any convincing evidence.

The term Azoic[1] applied by some writers to these oldest rocks suggests the absence of life during the period in which they were formed. Life there must have been, though we are unable to discover its records. The period of time represented by the Archaean or Pre-Cambrian rocks must be enormous, and it was in that earliest era that the first links in the chain of life were forged.

II. Cambrian.

The term Cambrian was adopted by Sedgwick for a series of sedimentary rocks in North Wales (*Cambria*). In that district, in South Wales, the Longmynd Hills, the Malverns, in Scotland, and other regions there occur more or less highly folded and contorted beds of pebbly conglomerate, sandstones, shales and slates resting on the uneven surface of an Archaean foundation.

It is in these Cambrian rocks that trustworthy records of organic life are first met with. Among the most constant and characteristic fossils of this period are the extinct and aberrant members of the crustacea, the trilobites; these with some brachiopods, sponges, and other fossils comprise the

---

[1] Whitney and Wadsworth (84).

oldest fauna, of which the ancestral types have yet to be discovered. During the last few decades the number of Cambrian fossils has been considerably increased, and in certain regions of North America and China there are found many thousand feet of strata above the typical Archaean rocks and below the newer fossiliferous beds of Cambrian age. It is reasonable to suppose that future research may extend the present limits of fossil-bearing rocks below the horizon, which is marked by the occurrence of the widely distributed and oldest known trilobite, the genus *Olenellus*.

The vast thickness of Cambrian strata was for the most part laid down on the floor of a comparatively deep sea; other members of the series represent the shingle beaches and coast deposits accumulated on the slopes of Archaean islands. There have been many conjectures as to the distribution of land and sea during the deposition of these rocks; but the data are too imperfect to enable us to restore with any degree of confidence the physical geography of this Palaeozoic epoch, of which the sediments stood out as islands of Cambrian land during many succeeding ages.

III. Ordovician.

Since the days when Sedgwick and Murchison first worked out the succession of Palaeozoic strata in North Wales, there has always existed a considerable difference of opinion as to the best method of subdividing the Cambrian-Silurian strata. Later research has shown that the rocks included by Sedgwick in his Cambrian system, fall naturally into two groups; for the upper of these Prof. Lapworth has suggested the term Ordovician, from the name of the Ordovices, who inhabited a part of northern Wales. At the base of the system we have a series of volcanic and sedimentary rocks to which Sedgwick gave the name Arenig; above these there occur the Llandeilo Flags, succeeded by a considerable thickness of rocks known as the Bala series. The rocks making up these Ordovician sediments consist for the most part of slates, sandstones and limestones with

volcanic ashes and lavas. Much of the typical Welsh scenery
owes its character to the folded and weathered rocks laid down
on the floor of the Ordovician sea, on which from many centres
of volcanic activity lava streams and showers of ash were
spread out between sheets of marine sediment. The Arenig
Hills, Snowdonia, and many other parts of North and South
Wales, parts of Shropshire, Scotland, Sweden, Russia, Bohemia,
North America and other regions consist of great thicknesses
of Ordovician strata.

IV. Silurian.

Passing up a stage higher in the geologic series, we
have a succession of conglomerates, sandstones, shales, and
limestones; in other words, a series of beds which represent
pebbly shore deposits, the sands and muds of deeper water,
and the accumulated débris of calcareous skeletons of animals
which lived in the clear water of the Silurian sea. The term
Silurian (Siluria was the country of Caractacus and the old
Britons known as Silures[1]) was first applied by Murchison in
1835 to a more comprehensive series of rocks than are now
included in the Silurian system. The rocks of this period occur
in Wales, Shropshire, parts of Scotland, Ireland, Scandinavia,
Russia, the United States and other countries. After the
accumulation of the thick Ordovician sediments, the sea-floor
was upraised and in places converted into ridges or islands of
land, of which the detritus formed part of the material of
Silurian deposits. The limestones of the Wenlock ridge have
yielded an abundant fauna, consisting of corals, crinoids,
molluscs and other invertebrates. In this period we have
the first representatives of the Vertebrata, discovered in the
rocks of Ludlow. In fact, in the Silurian period, "all the
great divisions of the Animal Kingdom were already represented[2]."

[1] Murchison (72), p. 5.   [2] Kayser and Lake (95), p. 88.

V. Devonian.

By the continued elevation of the Silurian sea-floor, large portions became dry land, and during the succeeding period most of the British area formed part of a continental mass. Over the southern part of England, there still lay an arm of the sea, and in this were laid down the marine sediments which now form part of Devon, and from which the name Devonian has been taken as a convenient designation for the strata of this period. In parts of the northern land, in the region now occupied by Scotland, there were large inland lakes, on the floor of which vast thicknesses of shingle beds and coarse sands ("Old Red Sandstone") were slowly accumulated; and it has been shown by Sir Archibald Geikie and others that during this epoch there were considerable outpourings of volcanic material in the Scotch area.

Farther to the West and South-west there was another large lake in which the so-called Kiltorkan beds of Ireland were deposited. In these Irish sediments, and others of the same age in Belgium and elsewhere a few forms of land plants have been discovered; but it is from the Devonian rocks of North America that most of our knowledge of the flora of this period has been obtained.

VI. Carboniferous.

From the point of view of palaeobotany, the shales, sandstones, and seams of coal included in the Carboniferous system are of special interest. It is from the relics of this Palaeozoic vegetation that the most important botanical lessons have been learnt.

The following classification of Carboniferous rocks shows the order of succession of the various beds, and the nature of the rocks which were formed at this stage in the Earth's history.

CARBONIFEROUS

Coal-Measures [1]
{
Upper Coal-Measures
Transition Series.
Middle Coal-Measures.
Lower Coal-Measures.
}

Millstone Grit.

Carboniferous limestone series
{
Upper limestone shales and
    Yoredale rocks.
Carboniferous or Mountain
    limestone.
Lower limestone shales.
}

Basement conglomerate.

In the classification of Carboniferous rocks adopted in Geikie's text-book of Geology the following arrangement is followed for the Carboniferous limestone series [2] :—

Carboniferous limestone series
{
Yoredale group of shales and grits passing down into dark shales and limestones.

Thick (Scaur or Main) limestone in the south and centre of England and Ireland, passing northwards into sandstones, shales and coals with limestones.

Lower limestone shale of the south and centre of England. The Calciferous sandstone group of Scotland (marine, estuarine, and terrestrial organisms) probably represents the Scaur limestone and lower limestone shale, and graduates downwards insensibly into the Upper Old Red Sandstone.
}

The thick beds of mountain limestone, with their characteristic marine fossil shells and corals play an important part in English scenery. In Derbyshire, West Yorkshire, and other places, the limestone crags and hills are made up of the raised floor of a comparatively deep Carboniferous sea, which covered a considerable portion of the British Isles at the beginning of this epoch.

The accumulation of the calcareous skeletons of marine animals, with masses of coral, veritable shell-banks of extinct oyster-like lamellibranchs, built up during the lapse of a long period of time, formed widespread deposits of calcareous

[1] Kidston (94).   [2] Geikie (93), p. 825.

sediments. These were eventually succeeded by less pure calcareous deposits, the sea became shallower, and land detritus found its way over an area formerly occupied by the clear waters of an open sea. The shallowing process was gradually continued, and the sea was by some means converted into a more confined fresh-water or brackish area, in which were laid down many hundred feet of coarse sandy sediments derived from the waste of granitic highlands. Finally the conditions became less constant; the continuous deposition of sandy detritus being interrupted by the more or less complete filling up of the area of sedimentation, and the formation of a land surface which supported a luxuriant vegetation, of which the débris was subsequently converted into beds of coal. By further subsidence the land was again submerged, and the forest-covered area became overspread with sands and muds.

Such are the imperfect outlines of the general physical conditions which are represented by the series of sedimentary strata included in the Carboniferous system. At the close of this period, the Earth's surface in Western Europe was subjected to crust-foldings on a large scale, along lines running approximately North and South and East and West, the two sets of movements resulting in the formation of ridges of Carboniferous rocks. The uppermost series of grits, sandstones and coal-seams were in great part removed by denudation from the crests of the elevated ridges, but remained in the intervening troughs or basins where they were less exposed to denudation. It is the direct consequence of this, that we have our Coal-Measures preserved in the form of detached basins of upper Carboniferous beds.

A closer examination of the comparative thickness and succession of Carboniferous rocks in different parts of Britain shows very clearly that in the northern area of Scotland and in the North of England the conditions were different from those which obtained further South. Seeing how much palaeo-botanical interest attaches to these rocks, it is important to treat a little more fully of their geology.

In parts of Devon, Cornwall and West Somerset, the Devonian strata are succeeded by a series of folded and contorted

rocks which have yielded a comparatively small number of
Carboniferous fossils. To this succession of limestones, shales
and grits the term *Culm-Measures* was applied by Sedgwick
and Murchison in 1837. The rocks of this series occupy a
trough between the Devonian rocks of North and South
Devon. While some authorities have correlated the Culm-
Measures with the Millstone Grit, others regard them as representing a portion of the true Coal-Measures, as well as the
Carboniferous and Lower Limestone Shale[1]. It has recently
been shown that among the lower Culm strata there occur
bands of ancient deep-sea sediments, consisting of beds of
chert containing siliceous casts of various species of Radiolaria.
There can be no doubt that the discovery of deep-sea fossils
in this particular development of the British Carboniferous
system leads to the conclusion that "while the massive
deposits of the Carboniferous limestone—formed of the skeletons of calcareous organisms—were in process of growth in
the seas to the North, there existed to the South-west a
deeper ocean in which siliceous organisms predominated and
formed these siliceous radiolarian rocks[2]."

The Upper Culm-Measures consist of conglomerates, grits,
sandstones and shales with some plant remains and other
fossils, and constitute a typical set of shallow water sediments.
In Westphalia, the Harz region, Thuringia, Silesia and Moravia
there are rocks corresponding to the Culm-Measures of Devon,
and some of these have also afforded evidence of deep water
conditions.

S. W. England, S. Wales, Derbyshire and Yorkshire. In
these districts the Carboniferous limestone reaches a considerable thickness; in the Mendips it has a thickness of
3000 feet, and in the Pennine chain of 4000 feet. At the base
of this limestone series there occurs in the southern districts
the so-called lower limestone shale, consisting of clays, shales
and sandy beds. Above the limestone we have the Millstone
grit and Coal-Measures; but in the Pennine district there is
a series of rocks consisting of impure limestones and shales,

[1] Woodward, H. B. (87), p. 197.
[2] Hinde and Fox (95), p. 662

intercalated between the Millstone grit and Carboniferous limestone; for this group of rocks the term *Yoredale series* has been proposed. In the Isle of Man and Derbyshire sheets of lava are interbedded with the calcareous sediments, affording clear proof of submarine volcanic eruptions.

N. England and Scotland. In the Carboniferous rocks of Northumberland we have distinct indications of a shallower sea. The regular succession of limestone strata in West Yorkshire and other districts, gives place to a series of thinner beds of limestones, interstratified with shales and impure calcareous rocks. We have come within the range of land detritus which was spread out on the floor of a shallow sea. The lowest portion of the Mountain limestone is here represented by about 200 feet of shales and other rocks grouped together in the *Tuedian series.* The Upper Carboniferous limestone and Yoredale rocks of Yorkshire are represented by sandstones, carbonaceous limestones and some seams of coal, included in the *Bernician series.* Further north, again, another classification has been proposed for the still more aberrant succession of rocks; the lowest being spoken of as the *Calciferous sandstone,* and the upper as the *Carboniferous limestone.* The calciferous sandstone may be compared with the lower limestone shale and part of the Carboniferous limestone of England. The Carboniferous limestone of Scotland probably represents the upper part of the limestone of England and the Yoredale rocks of the Pennine and other areas.

Turning to the upper members of the Carboniferous system—in the Coal-Measures, as they were called in 1817 by William Smith,—we have a series of coal seams, sandstones, shales, and ironstones occurring for the most part in basin-shaped areas. As a general rule, each seam of coal, which varies in thickness from one inch to thirty feet, rests on a characteristic unstratified argillaceous rock known as Underclay.

The accompanying diagram (Fig. 4) illustrates the frequent intercalation of small bands of argillaceous and sandy rocks associated with the seams of coal.

The usual classification adopted for the British Coal-Measures is that of Upper, Middle, and Lower Coal-Measures; between the Upper and Middle divisions there occur certain transition or passage beds which are known as the Transition series. Continental writers, and more recently Mr Kidston of Stirling, have attempted with considerable success to correlate the Coal-producing strata by means of fossil plants[1].

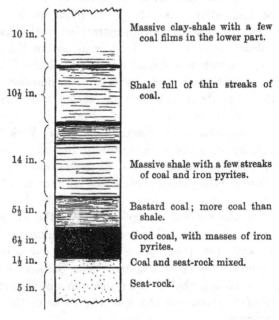

10 in.	Massive clay-shale with a few coal films in the lower part.
10½ in.	Shale full of thin streaks of coal.
14 in.	Massive shale with a few streaks of coal and iron pyrites.
5½ in.	Bastard coal; more coal than shale.
6½ in.	Good coal, with masses of iron pyrites.
1½ in.	Coal and seat-rock mixed.
5 in.	Seat-rock.

Fig. 4.

Vertical section of the Bassey or Salts Coal seam, Rushton Colliery, Blackburn (Lower Coal-Measures). From a specimen 4 feet 4 inches in height, presented by Mr P. W. Pickup to the Manchester Museum, Owens College.

Finally, some reference must be made to the occurrence of Carboniferous rocks underneath more recent strata. In a geological map, or bird's-eye view of a country, we see such rocks as appear at the surface; by means of deep borings, however, we are occasionally enabled to follow the course of older beds a considerable distance below the usually accessible

[1] Kidston (94).

part of the Earth's crust. In the neighbourhood of London, Dover, and other places we have Tertiary and Mesozoic strata forming the surface of the country, but below these comparatively recent formations, the sinking of deep wells and other borings have proved the existence of a ridge of Palaeozoic rocks. stretching from the South Wales Coal-field through the South-east of England to northern France, Belgium and Westphalia. It is from rocks forming part of this old ridge that characteristic Coal-Measure plants have been obtained from the Dover boring. In Fig. 5 is shown an almost complete pinnule of *Neuropteris Scheuchzeri* Hoffm., a well-known fern, marking a definite horizon of Upper Carboniferous rocks[1]. The small hairs on the pinnules, shown in the figure as fine lines lying more or less parallel to the midrib and across the lateral veins, are a characteristic feature of this species.

Fig. 5.

Imperfect pinnule of *Neuropteris Scheuchzeri* Hoffm., showing the character-istic hairs as fine lines traversing the lateral veins. From a specimen obtained from the Dover boring and now in the British Museum. Nat. size.

VII. Permian.

Reference has already been made to the earth-foldings. which marked the close of Carboniferous times; "the open Mediterranean sea of the Carboniferous period in Europe was converted into a large inland sea, like the Caspian of the present day, surrounded by a rocky and hilly continent, on which grew trees and plants of various kinds[2]." In parts of

[1] *Vide* Zeiller (92) for a list of species of Coal-Measure plants found in the pieces of shale included in the core brought up by the borer.

[2] Jukes-Browne (86), p. 252.

Lancashire, Westmoreland, the Eden Valley, and in the East of England from Sunderland to Nottingham, there occurs a succession of limestones, sandstones, clays and other rocks with occasional beds of rock-salt and gypsum, which represent the various forms of sediment and chemical precipitates formed on the floor of Permian lakes. The poverty of the fauna and flora of Permian strata points to conditions unfavourable to life; and there can be little doubt that the characteristic red rocks of St Bees Head, and the creamy limestones of the Durham coast are the upraised sediments of an inland salt-water lake. The term Dyas was proposed by Marcou for this series of strata as represented in Germany, where the rocks are conveniently grouped in two series, the *Magnesian limestone* or *Zechstein* and the red sandstones or *Rotheliegendes*. The older and better known name of Permian was instituted by Murchison for the rocks of this age, from their extreme development in the old kingdom of Permia in Russia. Unfortunately considerable confusion has arisen from the employment of different names for rocks of the same geological period; and the grouping of the beds varies in different parts of the world. It is of interest to note, that in the Tyrol, Carinthia, and other places there are found patches of old marine beds which were originally laid down in an open sea, which extended over the site of the Mediterranean, into Russia and Asia. In Bohemia, the Harz district, Autun in Burgundy, and other regions, there are seams of Permian coal interstratified with the marls and sands. From these last named beds many fossil plants have been obtained, and important palaeobotanical facts brought to light by the investigations of continental workers. Volcanic eruptions, accompanied by lava streams and showers of ash, have been recognised in the Permian rocks of Scotland, and elsewhere.

In North America, Australia, and India the term Permo-Carboniferous is often made use of in reference to the continuous and regular sequence of beds which were formed towards the close of the Carboniferous and into the succeeding Permian epoch. The enormous series of freshwater Indian rocks, to which geologists have given the name of the GONDWANA

SYSTEM, includes the sediments of more than one geological period, some of the older members being regarded as Permo-Carboniferous in age. These Indian beds, with others in Australia, South Africa, and South America, are of special interest on account of the characteristic southern hemisphere plants which they have afforded, and from the association with the fossiliferous strata of extensive boulder beds pointing to widespread glacial conditions.

VIII. Trias.

As we ascend the geologic series, and pass up to the rocks overlying the Permian deposits, there are found many indications of a marked change in the records of animal and plant life. Many of the characteristic Palaeozoic fossils are no longer represented, and in their place we meet with fresh and in many cases more highly differentiated organisms. The threefold division of the rocks of this period which suggested the term Trias to those who first worked out the succession of the strata, is typically illustrated over a wide area in Germany, in which the lowest or *Bunter* series is followed by the calcareous *Muschelkalk*, and this again by the clays, rock-salt, and sandstones of the *Keuper* series. In the Cheshire plain and in the low ground of the Midlands, we have a succession of red sandstones, conglomerates, and layers of rock-salt which correspond to the Bunter and Keuper beds of German geologists. These Triassic rocks were obviously formed in salt-water lakes, in which from time to time long continued evaporation gave rise to extensive deposit of rock-salt and other minerals. From the fact that it is this type of Triassic sediments which was first made known, it is often forgotten that the British and German rocks are not the typical representatives of this geological period. The 'Alpine' Trias of the Mediterranean region, in Asia, North America, and other countries, has a totally different facies, and includes limestones and dolomites of deep-sea origin. "The widespread Alpine Trias is the pelagic facies of the

formation; the more restricted German Trias, on the other
hand, is a shallow shore, bay or inland sea formation[1]."

In the Keuper beds of southern Sweden there are found
workable seams of coal, and the beds of this district have
yielded numerous well-preserved examples of the Triassic flora.
A more impure coal occurs in the lower Keuper of Thuringia
and S.-W. Germany, and to this group of rocks the term
Lettenkohle is occasionally applied.

In the Rhaetic Alps of Lombardy, in the Tyrol, and in
England, from Yorkshire to Lyme Regis, Devonshire, Somer-
setshire, and other districts there are certain strata at the top
of the Triassic system known as the *Rhaetic* or *Penarth* beds.
The uppermost Rhaetic beds, often described as the White
Lias, afford evidence of a change from the salt lakes of the
Trias to the open sea of the succeeding Jurassic period.
Passing beyond this period of salt lakes and wind-swept barren
tracts of land, we enter on another phase of the earth's history.

IX. Jurassic.

The Jura mountains of western Switzerland consist in
great part of folded and contorted rocks which were originally
deposited on the floor of a Jurassic sea. In England the
Jurassic rocks are of special interest, both for geological and
historical reasons, as it is in them that we find a rich fauna and
flora of Mesozoic age, and it was the classification of these beds
by means of their fossil contents that gained for William Smith
the title of the Father of English Geology. A glance at a
geological map of England shows a band of Jurassic rocks
stretching across from the Yorkshire coast to Dorset. These are
in a large measure calcareous, argillaceous, and arenaceous
sediments of an open sea; but towards the upper limit of the
series, both freshwater and terrestrial beds are met with. Numerous fragments of old coral reefs, sea-urchins, crinoids, and
other marine fossils are especially abundant; in the freshwater
beds and old surface-soils, as well as in the marine sandstones

[1] Kayser and Lake (95), p. 196.

and shales, we have remnants of an exceedingly rich and apparently tropical vegetation. This was an age of Reptiles as well as an age of Cycads. An interesting feature of these widely distributed Jurassic strata is the evidence they afford of distinct climatal zones; there are clear indications, according to the late Dr Neumayr, of a Mediterranean, a middle European, and a Boreal or Russian province[1]. The subdivisions of the English Jurassic rocks are as follows[2]:—

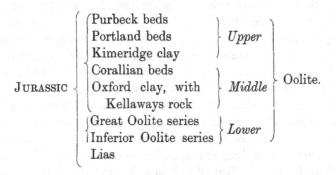

In tracing the several groups across England, and into other parts of Europe, their characters are naturally found to vary considerably; in one area a series is made up of typical clear water or comparatively deep sea sediments, and in another we have shallow water and shore deposits of the same age. The Lias rocks have been further subdivided into zones by means of the species of Ammonites which form so characteristic a feature of the Jurassic fauna. In the lower Oolite strata there are shelly limestones, clays, sandstones, and beds of lignite and ironstone. Without discussing the other subdivisions of the Jurassic period, we may note that in the uppermost members there are preserved patches of old surface-soils exposed in the face of the cliffs of the Dorset coast and of the Isle of Portland.

[1] Neumayr (83).   [2] Woodward, H. B. (87), p. 255.

X. Cretaceous.

In the south of England, and in some other districts, it is difficult to draw any definite line between the uppermost strata of the Jurassic and the lowest of the Cretaceous period. The rocks of the so-called *Wealden* series of Kent, Surrey, Sussex, and the Isle of Wight, are usually classed as Lower Cretaceous, but there is strong evidence in favour of regarding them as sediments of the Jurassic period. The Cretaceous rocks of England are generally speaking parallel to the Jurassic strata, and occupy a stretch of country from the east of Yorkshire and the Norfolk coast to Dorset in the south-west. The Chalk downs and cliffs represent the most familiar type of Cretaceous strata. In the white chalk with its numerous flints, we have part of the elevated floor of a comparatively deep sea, which extended in Cretaceous times over a large portion of the east and south-east of England and other portions of the European continent. On the bed of this sea, beyond the reach of any river-borne detritus, there accumulated through long ages the calcareous and siliceous remains of marine animals, to be afterwards converted into chalk and flints. At the beginning of the period, however, other conditions obtained, and there extended over the south-east of England, and parts of north and north-west Germany and Belgium, a lake or estuary in which were built up deposits of clay, sand and other material, forming the delta of one or more large rivers. For these sediments the name *Wealden* was suggested in 1828. Eventually the gradual subsidence of this area led to an incursion of the sea, and the delta became overflowed by the waters of a large Cretaceous sea. At first the sea was shallow, and in it were laid down coarse sands and other sediments known as the *Lower Greensand* rocks. By degrees, as the subsidence continued, the shallows became deep water, and calcareous material slowly accumulated, to be at last upraised as beds of white chalk. The distribution of fossils in the Cretaceous rocks of north and south Europe distinctly points to the existence of two fairly well-marked sets of organisms in the two regions; no doubt the expression of climatal zones similar to those recognised in Jurassic times. In North America, Cretaceous rocks are spread over a wide area, also in North Africa, India, South Africa, and other parts of the world. Within the Arctic Circle strata of this age have become famous, chiefly on account of the rich flora described from them by the Swiss palaeobotanist Heer. The fauna and flora of this epoch are alike in their advanced state of development and in the great variety of specific types; the highest class of plants is first met with at the base of the Cretaceous system.

XI. Tertiary.

"At the close of the Chalk age a change took place both in the distribution of land and water, and also in the development of organic life, so great and universal, that it has scarcely been equalled at any other period of the earth's geological history[1]." The Tertiary period seems to bring us suddenly to the threshold of our own times. In England at least, the deposits of this age are of the nature of loose sands, clays and other materials containing shells, bones, and fossil plants bearing a close resemblance to organisms of the present era. The chalk rocks, upheaved from the Cretaceous sea, stood out as dry land over a large part of Britain; much of their material was in time removed by the action of denuding agents, and the rest gradually sank again beneath the waters of Tertiary lakes and estuaries. In the south of England, and in north Europe generally, the Tertiary rocks have suffered but little disturbance or folding, but in southern Europe and other parts of the world, the Tertiary sands have been compacted and hardened into sandstones, and involved in the gigantic crust-movements which gave birth to many of our highest mountain chains. The Alps, Carpathians, Apennines, Himalayas, and other ranges consist to a large extent of piled up and strangely folded layers of old Tertiary sediments. The volcanic activity of this age was responsible for the basaltic lavas of the Giants' Causeway, the Isle of Staffa, and other parts of western Scotland.

[1] Kayser and Lake (95), p. 326.

During the succeeding phases of this period, the distribution of land and sea was continually changing, climatic conditions varied within wide limits; and in short wherever Tertiary fossiliferous beds occur, we find distinct evidence of an age characterised by striking activity both as regards the action of dynamical as well as of organic forces. Sir Charles Lyell proposed a subdivision of the strata of this period into Eocene, Miocene, and Pliocene, founding his classification on the percentage of recent species of molluscs contained in the various sets of rocks. His divisions have been generally adopted. In 1854 Prof. Beyrich proposed to include another subdivision in the Tertiary system, and to this he gave the name *Oligocene.*

Occupying a basin-shaped area around London and Paris there are beds of Eocene sands and clays which were originally deposited as continuous sheets of sediment in water at first salt, afterwards brackish and to a certain extent fresh. In the Hampshire cliffs and in some parts of the Isle of Wight, we have other patches of these oldest Tertiary sediments. Across the south of Europe, North Africa, Arabia, Persia, the Himalayas, to Java and the Philippine islands, there existed in early Tertiary times a wide sea connecting the Atlantic and Pacific oceans; and it may be that in the Mediterranean of to-day we have a remnant of this large Eocene ocean. Later in the Tertiary period a similar series of beds was deposited which we now refer to as the Oligocene strata; such occurs in the cliffs of Headon hill in the Isle of Wight, containing bones of crocodiles, and turtles, with the relics of a rich flora preserved in the delta deposits of an Oligocene river. At a still later stage the British area was probably dry land, and an open sea existed over the Mediterranean region. In the neighbourhood of Vienna we have beds of this age represented by a succession of sediments, at first marine and afterwards freshwater. Miocene beds occur over a considerable area in Switzerland and the Arctic regions, and they have yielded a rich harvest to palaeobotanical investigators.

On the coast of Essex, Suffolk, Norfolk, the south of Cornwall, and other districts there occur beds of shelly sand

and gravel long known under the name of 'Crag.' The beds
have a very modern aspect; the sands have not been converted
into sandstones, and the shells have undergone but little change.
These materials were for the most part accumulated on the
bed of a shallow sea which swept over a portion of East Anglia
in Pliocene times. In the sediments of this age northern forms
of shells and other organisms make their appearance, and in
the Cromer forest-bed there occur portions of drifted trees with
sands, clays and gravels, representing in all probability the
débris thrown down on the banks of an ancient river. At this
time the greater part of the North Sea was probably a low-
lying forest-covered region, through which flowed the waters of
a large river, of which part still exists in the modern Rhine.
The lowering of temperature which became distinctly pro-
nounced in the Pliocene age, continued until the greater part
of Britain and north Europe experienced a glacial period, and
such conditions obtained as we find to-day in ice-covered
Greenland. Finally the ice-sheet melted, the local glaciers of
North Wales, the English Lake district and other hilly regions,
retreated, and after repeated alterations in level, the land of
Great Britain assumed its modern form. The submerged
forests and peat beds familiar in many parts of the coast, the
diatomaceous deposits of dried up lakes, "remain as the very
finger touches of the last geological change."

The agents of change and geological evolution, which we
have passed in brief review, are still constantly at work carrying
one step further the history of the earth. A superficial review
of geological history gives us an impression of recurring and
wide-spread convulsions, and rapidly effected revolutions in
organic life and geographical conditions; on the other hand a
closer comparison of the past and present, with due allowance
for the enormous period of time represented by the records
of the rocks, helps us to realise the continuity of geological
evolution. "So that within the whole of the immense period
indicated by the fossiliferous stratified rocks, there is assuredly
not the slightest proof of any break in the uniformity of
Nature's operations, no indication that events have followed
other than a clear and orderly sequence[1]."

[1] Huxley (93), p. 27.

CHAPTER IV.

THE PRESERVATION OF PLANTS AS FOSSILS.

"The things, we know, are neither rich nor rare,
But wonder how the devil they got there."
PsALM, *Prologue to the Satires.*

THE discovery of a fossil, whether as an impression on the
surface of a slab of rock or as a piece of petrified wood,
naturally leads us back to the living plant, and invites speculation as to the circumstances which led to the preservation
of the plant fragment. There is a certain fascination in
endeavouring, with more or less success, to picture the exact
conditions which obtained when the leaf or stem was carried
along by running water and finally sealed up in a sedimentary
matrix. Attempts to answer the question—How came the
plant remains to be preserved as fossils?—are not merely of
abstract interest appealing to the imagination, but are of
considerable importance in the correct interpretation of the
facts which are to be gleaned from the records of plant-bearing
strata.

Before describing any specific examples of the commoner
methods of fossilisation; we shall do well to briefly consider
how plants are now supplying material for the fossils of a
future age. In the great majority of cases, an appreciation
of the conditions of sedimentation, and of the varied circumstances attending the transport and accumulation of vegetable
débris, supplies the solution of a problem akin to that of the
fly in amber and the manner in which it came there.

Seeing that the greater part of the sedimentary strata have
been formed in the sea, and as the sea rather than the land has
been for the most part the scene of rock-building in the past, it
is not surprising that fossil plants are far less numerous than
fossil animals. With the exception of the algae and a few
representatives of other classes of plants, which live in the
shallow-water belt round the coast, or in inland lakes and seas,
plants are confined to land-surfaces; and unless their remains
are swept along by streams and embedded in sediments which
are accumulating on the sea floor, the chance of their preservation is but small. The strata richest in fossil plants are often
those which have been laid down on the floor of an inland lake
or spread out as river-borne sediment under the waters of an
estuary. Unlike the hard endo- and exo-skeletons of animals,
the majority of plants are composed of comparatively soft
material, and are less likely to be preserved or to retain their
original form when exposed to the wear and tear which must
often accompany the process of fossilisation.

The Coal-Measure rocks have furnished numberless relics
of a Palaeozoic vegetation, and these occur in various forms of
preservation in rocks laid down in shallow water on the edge
of a forest-covered land. The underclays or unstratified
argillaceous beds which nearly always underlie each seam of
coal have often been described as old surface-soils, containing
numerous remains of roots and creeping underground stems of
forest trees. The overlying coal has been regarded as a mass
of the carbonised and compressed débris of luxuriant forests
which grew on the actual spot now occupied by the beds of coal.
There are, however, many arguments in favour of regarding the
coal seams as beds of altered vegetable material which was
spread out on the floor of a lagoon or lake, while the underclay
was an old soil covered by shallow water or possibly a swampy
surface tenanted by marsh-loving plants[1].

The Jurassic beds of the Yorkshire Coast, long famous as
some of the richest plant-bearing strata in Britain, and the
Wealden rocks of the south coast afford examples of Mesozoic
sediments which were laid down on the floor of an estuary or

[1] Discussed at greater length in vol. II.

large lake. Circumstances have occasionally rendered possible the preservation of old land-surfaces with the stumps of trees still in their position of growth. One of the best examples of this in Britain are the so-called dirt-beds or black bands of Portland and the Dorset Coast. On the cliffs immediately east of Lulworth Cove, the surface of a ledge of Purbeck limestone which juts out near the top of the cliffs, is seen to have the form here and there of rounded projecting bosses or 'Burrs' several feet in diameter. In the centre of each boss there is either an empty depression, or the remnants of a silicified stem of a coniferous tree. Blocks of limestone 3 to 5 feet long and of about equal thickness may be found lying on the rocky ledge presenting the appearance of massive sarcophagi in which the central trough still contains the silicified remains of an entombed tree. The calcareous sediment no doubt oozed up to envelope the thick stem as it sank into the soft mud. An examination of the rock just below the bed bearing these curious circular elevations reveals the existence of a comparatively narrow band of softer material, which has been worn away by denuding agents more rapidly than the over-lying limestone. This band consists of partially rounded or subangular stones associated with carbonaceous material, and probably marks the site of an old surface-soil. This old soil is well shown in the cliffs and quarries of Portland, and similar dirt-beds occur at various horizons in the Lower and Middle Purbeck Series[1]. In this case, then, we have intercalated in a series of limestone beds containing marine and freshwater shells two or three plant beds containing numerous and fre-quently large specimens of cycadean and coniferous stems, lying horizontally or standing in their original position of growth. These are vestiges of an ancient forest which spread over a considerable extent of country towards the close of the Jurassic period. The trunks of cycads, long familiar in the Isle of Portland as fossil crows' nests, have usually the form of round depressed stems with the central portion somewhat hol-lowed out. It was supposed by the quarrymen that they were petrified birds' nests which had been built in the forks of the

[1] Woodward, H. B. (95), Figs. 124 and 133 from photographs by Mr Strahan.

trees which grew in the Portland forest. The beds separating
the surface-soils of the Purbeck Series, as seen in the sections
exposed on the cliffs or quarries, point to the subsidence of a
forest-covered area over which beds of water-borne sediment
were gradually deposited, until in time the area became dry
land and was again taken possession of by a subtropical vegetation, to be once more depressed and sealed up under layers of
sediment[1].

A still more striking example of the preservation of forest
trees rooted in an old surface-soil is afforded by the so-called
fossil-grove in Victoria Park, Glasgow, (Frontispiece). The
stumps of several trees, varying in diameter from about one to
three feet, are fixed by long forking 'roots' in a bed of shale.
In some cases the spreading 'roots,' which bear the surface
features of *Stigmaria*, extend for a distance of more than ten
feet from the base of the trunk. The stem surface is marked
by irregular wrinklings which suggest a fissured bark; but the
superficial characters are very imperfectly preserved. In one
place a flattened *Lepidodendron* stem, about 30 feet long, lies
prone on the shale. Each of the rooted stumps is oval or
elliptical in section, and the long axes of the several stems are
approximately parallel, pointing to some cause operating in a
definite direction which gave to the stems their present form.
Near one of the trees, and at a somewhat higher level than its
base, the surface of the rock is clearly ripple-marked, and takes
us back to the time when the sinking forest trees were washed
by waves which left an impress in the soft mud laid down over
the submerged area. The stumps appear to be those of Lepidodendron trees, rooted in Lower Carboniferous rocks. From
their manner of occurrence it would seem that we have in
them a corner of a Palaeozoic forest in which Lepidodendra
played a conspicuous part. The shales and sandstones containing the fossil trees were originally overlain by a bed of
igneous rock which had been forced up as a sheet of lava into
the hardened sands and clays[2].

Other examples of old surface-soils occur in different parts
of the world and in rocks of various ages. As an instance of a

[1] Buckland (37) Pl. LVII.   [2] Young, Glen, and Kidston (88).

land surface preserved in a different manner, reference may be
made to the thin bands of reddish or brown material as well
as clays and shale which occasionally occur between the sheets
of Tertiary lava in the Western Isles of Scotland and the
north-east of Ireland. In the intervals between successive
outpourings of basaltic lava in the north-west of Europe during
the early part of the Tertiary period, the heated rocks became
gradually cooler, and under the influence of weathering agents
a surface-soil was produced fit for the growth of plants. In
some places, too, shallow lakes were formed, and leaves, fruits
and twigs became embedded in lacustrine sediments, to be
afterwards sealed up by later streams of lava. In the face of
the cliff at Ardtun Head on the coast of Mull a leaf-bed is
exposed between two masses of gravel underlying a basaltic
lava flow; the impressions of the leaves of *Gingko* and other
plants from the Tertiary sediments of this district are exceptionally beautiful and well preserved[1]. A large collection
obtained by Mr Starkie Gardner may be seen in the British
Museum.

In 1883 the Malayan island of Krakatoa, 20 miles from
Sumatra and Java, was the scene of an exceptionally violent
volcanic explosion. Two-thirds of the island were blown away,
and the remnant was left absolutely bare of organic life. In
1886 it was found that several plants had already established
themselves on the hardened and weathered crust of the Kraka-
toan rocks, the surface of the lavas having been to a large
extent prepared for the growth of the higher plants by the
action of certain blue-green algae which represent some of the
lowest types of plant life[2]. We may perhaps assume a somewhat similar state of things to have existed in the volcanic
area in north-west Europe, where the intervals between successive outpourings of lava are represented by the thin bands of
leaf-beds and old surface-soils.

On the Cheshire Coast at Leasowe[3] and other localities,
there is exposed at low water a tract of black peaty ground
studded with old rooted stumps of conifers and other trees

[1] Gardner (87), p. 279.   [2] Treub (88).
[3] Morton (91), p. 228.

Fig. 6. Part of a submerged Forest seen at low water on the Cheshire Coast at Leasowe. Drawn from a photograph.

(fig. 6). There is little reason to doubt that at all events the majority of the trees are in their natural place of growth. The peaty soil on which they rest contains numerous flattened stems of reeds and other plants, and is penetrated by roots, probably of some aquatic or marshy plants which spread over the site of the forest as it became gradually submerged. A lower forest-bed rests directly on a foundation of boulder clay. Such submerged forests are by no means uncommon around the British coast; many of them belong to a comparatively recent period, posterior to the glacial age. In many cases, however, the tree stumps have been drifted from the places where they grew and eventually deposited in their natural position, the roots of the trees, in some cases aided by stones entangled in their branches, being heavier than the stem portion. There is a promising field for botanical investigation in the careful analysis of the floras of submerged forests; the work of Clement Reid, Nathorst, Andersson and others, serves to illustrate the value of such research in the hands of competent students.

The following description by Lyell, taken from his American travels, is of interest as affording an example of the preservation of a surface-soil :

"On our way home from Charleston, by the railway from Orangeburg, I observed a thin black line of charred vegetable matter exposed in the perpendicular section of the bank. The sand cast out in digging the railway had been thrown up on the original soil, on which the pine forest grew ; and farther excavations had laid open the junction of the rubbish and the soil. As geologists, we may learn from this fact how a thin seam of vegetable matter, an inch or two thick, is often the only monument to be looked for of an ancient surface of dry land, on which a luxuriant forest may have grown for thousands of years. Even this seam of friable matter may be washed away when the region is submerged, and, if not, rain water percolating freely through the sand may, in the course of ages, gradually carry away the carbon[1]."

In addition to the remnants of ancient soils, and the preservation of plant fragments in rocks which have been formed on the floor of an inland lake or an estuary, it is by no means rare to find fossil plants in obviously marine sediments. In fig. 7 we

----

[1] Lyell (45), vol. I. p. 180.

have a piece of coniferous wood with the shell of an Ammonite (*Aegoceras planicosta* Sow.) lying on it; the specimen was found in the Lower Lias clay at Lyme Regis, and illustrates the accidental association of a drifted piece of a forest tree with a

FIG. 17. *Aegoceras planicosta* Sow. on a piece of coniferous wood, Lower Lias, Lyme Regis. From a specimen in the British Museum. Slightly reduced.

shell which marks at once the age and the marine character of the beds. Again in fig. 8 we have a block of flint partially enclosing a piece of coniferous wood in which the internal structure has been clearly preserved in silica. This specimen was found in the chalk, a deposit laid down in the clear and deep water of the Cretaceous sea. The wood must have floated for some time before it became water-logged and sank to the sea-floor. In the light coloured wood there occur here and there dark spots which mark the position of siliceous plugs *b, b* filling up clean cut holes bored by Teredos in the woody tissue. The wood became at last enclosed by siliceous sediment and its tissues penetrated by silica in solution, which gradually replaced and preserved in wonderful perfection the form of the original tissue. A similar instance of wood enclosed in flint was figured by Mantell in 1844 in his *Medals of Creation*[1].

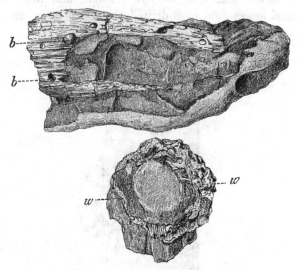

FIG. 8. Piece of coniferous wood in flint, from the Chalk, Croydon. Drawn from a specimen presented to the British Museum by Mr Murton Holmes. In the side view, shown above in the figure, the position of the wood is shown by the lighter portion, with holes, *b, b*, bored by Teredos or some other wood-eating animal. In the end view, below, the wood is seen as an irregular cylinder *w, w*, embedded in a matrix of flint. ⅓ Nat. size.

The specimen represented in fig. 9 illustrates the almost complete destruction of a piece of wood by some boring animal. The circular and oval dotted patches represent the filled up cavities made by a Teredo or some similar wood-boring animal.

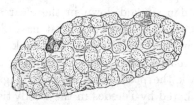

FIG. 9. Piece of wood from the Red Crag of Suffolk, riddled with holes filled in with mud. From a specimen in the York Museum. ⅓ Nat. size.

[1] Mantell (44), vol. I. p. 168.

Before discussing a few more examples of fossils illustrating different methods of fossilisation, it may not be out of place to quote a few extracts from travellers' narratives which enable us to realise more readily the circumstances and conditions under which plant remains have been preserved in the Earth's crust. In an account of a journey down the Rawas river in Sumatra, Forbes thus describes the flooded country :—

"The whole surface of the water was covered, absolutely in a close sheet, with petals, fruits and leaves, of innumerable species. In placid corners sometimes I noted a collected mass nearly half a foot deep, among which, on examination, I could scarcely find a leaf that was perfect, or that remained attached to its rightful neighbour, so that were they to become imbedded in some soft muddy spot, and in after ages to reappear in a fossil form they would afford a few difficult puzzles to the palaeontologist, both to separate and to put together[1]."

An interesting example of the mixture of plants and animals in sedimentary deposits is described by Hooker in his Himalayan Journals :—

"To the geologist the Jheels and Sunderbunds are a most instructive region, as whatever may be the mean elevation of their waters, a permanent depression of ten to fifteen feet would submerge an immense tract, which the Ganges, Burrampooter, and Soormah would soon cover with beds of silt and sand.

"There would be extremely few shells in the beds thus formed, the southern and northern divisions of which would present two very different floras and faunas, and would in all probability be referred by future geologists to widely different epochs. To the north, beds of peat would be formed by grasses, and in other parts temperate and tropical forms of plants and animals would be preserved in such equally balanced proportions as to confound the palaeontologist ; with the bones of the long-snouted alligator, Gangetic porpoise, Indian cow, buffalo, rhinoceros, elephant, tiger, deer, bear, and a host of other animals, he would meet with acorns of several species of oak, pine-cones and magnolia fruits, rose seeds, and *Cycas* nuts, with palm nuts, screw-pines, and other tropical productions[2]."

In another place the same author writes :

"On the 12th of January, 1848, the *Moozuffer* was steaming amongst the low, swampy islands of the Sunderbunds...... Every now and then the paddles of the steamer tossed up the large fruits of *Nipa fruticans*,

---

[1] Forbes, H. O. (85), p. 254.   [2] Hooker, J. D. (91), p. 477.

Thunb., a low stemless palm that grows in the tidal waters of the Indian Ocean, and bears a large head of nuts. It is a plant of no interest to the common observer, but of much to the geologist, from the nuts of a similar plant abounding in the Tertiary formations at the mouth of the Thames, having floated about there in as great profusion as here, till buried deep in the silt and mud that now forms the island of Sheppey[1]."

Of the drifting of timber, fruits, &c., we find numerous accounts in the writings of travellers. Rodway thus describes the formation of vegetable rafts in the rivers of Northern British Guiana :—

"Sometimes a great tree, whose timber is light enough to float, gets entangled in the grass, and becomes the nucleus of an immense raft, which is continually increasing in size as it gathers up everything that comes floating down the river[2]."

The undermining of river banks in times of flood, and the transport of the drifted trees to be eventually deposited in the delta is a familiar occurrence in many parts of the world. The more striking instances of such wholesale carrying along of trees are supplied by Bates, Lyell and other writers. In his description of the Amazon the former writes:

"The currents ran with great force close to the bank, especially when these receded to form long bays or *enseadas*, as they are called, and then we made very little headway. In such places the banks consist of loose earth, a rich crumbling vegetable mould, supporting a growth of most luxuriant forest, of which the currents almost daily carry away large portions, so that the stream for several yards out is encumbered with fallen trees, whose branches quiver in the current[3]."

In another place, Bates writes:

"The rainy season had now set in over the region through which the great river flows ; the sand-banks and all the lower lands were already under water, and the tearing current, two or three miles in breadth, bore along a continuous line of uprooted trees and islets of floating plants[4]."

The rafts of the Mississippi and other rivers described by Lyell afford instructive examples of the distant transport of

---

[1] Hooker, J. D. (91), p. 1. There are several good specimens of the black pyritised nipadite fruits in the British Museum and other collections.

[2] Rodway (95), p. 106.   [3] Bates (63), p. 139.

[4] Bates (63), p. 239.

vegetable material. The following passage is taken from the *Principles of Geology*;

"Within the tropics there are no ice-floes ; but, as if to compensate for that mode of transportation, there are floating islets of matted trees, which are often borne along through considerable spaces. These are sometimes seen sailing at the distance of fifty or one hundred miles from the mouth of the Ganges, with living trees standing erect upon them. The Amazons, the Orinoco, and the Congo also produce these verdant rafts[1]."

After describing the enormous natural rafts of the Atchafalaya, an arm of the Mississippi, and of the Red river, Lyell goes on to say:

"The prodigious quantity of wood annually drifted down by the Mississippi and its tributaries is a subject of geological interest, not merely as illustrating the manner in which abundance of vegetable matter becomes, in the ordinary course of nature, imbedded in submarine and estuary deposits, but as attesting the constant destruction of soil and transportation of matter to lower levels by the tendency of rivers to shift their courses.... It is also found in excavating at New Orleans, even at the depth of several yards below the level of the sea, that the soil of the delta contains innumerable trunks of trees, layer above layer, some prostrate as if drifted, others broken off near the bottom, but remaining still erect, and with their roots spreading on all sides, as if in their natural position[2]."

The drifting of trees in the ocean is recorded by Darwin in his description of Keeling Island, and their action as vehicles for the transport of boulders is illustrated by the same account.

"In the channels of Tierra del Fuego large quantities of drift timber are cast upon the beach, yet it is extremely rare to meet a tree swimming in the water. These facts may possibly throw light on single stones, whether angular or rounded, occasionally found embedded in fine sedimentary masses[3]."

Fruits may often be carried long distances from land, and preserved in beds far from their original source. Whilst cruising amongst the Solomon Islands, the Challenger met with fruits of *Barringtonia speciosa* &c., 130—150 miles from the coast. Off the coast of New Guinea long lines of drift

[1] Lyell (67) vol. II. p. 361.   [2] Lyell (67) vol. I. p. 445.
[3] Darwin (90) p. 443.

wood were seen at right angles to the direction of the river; uprooted trees, logs, branches, and bark, often floating separately.

"The midribs of the leaves of a pinnate-leaved palm were abundant, and also the stems of a large cane grass (*Saccharum*), like that so abundant on the shores of the great river in Fiji. Various fruits of trees and other fragments were abundant, usually floating confined in the midst of the small aggregations into which the floating timber was everywhere gathered.... Leaves were absent except those of the Palm, on the midrib of which some of the pinnae were still present. The leaves evidently drop first to the bottom, whilst vegetable drift is floating from a shore; thus, as the débris sinks in the sea water, a deposit abounding in leaves, but with few fruits and little or no wood, will be formed near shore, whilst the wood and fruits will sink to the bottom farther off the land. Much of the wood was floating suspended vertically in the water, and most curiously, logs and short branch pieces thus floating often occurred in separate groups apart from the horizontally floating timber. The sunken ends of the wood were not weighted by any attached masses of soil or other load of any kind; possibly the water penetrates certain kinds of wood more easily in one direction with regard to its growth than the other, hence one end becomes water-logged before the other.... The wood which had been longest in the water was bored by a *Pholas*[1]."

The bearing of this account on the manner of preservation of fossils, and the differential sorting so frequently seen in plant beds, is sufficiently obvious.

As another instance of the great distance to which land plants may be carried out to sea and finally buried in marine strata, an observation by Bates may be cited. When 400 miles from the mouth of the main Amazons, he writes:

"We passed numerous patches of floating grass mingled with tree trunks and withered foliage. Amongst these masses I espied many fruits of that peculiar Amazonian tree the Ubussú Palm; this was the last I saw of the great river[2]."

The following additional extract from the narrative of the Cruise of H.M.S. Challenger illustrates in a striking degree the conflicting evidence which the contents of fossiliferous beds may occasionally afford; it describes what was observed in an excursion from Sydney to Browera Creek, a branch of the main estuary or inlet into which flows the Hawkesbury river. It

---

[1] Challenger (85), Narrative, vol. I. Pt. ii. p. 679.
[2] Bates (63) p. 389.

was impossible to say where the river came to an end and the sea began. The Creek is described as a long tortuous arm of the sea, 10 to 15 miles long, with the side walls covered with orchids and *Platycerium.* The ferns and palms were abundant in the lateral shady glens; marine and inland animals lived in close proximity.

"Here is a narrow strip of the sea water, twenty miles distant from the open sea ; on a sandy shallow flat close to its head are to be seen basking in the sun numbers of sting-rays.... All over these flats, and throughout the whole stretch of the creek, shoals of Grey Mullet are to be met with ; numerous other marine fish inhabit the creek. Porpoises chase the mullet right up to the commencement of the sand-flat. At the shores of the creek the rocks are covered with masses of excellent oysters and mussel, and other shell-bearing molluscs are abundant, whilst a small crab is to be found in numbers in every crevice. On the other hand the water is overhung by numerous species of forest trees, by orchids and ferns, and other vegetation of all kinds ; mangroves grow only in the shallow bays. The gum trees lean over the water in which swim the *Trygon* and mullet, just as willows hang over a pool of carp. The sandy bottom is full of branches and stems of trees, and is covered in patches here and there by their leaves. Insects constantly fall in the water, and are devoured by the mullet. Land birds of all kinds fly to and fro across the creek, and when wounded may easily be drowned in it. Wallabies swim across occasionally, and may add their bones to the débris at the bottom. Hence here is being formed a sandy deposit, in which may be found cetacean, marsupial, bird, fish, and insect remains, together with land and sea shells, and fragments of a vast land flora ; yet how restricted is the area occupied by this deposit, and how easily might surviving fragments of such a record be missed by future geological explorers![1]"

The term 'fossil' suggests to the lay mind a petrifaction or a replacement by mineral matter of the plant tissues. In the scientific sense, a fossil plant, that is a plant or part of a plant whether in the form of a true petrifaction or a structureless mould or cast, which has been buried in the earth by natural causes, may be indistinguishable from a piece of recent wood lately fallen from the parent tree. In the geologically recent peat beds such little altered fossils (or sub-fossils) are common enough, and even in older rocks the more resistant parts of plant fragments are often found in a practically unaltered state. In the leaf impressions on an impervious clay, the brown-walled

[1] Challenger (85), Narrative, vol. i. p. 459.

epidermis shows scarcely any indication of alteration since it
was deposited in the soft mud of a river's delta. Such fossil
leaves are common in the English Tertiary beds, and even in
Palæozoic rocks it is not uncommon to find an impression of a
plant on a bed of shale from which the thin brown epidermis
may be peeled off the rock, and if microscopically examined it
will be found to have retained intact the contours of the
cuticularised epidermal cells. A striking example of a similar
method of preservation is afforded by the so-called paper-coal
of Culm age from the Province of Toula in Russia[1]. In the
Russian area the Carboniferous or Permian rocks have been
subjected to little lateral pressure, and unlike the beds of the
same age in Western Europe, they have not been folded and compressed by widespread and extensive crust-foldings. Instead of
the hard seams of coal there occur beds of a dark brown
laminated material, made up very largely of the cuticles of
Lepidodendroid plants.

From such examples we may naturally pass to fossils in
which the plant structure has been converted into carbonaceous matter or even pure coal. This form of preservation is
especially common in plant-bearing beds at various geological
horizons. In other cases, again, some mineral solution, oxide of
iron, talc, and other substances, has replaced the plant tissues.
From the Coal-Measures of Switzerland Heer has figured numerous specimens of fern fronds and other plants in which the leaf
form has been left on the dark coloured rock surface as a thin
layer of white talcose material[2]. In the Buntersandstone of
the Vosges and other districts the red imperfectly preserved
impressions of plant stems and leaves are familiar fossils[3];
the carbonaceous substance of the tissues has been replaced by
a brown or red oxide of iron.

Plants frequently occur in the form of incrustations; and
in fact incrustations, which may assume a variety of forms, are
the commonest kind of fossil. The action of incrusting springs,
or as they are often termed petrifying springs, is illustrated at
Knaresborough, in Yorkshire, and many other places where

---

[1] Zeiller (82) and Renault (95).   [2] Heer (76).
[3] Schimper and Mougeot (44).

water highly charged with carbonate of lime readily deposits calcium carbonate on objects placed in the path of the stream.

The travertine deposited in this manner forms an incrustation on plant fragments, and if the vegetable substance is subsequently removed by the action of water or decay, a mould of the embedded fragment is left in the calcareous matrix. An instructive example of this form of preservation was described in 1868[1] by Sharpe from an old gravel pit near Northampton. He found in a section eight feet high (fig. 10), a mass of incrusted plants of *Chara* (*a*) resting on and overlain by a calcareous paste (*c*) and (*d*) made up of the decomposed material of the overlying rock, and this again resting on sand. The place where the section occurred was originally the site of a pool in which Stoneworts

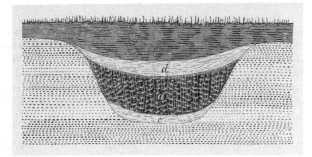

Fɪɢ. 10. Section of an old pool filled up with a mass of *Chara*. (From the *Geol. Mag.* vol. v. 1868, p. 563.)

grew in abundance. Large blocks of these incrusted Charas may be seen in the fossil-plant gallery of the British Museum.

In the Natural History Museum in the Jardin des Plantes, Paris, one of the table-cases contains what appear to be small models of flowers in green wax. These are in reality casts in wax of the moulds or cavities left in a mass of calcareous travertine, on the decay and disappearance of the encrusted flowers and other plant fragments[2]. This porous calcareous

---

[1] Sharpe, S. (68) p. 563.

[2] There are still more perfect casts from Sézanne in Prof. Munier-Chalmas' Geological collection in the Sorbonne. The best examples have not yet been figured.

rock occurs near Sézanne in Southern France, and is of Eocene age[1]. The plants were probably blown on to the freshly deposited carbonate of lime, or they may have simply fallen from the tree on to the incrusting matrix; more material was afterwards deposited and the flowers were completely enclosed. Eventually the plant substance decayed, and as the matrix hardened moulds were left of the vegetable fragments. Wax was artificially forced into these cavities and the surrounding substance removed by the action of an acid, and thus perfect casts were obtained of Tertiary flowers.

Darwin has described the preservation of trees in Van Diemen's land by means of calcareous substances. In speaking of beds of blown sand containing branches and roots of trees he says:

"The whole became consolidated by the percolation of calcareous matter; and the cylindrical cavities left by the decaying of the wood were thus also filled up with a hard pseudo-stalactitical stone. The weather is now wearing away the softer parts, and in consequence the hard casts of the roots and branches of the trees project above the surface, and, in a singularly deceptive manner, resemble the stumps of a dead thicket[2]."

As a somewhat analogous method of preservation to that in travertine, the occurrence of plants in amber should be mentioned. In Eocene times there existed over a region, part of which is now the North-east German coast, an extensive forest of conifers and other trees. Some of the conifers were rich in resinous secretions which were poured out from wounded surfaces or from scars left by falling branches. As these flowed as a sticky mass over the stem or collected on the ground, flowers, leaves, and twigs blown by the wind or falling from the trees, became embedded in the exuded resin. Evaporation gradually hardened the resinous substance until the plant fragments became sealed up in a mass of amber, in precisely the same manner in which objects are artificially preserved in Canada balsam. In many cases the amber acts as a petrifying agent, and by penetrating the tissues of a piece of wood it preserves the minute structural details in wonderful

¹ Saporta (68).   ² Darwin (90) p. 432.

perfection[1]. Dr Thomas in an account of the amber beds of East Prussia in 1848, refers to the occurrence of large fossil trees; he writes:

"The continuous changes to which the coast is exposed, often bring to light enormous trunks of trees, which the common people had long regarded as the trunks of the amber tree, before the learned declared that they were the stems of palm trees, and in consequence determined the position of Paradise to be on the coast of East Prussia[2]."

In 1887 an enormous fossil plant was discovered in a sandstone quarry at Clayton near Bradford[3]. The fossil was in the form of a sandstone cast of a large and repeatedly branched *Stigmaria,* and it is now in the Owens College Museum, where it was placed through the instrumentality of Prof. Williamson. The plant was found spread out in its natural position on the surface of an arenaceous shale, and overlain by a bed of hard sandstone identical with the material of which the cast is composed. Williamson has thus described the manner of formation of the fossil:

"It is obvious that the entire base of the tree became encased in a plastic material, which was firmly moulded upon these roots whilst the latter retained their organisation sufficiently unaltered to enable them to resist all superincumbent pressure. This external mould then hardened firmly, and as the organic materials decayed they were floated out by water which entered the branching cavity; at a still later period the same water was instrumental in replacing the carbonaceous elements by the sand of which the entire structure now consists[4]."

Although the branches have not been preserved for their whole length, they extend a distance of 29 feet 6 inches from right to left, and 28 feet in the opposite direction.

The fossil represented in fig. 1 (p. 10), from the collection of Dr John Woodward, affords a good example of a well-defined impression. The surface of the specimen, of which a cast is represented in fig. 1, shows very clearly the characteristic

[1] For figures of fossil plants in amber, *vide* Göppert and Berendt (45), Conwentz (90), Conwentz (96) &c.

[2] Thomas (48).   [3] Adamson (88).

[4] Williamson (87) Pl. xv. p. 45. A very fine specimen, similar to that in the Manchester Museum, has recently been added to the School of Mines Museum in Berlin; Potonié (90).

leaf-cushions and leaf-scars of a *Lepidodendron.* The stem
was embedded in soft sand, and as the latter became hard and
set, an impression was obtained of the external markings of
the *Lepidodendron.* Decay subsequently removed the substance
of the plant.

FIG. 11. *Equisetites columnaris* Brongn. From a specimen in the Woodwardian
Museum, Cambridge. ¼ nat. size.

In fig. 11 some upright stems of a fossil Horse-tail (*Equisetites columnaris*) from the Lower Oolite rocks near Scarborough,
are seen in a vertical position in sandstone.   On the surface of
the fossils there is a thin film of carbonaceous matter, which is
all that remains of the original plant substance ; the stems were
probably floated into their present position and embedded vertically in an arenaceous matrix.   The hollow pith-cavity was
filled with sand, and as the tissues decayed they became in part
converted into a thin coaly layer.   The vertical position of
such stems as those in fig. 11 naturally suggests their preservation *in situ,* but in this as in many other cases the erect
manner of occurrence is due to the settling down of the drifted
plants in this particular position.

An example of *Stigmaria* drawn in fig. 12 further illustrates the formation of casts[1]. The outer surface with the characteristic spirally arranged circular depressions, represents the wrinkled bark of the dried plant; the smaller cylinder, on the left side of the upper end (fig. 12, 2, *p*), marks the position of the pith

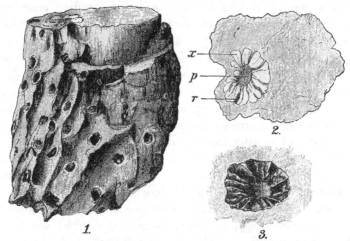

Fig. 12. *Stigmaria ficoides* Brongn. 1. Side view, showing wrinkled surface and the scars of appendages. 2. End view (upper) showing the displaced central cylinder; *p*, pith, *x*, xylem, *r*, medullary rays. 3. End view (lower). From a specimen in the Woodwardian Museum. ½ nat. size.

surrounded by the secondary wood, which has been displaced from its axial position. The pith decayed first, and the space was filled in with mud; somewhat later the wood and cortex were partially destroyed, and the rod of material which had been introduced into the pith-cavity dropped towards one side of the decaying shell of bark.

As the parenchymatous medullary rays readily decayed, the mud in the pith extended outwards between the segments of wood which still remained intact, and so spokes of argillaceous material were formed which filled the medullary ray cavities. The cortical tissues were decomposed, and their place taken by more argillaceous material. At one end of the specimen (fig. 12, 3) we find the wood has decayed without its place being afterwards filled up with foreign material. At the opposite

---

[1] The British Museum collection contains a specimen of *Stigmaria* preserved in the same manner as the example shown in fig. 12.

end of the specimen, the woody tissue has been partially preserved by the infiltration of a solution containing carbonate of lime (fig. 12, 2).

Numerous instances have been recorded from rocks of various geological ages of casts of stems standing erect and at right angles to the bedding of the surrounding rock. These vertical trees occasionally attain a considerable length, and have been formed by the filling in by sand or mud of a pipe left by the decay of the stem. It is frequently a matter of some difficulty to decide how far such fossils are in the position of growth of the tree, or whether they are merely casts of drifted stems, which happen to have been deposited in an erect position. The weighting of floating trees by stones held in the roots, added to the greater density of the root wood, has no doubt often been the cause of this vertical position. In attempting to determine if an erect cast is in the original place of growth of the tree, it is important to bear in mind the great length of time that wood is able to resist decay, especially under water. The wonderful state of preservation of old piles found in the bed of a river, and the preservation of wooden portions of anchors of which the iron has been completely removed by disintegration, illustrate this power of resistance. In this connection, the following passage from Lyell's travels in America is of interest. In describing the site of an old forest, he writes[1]:

"Some of the stumps, especially those of the fir tribe, take fifty years to rot away, though exposed in the air to alternations of rain and sunshine, a fact on which every geologist will do well to reflect, for it is clear that the trees of a forest submerged beneath the water, or still more, if entirely excluded from the air, by becoming imbedded in sediment, may endure for centuries without decay, so that there may have been ample time for the slow petrifaction of erect fossil trees in the Carboniferous and other formations, or for the slow accumulation around them of a great succession of strata."

In another place, in speaking of the trees in the Great Dismal Swamp, Lyell writes:—"When thrown down, they are soon covered by water, and keeping wet they never decompose, except the sap wood, which is less than an inch thick[2]." We

---

[1] Lyell (45) vol. I. p. 60.   [2] Lyell (45) vol. I. p. 147.

see, then, that trees may have resisted decay for a sufficiently long time to allow of a considerable deposition of sediment. It is very difficult to make any computation of the rate of deposition of a particular set of sedimentary strata, and, therefore, to estimate the length of time during which the fossil stems must have resisted decay.

The protective qualities of humus acids, apart from the almost complete absence of Bacteria[1] from the waters of Moor- or Peat-land, is a factor of great importance in the preservation of plants against decay for many thousands of years.

From examples of fossil stems or leaves in which the organic material has been either wholly or in part replaced by coal, we may pass by a gradual transition to a mass of opaque coal in which no plant structure can be detected. It is by no means uncommon to notice on the face of a piece of coal a distinct impression of a plant stem, and in some cases the coal is obviously made up of a number of flattened and compressed branches or leaves of which the original tissues have been thoroughly carbonised. A block of French coal, represented in fig. 13, consists very largely of laminated bands composed of the long parallel veined leaves of the genus *Cordaites* and of the bark of *Lepidodendron*, *Sigillaria*, and other Coal-Measure genera. The long rhizomes and roots below the coal are preserved as casts in the underclay.

In examining thin sections of coal, pieces of pitted tracheids or crushed spores are frequently met with as fragments of plant structures which have withstood decay more effectually than the bulk of the vegetable débris from which the coal was formed.

The coaly layer on a fossil leaf is often found to be without any trace of the plant tissues, but not infrequently such carbonised leaves, if treated with certain reagents and examined microscopically, are seen to retain the outlines of the epidermal cells of the leaf surface. If a piece of the Carbonaceous film detached from a fossil leaf is left for some days in a small quantity of nitric acid containing a crystal of chlorate of potash, and, after washing with water, is transferred to ammonia,

[1] Warming (96) p. 170.

transparent film often shows very clearly the outlines of the epidermal cell and the form of the stomata. Such treatment has been found useful in many cases as an aid to determination[1].

FIG. 15. Part of a coal seam largely made up of *Cordaites* leaves. *Stigmaria* and *Stigmariopsis* shown in the rock (underclay) underlying the coal. (After Grand'Eury [82] Pl. I. fig. 3.)

Prof. Zeiller informs me that he has found it particularly satisfactory in the case of cycadean leaves.

It is sometimes possible to detach the thin lamina repre-

[1] Bornemann (56), Schenk (67), Zeiller (82).

senting the carbonised leaf or other plant fragment from the rock on which it lies and to mount it whole on a slide. Good examples of plants treated in this way may be seen in the Edinburgh and British Museums, especially *Sphenopteris* fronds from the Carboniferous oil shales of Scotland. In the excellent collection of fossil plants in Stockholm there are still finer examples of such specimens, obtained by Dr Nathorst from some of the Triassic plants of Southern Sweden. In a few instances the tissues of a plant have been converted into coal in such a manner as to retain the form of the individual cells, which appear in section as a black framework in a lighter coloured matrix. Examples of such carbonised tissues were figured by some of the older writers, and Solms-Laubach has recently[1] described sections of Palaeozoic plants preserved in this manner. The section represented in fig. 70 is that of a Calamite stem (8 × 9·5 cm.) in which the wood has been converted into carbonaceous material, but the more delicate tissues have been almost completely destroyed. The thin and irregular black line a little distance outside the ring of wood, and forming the limit of the drawing, probably represents the cuticle. The whole section is embedded in a homogeneous matrix of calcareous rock, in which the more resistant tissues of the plant have been left as black patches and faint lines.

Mention should be made of a special form of preservation which has been described as fossilisation in half-relief. If a stem is imbedded in sand or mud, the matrix receives an impression of the plant surface, and if the hollow pith-cavity is filled with the surrounding sediment, the surface of the medullary cast will exhibit markings different from those seen on the surface in contact with the outside of the stem. The space separating the pith-cast from the mould bearing the impression of the stem surface may remain empty, or it may be filled with sedimentary material. In half-relief fossils, on the other hand, we have projecting from the under surface of a bed a more or less rounded and prominent ridge with certain surface markings, and fitting into a corresponding groove in the underlying rock on which the same markings have been impressed. It is

[1] Solms-Laubach (95²).

conceivable that such a cast might be obtained if soft plant
fragments were lying on a bed of sand, and were pressed
into it by the weight of superincumbent material.   The plant
fragment would be squeezed into a depression, and its substance
might eventually be removed and leave no other trace than the
half-relief cast and hollow mould.   A twig lying on sand would
by its own weight gradually sink a little below the surface; if
it were then blown away or in some manner removed, the
depression would show the surface features of the twig.   When
more sand came to be spread out over the depression, it would
find its way into the pattern of the mould, and so produce a
cast.   If at a later period when the sand had hardened, the
upper portion were separated from the lower, from the former
there would project a rounded cast of the hollow mould.
The preservation of soft algae as half-relief casts has been
doubted by Nathorst[1] and others as an unlikely occurrence in
nature.   They prefer to regard such ridges on a rock face as the
casts of the trails or burrows of animals.   This question of the
preservation of the two sides of a mould showing the same impression of a plant has long been a difficult problem; it is discussed
by Parkinson in his *Organic Remains*.   In one of the letters
(No. XLVI), he quotes the objection of a sceptical friend, who
refuses to believe such a manner of preservation possible,
"until," says Parkinson, "I can inform him if, by involving a
guinea in plaster of Paris, I could obtain two impressions of the
king's head, without any impression of the reverse[2]."

It would occupy too much space to attempt even a brief
reference to the various materials in which impressions of plants
have been preserved.   Carbonaceous matter is the most usual
substance, and in some cases it occurs in the form of graphite
which on dark grey or black rocks has the appearance of a plant
drawn in lead pencil.   The impressions of plants on the Jurassic
(Kimeridgian) slates of Solenhofen[3] in Bavaria, like those on
the Triassic sandstones of the Vosges, are usually marked out in
red iron oxide.

[1] Nathorst (86) p. 9.   See also Delgado (86).
[2] Parkinson (11) vol. I. p. 431.
[3] The British Museum collection contains many good examples of the Solenhofen plants.

So far we have chiefly considered examples of plants preserved in various ways by *incrustation*, that is, by having been enclosed in some medium which has received an impression of the surface of the plant in contact with it. By far the most valuable fossil specimens from a botanical point of view are however those in which the internal structure has been preserved; that is in which the preserving medium has not served merely as an encasing envelope or internal cast, but has penetrated into the body of the plant fragment and rendered permanent the organization of the tissues. In almost every Natural History or Geological Museum one meets with specimens of petrified trees or polished sections of fossil palm stems and other plants, in which the internal structure has been preserved in siliceous material, and admits of detailed investigation in thin sections under the microscope. Silica, calcium carbonate, with usually a certain amount of carbonate of iron and magnesium carbonate, iron pyrites, amber, and more rarely calcium fluoride or other substances have taken the place of the original cell-walls. Of silicified stems, those from Antigua, Egypt, Central France, Saxony, Brazil, Tasmania[1], and numerous other places afford good examples. Darwin records numerous silicified stems in Northern Chili, and the Uspallata Pass. In the central part of the Andes range, 7000 feet high, he describes the occurrence of "Snow-white projecting silicified columns...They must have grown," he adds, "in volcanic soil, and were subsequently submerged below sea-level, and covered with sedimentary beds and lava-flows[2]." A striking example of the occurrence of numerous petrified plant stems has been described by Holmes from the Tertiary forests of the Yellowstone Park. From the face of a cliff on the north side of Ameythryst mountain "rows of upright trunks stand out on the ledges like the columns of a ruined temple. On the more gentle slopes farther down, but where it is still too steep to support vegetation, save a few pines, the petrified trunks fairly cover the surface, and were at first supposed by us to be the shattered remains of a recent forest[3]." Marsh[4] and

---

[1] There is a splendid silicified tree stem from Tasmania of Tertiary age several feet in height in the National Museum.

[2] Darwin (90) p. 317.   [3] Holmes (80) p. 126, fig. 1.   [4] Marsh (71).

Conwentz[1] have described silicified trees more than fifty feet in
length from a locality in California where several large forest
trees of Tertiary age have been preserved in volcanic strata.
In South Africa on the Drakenberg hills there occur numerous
silicified trunks, occasionally erect and often lying on the
ground, probably of Triassic age[2]. In some instances the
specimens measure several feet in length and diameter. Some
of the coniferous stems seen in Portland, and occasionally met
with reared up against a house side, illustrate the silicification of
plant structure on a large scale. These are of Upper Jurassic
(Purbeck) age. From Grand'Croix in France a silicified stem
of *Cordaites* of Palaeozoic age has been recorded with a length
of twenty meters. The preservation of plants by siliceous infil-
trations has long been known. One of the earliest descriptions
of this form of petrifaction in the British Isles is that of stems
found in Lough Neagh, Ireland. In his lectures on Natural
Philosophy, published at Dublin in 1751, Barton gives several
figures of Irish silicified wood, and records the following
occurrence in illustration of the peculiar properties erroneously
attributed to the waters of Lough Neagh. Describing a certain
specimen (No. XXVI), he writes:—

"This is a whetstone, which as Mr Anthony Shane, apothecary, who
was born very near the lake, and is now alive, relates, he made by putting
a piece of holly in the water of the lake near his father's house, and fixing
it so as to withstand the motion of the water, and marking the place so as
to distinguish it, he went to Scotland to pursue his studies, and seven
years after took up a stone instead of holly, the metamorphosis having
been made in that time. This account he gave under his handwriting.
The shore thereabouts is altogether loose sand, and two rivers discharge
themselves into the lake very near that place[3]."

The well-known petrified trees from the neighbourhood of
Lough Neagh are probably of Pliocene age, but their exact
source has been a matter of dispute[4].

In 1836 Stokes described certain stems in which the tissues
had been partially mineralised. In describing a specimen of

[1] Conwentz (78).
[2] A large piece from one of these South African trees is in the Fossil-plant
Gallery of the British Museum.
[3] Barton (1751) p. 58.   [4] Gardner (84) p. 314.

beech from a Roman aqueduct at Eibsen in Lippe Bucheburg, he says :—

"The wood is, for the most part, in the state of very old dry wood, but there are several insulated portions, in which the place of the wood has been taken by carbonate of lime. These portions, as seen on the surface of the horizontal section, are irregularly circular, varying in size, but generally a little less or more than ⅛ of an inch in diameter, and they run through the whole thickness of the specimen in separate, perpendicular columns. The vessels of the wood are distinctly visible in the carbonate of lime, and are more perfect in their form and size in those portions of the specimen than in that which remains unchanged[1]."

This partial petrifaction of the structure in patches is often met with in fossil stems, and may be seriously misleading to those unfamiliar with the appearance presented by the crystallisation of silica from scattered centres in a mass of vegetable tissue. A good example of this is afforded by the gigantic stems discovered in 1829 in the Craigleith Quarry near Edinburgh[2]. Of those two large stems found in the Sandstone rock, the longest, originally 11 meters long and 3·3—3·9 meters in girth, is now set up in the grounds of the British Museum, and a large polished section (1 m. × 87 cm.) is exhibited in the

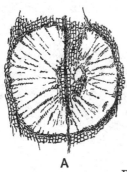 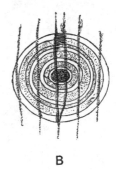

A   B

FIG. 14.

A. *Araucarioxylon Withami* (L. and H.). Radiating lines of crystallisation in secondary wood, as seen in transverse section.

B. *Lepidodendron* sp. Concentric lines of crystallisation, and scalariform tracheids, as seen in longitudinal section.

Fossil-plant Gallery. The other stem is in the Botanic Garden, Edinburgh. Transverse sections of the wood of the London

[1] Stokes (40) p. 207.   [2] Witham (31), Christison (76).

specimen show scattered circular patches (fig. 14 A) in the mineralised wood in which the tracheids are very clearly preserved; while in the other portion the preservation is much less perfect. The patch of tissue in fig. 14 A shows a portion of the wood of the Craigleith tree [*Araucarioxylon Withami* (L. and H.)] in which the mineral matter, consisting of dolomite with a little silica here and there, has crystallised in such a

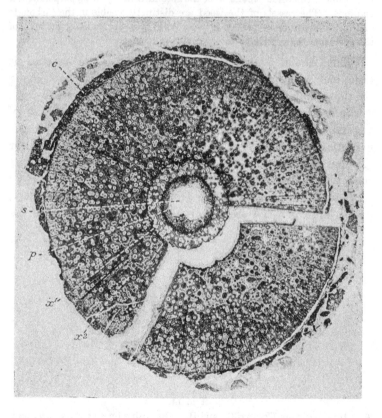

FIG. 15.   Transverse section of the central cylinder of a Carboniferous Lepido-dendroid stem in the collection of Mr Kidston. From Dalmeny, Scotland. *s*. Silica filling up the central portion of the pith. *p*. Remains of the pith tissue. x^1. Primary xylem. x^2. Secondary xylem. *c*. Innermost cortex.

manner as to produce what is practically a cone-in-cone structure on a small scale, which has partially obliterated the structural features. This minute cone-in-cone structure is not uncommon in petrified tissues; it is precisely similar in appearance to that described by Cole[1] in certain minerals. The crystallisation has been set up along lines radiating from different centres, and the particles of the tissue have been pushed as it were along these lines.

A somewhat different crystallisation phenomenon is illustrated by the extremely fine section of a Lepidodendroid plant shown in fig. 15. The tissues of the primary and secondary wood (x^1 and x^2) are well preserved throughout in silica, but scattered through the siliceous matrix there occur numerous circular patches, as seen in the figure. One of these is more clearly shown in fig. 14 B drawn from a longitudinal section through the secondary wood, x^2; it will be noticed that where the concentric lines of the circular patch occur, the scalariform thickenings of the tracheids are sharply defined, but immediately a tracheid is free of the patch these details are lost. It would appear that in this case silicification was first completed round definite isolated centres, and the secondary crystallisation in the matrix partially obliterated some of the more delicate structural features. The same phenomenon has been observed in oolitic rocks[2], in which the oolitic grains have resisted secondary crystallisation and so retained their original structure.

Among the most important examples of silicified plants are those from a few localities in Central France. In the neighbourhood of Autun there used to be found in abundance loose nodules of siliceous rock containing numerous fragments of seeds, twigs, and leaves of different plants. The rock of which the broken portions are found on the surface of the ground was formed about the close of the Carboniferous period.

At the hands of French investigators the microscopic examination of these fragments of a Palaeozoic vegetation have thrown a flood of light on the anatomical structure of many extinct types. Sometimes the silica has penetrated the cavities of the cells and vessels, and the walls have decayed without their substance being replaced by mineral material. Sections of tissues preserved in this manner, if soaked in a coloured

---

[1] Cole (94), figs. 1 and 3.   [2] Harker (95) p. 233, fig. 56.

solution assume an appearance almost identical with that of stained sections of recent plants. The spaces left by the decayed walls act as fine capillaries and suck up the coloured solution[1].

In the Coal-Measure sandstones of England large pieces of woody stems are occasionally met with in which the mineralisation has been incomplete. A brown piece of fossil stem lying in a bed of sandstone shows on the surface a distinct woody texture, and the lines of wood elements are clearly visible. The whole is, however, very friable and falls to pieces if an attempt is made to cut thin sections of it; the tracheids of the wood easily fall apart owing to the walls being imperfectly preserved, and the absence of a connecting framework such as would have been formed had the membranes been thoroughly silicified. It is occasionally possible to obtain from petrified plant stems perfect casts in silica or other substances of the cavity of a sclerenchymatous fibre, in which the mineral has been deposited not only in the cavity but in the fine pit-canals traversing the lignified walls. Such a cast is represented in fig. 16, the fine lateral projections are the delicate casts of the pit canals. Numerous instances of minute and delicate tissues preserved in silica are recorded in later chapters. A somewhat

FIG. 16.   Internal cast of a sclerenchymatous cell from the root of a Cretaceous fern (*Rhizodendron oppoliense* Göpp.). After Stenzel (86) Pl. III. fig. 29. × 240 and reduced to one-half.

unusual type of silicification is met with in some of the Gondwana rocks of India, in which cycadean fronds occur as white porcellaneous specimens showing a certain amount of internal structure in a siliceous matrix. Specimens of such leaves may be seen in the British Museum.

[1] I am indebted to Dr Renault of Paris for showing to me several preparations illustrating this method of petrifaction.

In the Coal-Measures of England, especially in the neighbourhood of Halifax in Yorkshire, and in South Lancashire, the seams of coal occasionally contain calcareous nodules varying in size from a nut to a man's head, and consisting of about 70 % of carbonate of calcium and magnesium, and 30 % of oxide of iron, sulphide of iron, &c.[1] The nodules, often spoken of by English writers as 'coal-balls,' contain numerous fragments of plants in which the minute cellular structure is preserved with remarkable perfection. It should be noted that the term coal-ball is also applied to rounded or subangular pieces of coal which are occasionally met with in coal seams, and especially in

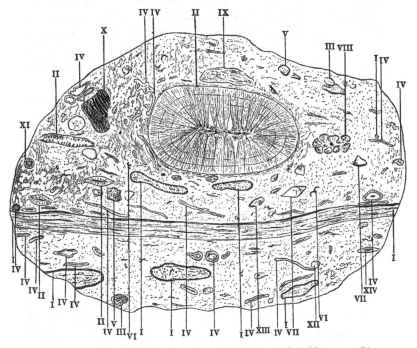

FIG. 17. A thin section of a calcareous nodule from the Coal-Measures. Binney collection, Woodwardian Museum, Cambridge. Very slightly reduced.

certain French coal fields. To avoid confusion it is better to speak of the plant-containing nodules as calcareous nodules, restricting the term coal-ball to true coal pebbles. A section

[1] Cash and Hick (78).

of a calcareous nodule, when seen under the microscope, presents
the appearance of a matrix of a crystalline calcareous substance
containing a heterogeneous mixture of all kinds of plant tissues,
usually in the form of broken pieces and in a confused mass.

A large section of one of these nodules (12·5 cm. × 8·5 cm.)
is shown in fig. 17. It illustrates the manner of occurrence of
various fragments of different plants in which the structure
has been more or less perfectly preserved. In this particular
example we see sections of *Myeloxylon* (I), *Calamites* (II), Fern
petioles (*Rachiopteris*) (III), Stigmarian appendages (IV),
Lepidodendroid leaves (V), *Myeloxylon* pinnules (VI), Gymnospermous seeds (VII), Twig of a *Lepidodendron*, showing the
central xylem cylinder and large leaf-bases on the outer cortex,
(VIII), Sporangia and spores of a strobilus (IX), Tangential
section of a *Myeloxylon* petiole (X), *Rachiopteris* sp. (XI),
Rachiopteris sp. (XII), Band of sclerenchymatous tissue (XIII),
Rachiopteris sp. (XIV).

The general appearance of a calcareous plant-nodule suggests
a soft pulpy mass of decaying vegetable débris, through which
roots were able to bore their way, as in a piece of peat or leafy
mould. Overlying this accumulation of soft material there
was spread out a bed of muddy sediment containing numerous
calcareous shells, which supplied the percolating water with
the material which was afterwards deposited in portions of
the vegetable débris. According to this view the calcareous
nodules of the coal seams represent local patches of a widespread mass of débris which were penetrated by a carbonated
solution, and so preserved as samples of a decaying mass of
vegetation, of which by far the greater portion became
eventually converted into coal[1].

In such nodules, we find that not only has the framework of
the tissues been preserved, but frequently the remains of cell
contents are clearly seen. In some cases the cells of a tissue may
contain in each cavity a darker coloured spot, which is probably
the mineralised cell nucleus. (Fig. 42, *A*, 1, p. 214.) The
contents of secretory sacs, such as those containing gum or resin,
are frequently found as black rods filling up the cavity of the cell

[1] Stur (85).

or canal. The contents of cells in some cases closely simulate starch grains, and such may have been actually present in the tissues of a piece of a fossil dicotyledonous stem described by Thiselton-Dyer from the Lower Eocene Thanet beds[1], and in the rhizome of a fossil *Osmunda* recorded by Carruthers[2] (Fig. 42, *B*, p. 214.)

Schultze in 1855[3] recorded the discovery of cellulose by microchemical tests applied to macerated tissue from Tertiary lignite and coal. With reference to the possibility of recognising cell contents in fossil tissue it is interesting to find that Dr Murray of Scarborough had attempted, and apparently with success, to apply chemical tests to the tissues of Jurassic leaves. In a letter written to Hutton in 1833 Murray speaks of his experiments as follows :—

"Reverting to the Oolitic plants, I have again and with better success been experimenting upon the thin transparent films of leaves, chiefly of *Taeniopteris vittata* and *Cyclopteris*, which from their tenuity offer fine objects for the microscope.... By many delicate trials I have ascertained the existence still in these leaves of resin and of tannin.... I am seeking among the filmy leaves of the *Fucoides* of A. Brongniart for iodine, but hitherto without success, and indeed can hardly expect it, as probably did iodine exist in them, it must have long ago entered into new combinations[4]."

Apart from this difficulty, it is not surprising that Dr Murray's search for iodine was unsuccessful, considering how little algal nature most of the so-called Fucoids possess.

Some of the most perfectly preserved tissues as regards the details of cell contents are those of gymnospermous seeds from Autun. In sections of one of these seeds which I recently had the opportunity of examining in Prof. Bertrand's collection, the parenchymatous cells contained very distinct nuclei and protoplasmic contents. In one portion of the tissue in the nucellus of *Sphaerospermum* the cell walls had disappeared, but the nuclei remained in a remarkable state of preservation. The cells shown in fig. 42 are from the ground tissue of a petiole of

---

[1] Thiselton-Dyer (72) Pl. vi.   [2] Carruthers (70).   [3] Schultze (55).

[4] I am indebted to Prof. Lebour of the Durham College of Science for the loan of this letter.

Cycadeoidea gigantea Sew.[1], a magnificent Cycadean stem from
Portland recently added to the British Museum collection; in
the cell *A*, 1, the nucleus is fairly distinct and in 2 and 4 the
contracted cell-contents is clearly seen. Other interesting
examples of fossil nuclei are seen in a *Lyginodendron* leaf
figured by Williamson and Scott in a recent Memoir on that
genus[2]. Each mesophyll cell contains a single dark nucleus.
The mineralisation of the most delicate tissues and the
preservation of the various forms of cell-contents are now
generally admitted by those at all conversant with the possibilities of plant petrifaction. If we consider what these facts
mean—the microscopic investigation of not only the finest
framework but even the very life-substance of Palaeozoic
plants—we feel that the aeons since the days when these
plants lived have been well-nigh obliterated.

Occasionally the plant tissues have assumed a black and
somewhat ragged appearance, giving the impression of charred
wood. A section of a recent burnt piece of wood resembles very
closely some of the fossil twigs from the coal seam nodules. It
is possible that in such cases we have portions of mineralised
tissues which were first burnt in a forest fire or by lightning
and then infiltrated with a petrifying solution. An example of
one of these black petrified plants is shown in fig. 74 B. Chap. x.
In many of the fossil plants there are distinct traces of fungus
or bacterial ravages, and occasionally the section of a piece of
mineralised wood shows circular spaces or canals which have the
appearance of being the work of some wood-eating animal, and
small oval bodies sometimes occur in such spaces which may
be the coprolites of the xylophagous intruder. (Fig. 24, p. 107.)

It is well known to geologists that during the Permian and
Carboniferous periods the southern portion of Scotland was the
scene of widespread volcanic activity. Forests were overwhelmed
by lava-streams or showers of ash, and in some districts tree
stems and broken plant fragments became sealed up in a volcanic
matrix. Laggan Bay in the north-east corner of the Isle of
Arran, and Petticur a short distance from Burntisland on the
north shore of the Firth of Forth, are two localities where

[1] Seward (97).   [2] Williamson and Scott (96) Pl. xxiv. fig. 16.

petrified plants of Carboniferous age occur in such preservation as allows of a minute investigation of their internal structure. The occurrence of plants in the former locality was first discovered by Mr Wünsch of Glasgow; the fossils occur in association with hardened shales and beds of ash, and are often exceedingly well preserved[1]. In fig. 18 is reproduced a sketch of a hollow tree trunk from Arran, probably a *Lepidodendron* stem, in which only the outer portion of the bark has been preserved, while the inner cortical tissues have been removed and the space occupied by volcanic detritus.

The smaller cylindrical structures in the interior of the hollow trunk are the central woody cylinders of Lepidodendroid trees; each consists of an axial pith surrounded by a band

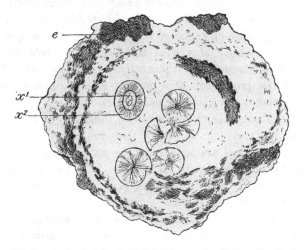

Fig. 18.   Diagrammatic sketch of a slab cut from a fossil stem (*Lepidodendron*?) from Laggan Bay. *e*, Imperfectly preserved bark of a large stem, extending in patches round the periphery of the specimen; the oval and circular bodies in the interior are the xylem portions of the central cylinders of *Lepidodendron* stems, x^1, primary wood, x^2, secondary wood. From a specimen in the Binney collection, Woodwardian Museum, Cambridge. ¼ nat. size.

of primary wood and a broader zone of secondary wood. One of the axes probably belonged to the stem of which only the shell has been preserved, the others must have come from other

[1] Bryce (72) p. 126, fig. 23.

trees and may have been floated in by water[1]. The microscopic
details of the wood and outer cortex have in this instance been
preserved in a calcareous material, which was no doubt derived
by water percolating through the volcanic ash. It is frequently
found that in fossil trees or twigs a separation of the tissues
has taken place along such natural lines of weakness as the
cambium or the phellogen, before the petrifying medium had
time to permeate the entire structure. Tree stems recently
killed by lava streams during volcanic eruptions at the present
day supply a parallel with the Palaeozoic forest trees of
Carboniferous times.

Guillemard in describing a volcanic crater in Celebes, speaks
of burnt trees still standing in the lava stream, " so charred at the
base of the trunk that we could easily push them down[2]." An
interesting case is quoted by Hooker in his *Himalayan Journals*,
illustrating the occurrence of a hollow shell of a tree, in which
the outer portions of a stem had been left while the inner
portions had disappeared, the wood being hollow and so favourable to the production of a current of air which accelerated
the destruction of the internal tissues.

On the coast near Burntisland on the Firth of Forth blocks
of rock are met with in which numerous plant fragments of
Carboniferous age are scattered in a confused mass through a
calcareous volcanic matrix. The twigs, leaves, spores, and other
portions are in small fragments, and their delicate cells are
often preserved in wonderful perfection.

The manner of occurrence of plants in sandstones, shales
or other rocks is often of considerable importance to the
botanist and geologist, as an aid to the correct interpretation
of the actual conditions which obtained at the time when the
plant remains were accumulating in beds of sediment. To
attempt to restore the conditions under which any set of plants
became preserved, we have to carefully consider each special
case. A nest of seeds preserved as internal casts in a mass of
sandstone, such as is represented by the block of Carboniferous
sandstone in fig. 19, suggests a quiet spot in an eddy where

[1] An erroneous interpretation of the Arran stems is given in Lyell's *Elements
of Geology* : Lyell (78) p. 547.   [2] Guillemard (86) p. 322.

seeds were deposited in the sandy sediment. Delicate leaf structures with sporangia still intact, point to quietly flowing water and a transport of no great distance. Occasionally the

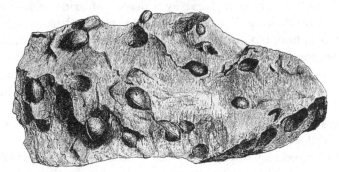

FIG. 19. Piece of Coal-Measures Sandstone with casts of *Trigonocarpon* seeds, from Peel Quarry near Wigan. From a specimen in the Manchester Museum, Owens College. ½ nat. size.

large number of delicate and light plant fragments, associated it may be with insect wings, may favour the idea of a wind storm which swept along the lighter pieces from a forest-clad slope and deposited them in the water of a lake. In some Tertiary plant-beds the manner of occurrence of leaves and flowers is such as to suggest a seasonal alternation, and the different layers of plant débris may be correlated with definite seasons of growth[1].

The predominance of certain classes of plants in a particular bed may be due to purely mechanical causes and to differential sorting by water, or it may be that the district traversed by the stream which carried down the fragments was occupied almost exclusively by one set of plants. The trees from higher ground may be deposited in a different part of a river's course to those growing in the plains or lowland marshes. It is obviously impossible to lay down any definite rules as to the reading of plant records, as aids to the elucidation of past physical and botanical conditions. Each case must be separately considered, and the various probabilities taken into account, judging by reference to the analogy of present day conditions.

Various attempts, more or less successful, have been made

[1] Heer (55).

to imitate the natural processes of plant mineralisation[1]. By soaking sections of wood for some time in different solutions, and then exposing them to heat, the organic substance of the cell walls has been replaced by a deposit of oxide of iron and other substances. Fern leaves heated to redness between pieces of shale have been reduced to a condition very similar to that of fossil fronds. Pieces of wood left for centuries in disused mines have been found in a state closely resembling lignite[2]. Attempts have also been made to reproduce the conditions under which vegetable tissues were converted into coal, but as yet these have not yielded results of much scientific value. The Geysers of Yellowstone Park have thrown some light on the manner in which wood may be petrified by the percolation of siliceous solutions; and it has been suggested that the silicification of plants may have been effected by the waters of hot springs holding silica in solution. Examples of wood in process of petrifaction in the Geyser district of North America have been recorded by Kuntze[3], and discussed by Schweinfurth[4], Solms-Laubach[5] and others[6]. The latter expresses the opinion that by a long continuance of such action as may now be observed in the neighbourhood of hot springs, the organic substance of wood might be replaced by siliceous material. The exact manner of replacement needs more thorough investigation. Kuntze describes the appearance of forest trees which have been reached by the waters of neighbouring Geysers. The siliceous solution rises in the wood by capillarity; the leaves, branches and bark are gradually lost, and the outer tissues of the wood become hardened and petrified as the result of evaporation from the exposed surface of the stem. The products of decay going on in the plant tissues must be taken into account, and the double decomposition which might result. There is no apparent reason why experiments undertaken with pieces of recent wood exposed to permeation by various calcareous and siliceous solutions under different conditions should not furnish useful results.

---

[1] Göppert (36), etc.   [2] Hirschwald (73).   [3] Kuntze (80) p. 8.
[4] Schweinfurth (82).   [5] Solms-Laubach (91), p. 29.
[6] Göppert (57). Some of the large silicified trees mentioned by Göppert may be seen in the Breslau Botanic gardens.

CHAPTER V.

DIFFICULTIES AND SOURCES OF ERROR IN THE DETERMINATION OF FOSSIL PLANTS.

"Robinson Crusoe did not feel bound to conclude, from the single human footprint which he saw in the sand, that the maker of the impression had only one leg."

HUXLEY'S *Hume*, p. 105, 1879.

THE student of palaeobotany has perhaps to face more than his due share of difficulties and fruitful sources of error; but on the other hand there is the compensating advantage that trustworthy conclusions arrived at possess a special value. While always on the alert for rational explanations of obscure phenomena by means of the analogy supplied by existing causes, and ready to draw from a wide knowledge of recent botany, in the interpretation of problems furnished by fossil plants, the palaeobotanist must be constantly alive to the necessity for cautious statement. That there is the greatest need of moderation and safe reasoning in dealing with the botanical problems of past ages, will be apparent to anyone possessing but a superficial acquaintance with fossil plant literature. The necessity for a botanical and geological training has already been referred to in a previous chapter.

It would serve no useful purpose, and would occupy no inconsiderable space, to refer at length to the numerous mistakes which have been committed by experienced writers on the subject of fossil plants. Laymen might find in such a list of blunders a mere comedy of errors, but the palaeobotanist must

see in them serious warnings against dogmatic conclusions or
expressions of opinion on imperfect data and insufficient evidence.
The description of a fragment of a handle of a Wedgewood
teapot as a curious form of Calamite[1] and similar instances of
unusual determinations need not detain us as examples of
instructive errors.   The late Prof. Williamson has on more than
one occasion expressed himself in no undecided manner as to
the futility of attempting to determine specific forms among
fossil plants, without the aid of internal structure[2]; and even in
the case of well-preserved petrifactions he always refused to
commit himself to definite specific diagnoses.   In his remarks in
this connection, Williamson no doubt allowed himself to express
a much needed warning in too sweeping language.   It is one of
the most serious drawbacks in palaeobotanical researches that
in the majority of cases the specimens of plants are both
fragmentary and without any trace of internal structure.
Specimens in which the anatomical characters have been preserved necessarily possess far greater value from the botanist's
point of view than those in which no such petrifaction has
occurred.   On the other hand, however, it is perfectly possible
with due care to obtain trustworthy and valuable results from
the examination of structureless casts and impressions.   In
dealing with the less promising forms of plant fossils, there is
in the first place the danger of trusting to superficial resemblance.
Hundreds of fossil plants have been described under the names
of existing genera on the strength of a supposed agreement in
external form; but such determinations are very frequently not
only valueless but dangerously misleading.   Unless the evidence is of the best, it is a serious mistake to make use
of recent generic designations.   If we consider the difficulties
which would attend an attempt to determine the leaves,
fragments of stems and other detached portions of various
recent genera, we can better appreciate the greater probability
of error in the case of imperfectly preserved fossil fragments.
    The portions of stems represented in figures 20 and 21, ex-
hibit a fairly close resemblance to one another; in the absence

---

[1] An example referred to by Carruthers (71) p. 444.
[2] Williamson (71) p. 507.

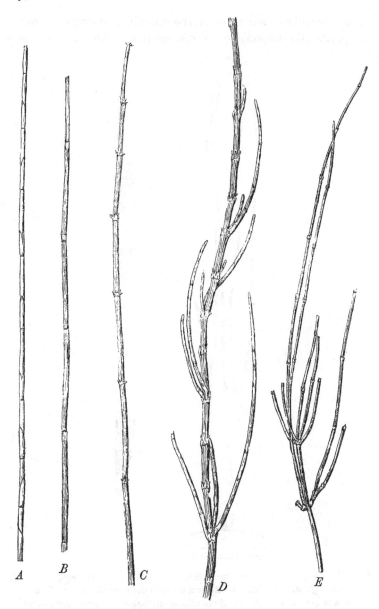

FIG. 20.

A. *Restio tetraphylla* Labill. (Monocotyledon).
B. *Equisetum variegatum* Schleich. ⎫ (Vascular Cryptogam).
C. *Equisetum debile* Roxb.   ⎬
D. *Casuarina stricta* Dryand. (Dicotyledon).
E. *Ephedra distachya* Linn. (Gymnosperm).   (*A—E* ½ nat. size).

of microscopical sections or of the reproductive organs it would be practically impossible to discriminate with any certainty

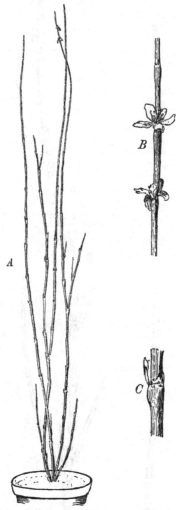

FIG. 21. *Polygonum Equisetiforme* Sibth. and Sm. *A*. Showing habit of plant. ½ nat. size. The two flowers towards the apex of one branch, drawn to a larger scale in *B*. *C*. Node with small leaf and ochrea characteristic of *Polygonaceæ*. From a plant in the Cambridge Botanic Garden.

between fossil specimens of the plants shown in the drawings. Examples such as these, and many others which might be cited,

serve to illustrate the possibility of confusion not merely between
different genera of the same family, but even between members
of different classes or groups.   The long slender branches of the
Polygonum represented in (fig. 21) would naturally be referred
to *Equisetum* in the absence of the flowers (fig. 20 *B*), or without
a careful examination of the insignificant scaly leaves borne at

Fig. 22.   *Kaulfussia æsculifolia* Blume.   From a specimen from Java in the
British Museum herbarium.   ½ nat. size.

the nodes.   The resemblance between *Casuarina* and *Ephedra*
and the British species of *Equisetum*, or such a tropical form as
E. debile, speaks for itself.

s.   7

Endless examples might be quoted illustrating the absolute futility, in many cases, of relying on external features even for the purpose of class distinction. An acquaintance with the general habit and appearance of only the better known members of a family, frequently leads to serious mistakes. The specimen shown in fig. 22 is a leaf of a tropical fern *Kaulfussia,* a genus now living in South-eastern Asia, and a member of one of the most important and interesting families of the Filicinæ, the Marrattiaceæ; its form is widely different from that which one is accustomed to associate with fern fronds. It is unlikely that the impression of a sterile leaf of *Kaulfussia* would be recognised as a portion of a fern plant.

Similarly in another exceedingly important group of plants, the Cycadaceæ[1], the examples usually met with in botanical gardens are quite insufficient as standards of comparison when we are dealing with fossil forms. Familiarity with a few commoner types leads us to regard them as typical for the whole family. In Mesozoic times cycadean plants were far more numerous and widely distributed than at the present time, and to adequately study the numerous fossil examples we need as thorough an acquaintance as possible with the comparatively small number of surviving genera and species. The less common and more isolated species of an existing family may often be of far greater importance to the palæobotanist than the common and more typical forms. This importance of rare and little known types will be more fully illustrated in the chapters dealing with the Cycadaceæ and other plant groups. Among Dicotyledons, the Natural Order Proteaceæ, at present characteristic of South Africa and Australia, and also represented in South America and the Pacific Islands, is of considerable interest to the student of fossil Angiosperms. In a valuable address delivered before the Linnean Society[2] in 1870 Bentham drew attention to the marked 'protean' character of the members of this family. He laid special stress on this particular division of the Dicotyledons in view of certain far-reaching conclusions, which had been based on the occurrence in different parts of Europe of fossil leaves supposed to be those of Proteaceous

[1] Dealt with more fully in vol. II.   [2] Bentham (70).

genera[1]. Speaking of detached leaves, Bentham says:—"I do not know of a single one which, in outline or venation, is exclusively characteristic of the order, or of any one of its genera." Species of *Grevillea*, *Hakea* and a few other genera are more or less familiar in plant houses, but the leaf-forms illustrated by the commoner members of the family convey no idea of the enormous variation which is met with not only in the family as a whole, but in the different species of the same genus. The striking diversity of leaf within the limits of a single genus will be dealt with more fully in volume II. under the head of Fossil Dicotyledons.

There is a common source of danger in attempting to carry too far the venation characters as tests of affinity. The parallel venation of Monocotyledons is by no means a safe guide to follow in all cases as a distinguishing feature of this class of plants. In addition to such leaves as those of the Gymnosperm *Cordaites* and detached pinnæ of Cycads, there are certain species of Dicotyledons which correspond in the character of their venation to Monocotyledonous leaves. *Eryngium montanum* Coult., *E. Lassauxi* Dcne., and other species of this genus of Umbelliferæ agree closely with such a plant as *Pandanus* or other Monocotyledons; similarly the long linear leaves of *Richea dracophylla*, R. Br., one of the Ericaceæ, are identical in form with many monocotyledonous leaves. Instances might also be quoted of monocotyledonous leaves, such as species of *Smilax* and others which Lindley included in his family of Dictyogens which correspond closely with some types of Dicotyledons[2]. Venation characters must be used with care even in determining classes or groups, and with still greater reserve if relied on as family or generic tests.

It is too frequently the case that while we are conversant with the most detailed histological structure of a fossil plant stem, its external form is a matter of conjecture. The conditions which have favoured the petrifaction of plant tissues have as a rule not been favourable for the preservation of good casts or impressions of the external features; and, on the other hand, in the best impressions of fern fronds or other plants, in which

---

[1] See also Bunbury (83) p. 309.   [2] Seward (96) p. 208.

the finest veins are clearly marked, there is no trace of internal
structure. It is, however, frequently the case that a knowledge
of the internal structure of a particular plant enables us to
interpret certain features in a structureless cast which could
not be understood without the help of histological facts. A
particularly interesting example of anatomical knowledge
affording a key to apparently abnormal peculiarities in a
specimen preserved by incrustation, is afforded by the fructification of the genus *Sphenophyllum*. Some few years ago
Williamson described in detail the structure of a fossil strobilus
(*i.e.* cone) from the Coal-Measures, but owing to the isolated
occurrence of the specimens he was unable to determine the
plant to which the strobilus belonged. On re-examining some
strobili of *Sphenophyllum*, preserved by incrustation, in the
light of Williamson's descriptions, Zeiller was able to explain
certain features in his specimens which had hitherto been a
puzzle, and he demonstrated that Williamson's cone was that of
a *Sphenophyllum*. Similar examples might be quoted, but
enough has been said to emphasize the importance of dealing
as far as possible with both petrifactions and incrustations.
The facts derived from a study of a plant in one form of preservation may enable us to interpret or to amplify the data
afforded by specimens preserved in another form.

The fact that plants usually occur in detached fragments,
and that they have often been sorted by water, and that portions
of the same plant have been embedded in sediment considerable
distances apart, is a constant source of difficulty. Deciduous leaves,
cones, or angiospermous flowers, and other portions of a plant
which become naturally separated from the parent tree, are met
with as detached specimens, and it is comparatively seldom that
we have the necessary data for reuniting the isolated members.
As the result of the partial decay and separation of portions of
the same stem or branch, the wood and bark may be separately
preserved. Darwin[1] describes how the bark often falls from
Eucalyptus trees, and hangs in long shreds, which swing about
in the wind, and give to the woods a desolate and untidy
appearance. In the passage already quoted from the narrative

[1] Darwin, (90) p. 416.

of the voyage of the Challenger, illustrations are afforded of the
manner in which detached portions of plants are likely to be
preserved in a fossil state. The epidermal layer of a leaf or the
surface tissues of a twig may be detached from the underlying
tissues and separately preserved[1]. It is exceedingly common
for a stem to be partially decorticated before preservation, and
the appearance presented by a cast or impression of the surface
of a woody cylinder, and by the same stem with a part or the
whole of its cortex intact is strikingly different. The late
Prof. Balfour[2] draws attention to this source of error in his
text-book of palaeobotany, and gives figures illustrating the
different appearance presented by a branch of *Araucaria imbricata* Pav. when seen with its bark intact and more or less
decorticated. Specimens that are now recognised as casts of
stems from which the cortex had been more or less completely
removed before preservation, were originally described under
distinct generic names, such as *Bergeria, Knorria* and others.
These are now known to be imperfect examples of Sigillarian or
Lepidodendroid plants. Grand'Eury[3] quotes the bark of *Lepidodendron Veltheimianum* Presl. as a fossil which has been
described under twenty-eight specific names, and placed in
several genera.

Since the microscopical examination of fossil plant-anatomy
was rendered possible, a more correct interpretation of decorticated and incomplete specimens has been considerably facilitated.
The examination of tangential sections taken at different levels
in the cortex of such a plant as *Lepidodendron* brings out the
distribution of thin and thick-walled tissue. Regularly placed
prominences on such a stem as the *Knorria* shown in fig. 23 are
due to the existence in the original stem of spirally disposed
areas of thin-walled and less resistant tissue; as decay pro-
ceeded, the thinner cells would be the first to disappear, and
depressions would thus be formed in the surrounding thicker
walled and stronger tissue. If the stem became embedded in
mud or sand before the more resistant tissue had time to decay,
but after the removal of the thin-walled cells, the surrounding

[1] Solms-Laubach (91) p. 9.   [2] Balfour (72) p. 5.
[3] Grand'Eury (77) Pt. i., p. 3.

sediment would fill up the depressions and finally, after the complete decay of the stem, the impression on the mould or on

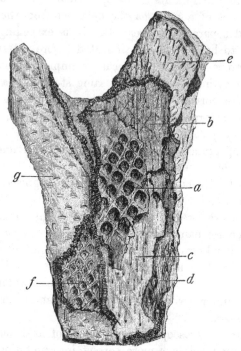

FIG. 23. A dichotomously branched Lepidodendroid stem (*Knorria mirabilis* Ren. and Zeill.). After Renault and Zeiller[1]. (¼ nat. size.) The original specimen is in the Natural History Museum, Paris.

the cast, formed by the filling up of the space left by the stem, would have the form of regularly disposed projections marking the position of the more delicate tissues. The specimen represented in the figure is an exceedingly interesting and well preserved example of a Coal-Measure stem combining in itself representatives of what were formerly spoken of as distinct genera.

The surface of the fossil as seen at *e* affords a typical example of the Knorria type of stem; the spirally disposed peg-like projections are the casts of cavities formed by the

[1] Renault and Zeiller (88) Pl. LX. fig. 1.

decay of the delicate cells surrounding each leaf-trace bundle on its way through the cortex of the stem. The surface *g* exhibits a somewhat different appearance, owing to the fact that we have the cast of the stem taken at a slightly different level. The surface of the thick layer of coal at *a* shows very clearly the outlines of the leaf-cushions; on the somewhat deeper surfaces *b*, *c* and *d* the leaf-cushions are but faintly indicated, and the long narrow lines on the coal at *c* represent the leaf-traces in the immediate neighbourhood of the leaf-cushions.

It is not uncommon among the older plant-bearing rocks to find a piece of sandstone or shale of which the surface exhibits a somewhat irregular reticulate pattern, the long and oval meshes having the form of slightly raised bosses. The size of such a reticulum may vary from one in which the pattern is barely visible to the unaided eye to one with meshes more than an inch in length. The generic name *Lyginodendron*[1] was proposed several years ago (1843) for a specimen having such a pattern on its surface, but without any clue having been found as to the meaning of the elongated raised areas separated from one another by a narrow groove. At a later date Williamson investigated the anatomy of some petrified fragments of a Carboniferous plant which suggested a possible explanation of the surface features in the structureless specimens. The name *Lyginodendron* was applied to this newly discovered plant, of which one characteristic was found to be the occurrence of a hypodermal band of strong thick-walled tissue arranged in the form of a network with the meshes occupied by thin-walled parenchyma. If such a stem were undergoing gradual decay, the more delicate tissue of the meshes would be destroyed first and the harder framework left. A cast of such a partially decayed stem would take the form, therefore, of projecting areas, corresponding to the hollowed out areas of decayed tissue, and intervening depressions corresponding to the projecting framework of the more resistant fibrous tissue. A precisely similar arrangement of hypodermal strengthening tissue occurs in various Palaeozoic and other plants, and casts presenting a

[1] Williamson (73) p. 393, Pl. xxvii. Described in detail in vol. ii. See also Solms-Laubach (91) p. 7, fig. 1.

corresponding appearance cannot be referred with certainty to one special genus; such casts are of no real scientific value[1]

The old generic terms *Artisia* and *Sternbergia* illustrate another source of error which can be avoided only by means of a knowledge of internal structure. The former name was proposed by Sternberg and the latter by Artis for precisely similar Carboniferous fossils, having the form of cylindrical bodies marked by numerous transverse annular ridges and grooves. These fossils are now known to be casts of the large discoid pith of the genus *Cordaites,* an extinct type of Palæozoic Gymnosperms. *Calamites* and *Tylodendron* afford other instances of plants in which the supposed surface characters have been shown to be those of the pith-cast. The former genus is described at length in a later chapter, but the latter may be briefly referred to   A cast, apparently of a stem, from the Permian rocks of Russia was figured in 1870 under the name *Tylodendron*; the surface being characterised by spirally arranged lozenge-shaped projections, described as leaf-scars. Specimens were eventually discovered in which the supposed stem was shown to be a cast of the large pith of a plant possessing secondary wood very like that of the recent genus *Araucaria.* The projecting portions, instead of being leaf-cushions, were found to be the casts of depressions in the inner face of the wood where strands of vascular tissue bent outwards on their way to the leaves. If a cast is made of the comparatively large pith of *Araucaria imbricata* the features of *Tylodendron* are fairly closely reproduced[2].

A dried Bracken frond lying on the ground in the Autumn presents a very different appearance as regards the form of the ultimate segments of the frond to that of a freshly cut leaf. In the former the edges of the pinnules are strongly recurved, and their shape is considerably altered. Immersed in water for some time fern fronds or other leaves undergo maceration, and the more delicate lamina of the leaf rots away much more rapidly than the scaffolding of veins. Among fossil fern fronds

[1] A good example is figured by Newberry (88) Pl. xxv. as a decorticated coniferous stem of Triassic age.

[2] Potonié (87).

differences in the form of the pinnules and in the shape and
extent of the lamina, to which a specific value is assigned, are
no doubt in many cases merely the expression either of
differences in the state of the leaves at the time of fossilisation
or of the different conditions under which they became embedded. Differential decay and disorganisation of plant tissues
are factors of considerable importance with regard to the fossilisation of plants. As Lindley[1] and later writers have suggested,
the absence or comparative scarcity of certain forms of plants
from a particular fossil flora may in some cases be due to their
rapid decay and non-preservation as fossils; it does not necessarily mean that such plants were unrepresented in the
vegetation of that period. The decayed rhizomes of the
Bracken fern often seen hanging from the roadside banks
on a heath or moorland, and consisting of flat dark coloured
bands of resistant sclerenchyma in a loose sheath of the hard
shrivelled tissue, are in striking contrast to the perfect stem.
A rotting Palm stem is gradually reduced to a loose stringy
mass consisting of vascular strands of which the connecting
parenchymatous tissue has been entirely removed. It must
frequently have happened that detached vascular bundles or
strands and plates of hard strengthening tissue have been preserved as fossils and mistaken for complete portions of plants.

Apart from the necessity of keeping in view the possible
differences in form due to the state of the plant fragments at
the time of preservation, and the marked contrast between the
same species preserved in different kinds of rock, there are
numerous sources of error which belong to an entirely different
category. The so-called moss-agates and the well-known
dendritic markings of black oxide of manganese, are among
the better known instances of purely inorganic structures
simulating plant forms.

An interesting example of this striking similarity between
a purely mineral deposit and the external form of a plant is
afforded by some specimens originally described as impressions
of the oldest known fern. The frontispiece to a well-known
work on fossil plants, *Le monde des plantes avant l'apparition de*

---

[1] Lindley and Hutton (31) vol. III. p. 4. See also Schenk (88) p. 202.

l'homme[1], represents a fern-like fossil on the surface of a piece of Silurian slate. The supposed plant was named *Eopteris Morierei* Sap., and it is occasionally referred to as the oldest land plant in books of comparatively recent date. In the Museum of the School of Mines, Berlin, there are some specimens of Angers slate on some of which the cleavage face shows a shallow longitudinal groove bearing on either side somewhat irregularly oblong and oval appendages of which the surface is traversed by fine vein-like markings. A careful examination of the slate reveals the fact that these apparent fern pinnules are merely films of iron pyrites deposited from a solution which was introduced along the rachis-like channel. Many of the extraordinary structures described as plants by Reinsch[2] in his Memoir on the minute structure of coal have been shown to be of purely mineral origin.

The innumerable casts of animal-burrows and trails as well as the casts of egg-cases and various other bodies, which have been described as fossil algæ, must be included among the most fruitful sources of error.

It requires but a short experience of microscopical investigation of fossil plant structures to discover numerous pitfalls in the appearance presented by sections of calcareous and siliceous nodules. The juxtaposition of tissues apparently parts of the same plant, and the penetration by growing roots of partially decayed plant débris, serve to mislead an unpractised observer. In sections of the English 'calcareous nodules' one very frequently finds the tissue of Stigmarian appendages occupying every conceivable position, and preserved in places admirably calculated to lead to false interpretations. The more minute investigation of tissues is often rendered difficult by deceptive appearances simulating original structures, but which are in reality the result of mineralisation. It is no easy matter in some cases to discover whether a particular cell in a fossil tissue was originally thick-walled, or whether its sclerous

---

[1] Saporta (79) (77). *Eopteris* is included among the ferns in Schimper and Schenk's volume of Zittel's *Handbuch der Palaeontologie* (p. 115), and in some other modern works.

[2] Reinsch (81).

appearance is due to the deposition of mineral matter on the inside of the thin cell-membrane. Examples of such sources of error as have been briefly referred to, and others, will be found in various parts of the descriptive portions of this book.

There is one other form of pitfall which should be briefly noticed. In sections of petrified plants one occasionally finds clean cut canals penetrating a mass of tissue, and differing in

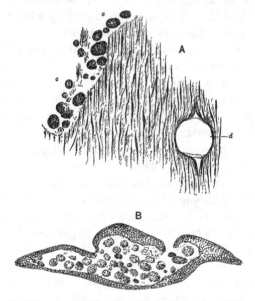

Fig. 24.   *A*.   Section of partially disorganised tissue attacked by some boring animal.   *c, c*, coprolites ;   *d*, a tunnel made by the borer through the plant tissue.

B.   Transverse section of a Lepidodendroid leaf, of which the inner tissues have been destroyed and the cavity filled with coprolites ; simulating a sporangium containing spores. (A and B from specimens in the Botanical Laboratory collection, Cambridge.)

their manner of occurrence and in their somewhat larger size from ordinary secretory ducts. Such tunnels or canals are probably the work of a wood-boring animal. An example is illustrated in fig. 24 *A*. Similarly it is not unusual to meet with groups or nests of spherical or elliptical bodies lying among plant tissues, and having the appearance of spores. Such spore-like bodies appear on close examination to be made up of finely comminuted particles of tissue, and in all probability they are the coprolites of some xylophagous animal. Examples of such coprolites are shown in fig. 24 A^1, and in fig. 24 B an interesting manner of occurrence of these misleading bodies is represented. The framework of cells enclosing the nest of coprolites in fig. 24 B, represents the outer tissues of a Lepidodendroid or a Sigillarian leaf; the inner tissues have been destroyed and the cavity is now occupied by what may possibly be the excreta of the wood-eating animal.

Some of the oval spore-like structures met with in plant tissues may, as Renault has suggested, be the eggs of an Arthropod[2]. In a section of a calcareous Coal-Measure nodule in the Williamson collection (British Museum)[3] there occur several fungal spores or possibly oogonia lying among imperfectly preserved Stigmarian appendages. Associated with these are numerous dark coloured and larger bodies consisting of a cavity bounded by a simple membrane; the larger bodies may well be the eggs of some Arthropod or other animal.

In looking through the collections of Coal-Measure plants in the Museums of Berlin, Vienna and other continental towns, one cannot fail to be struck with the larger size of many of the specimens as compared with those usually seen in English Museums. The facilities afforded in the State Collieries of Germany to the scientific investigator may account in part at least for the better specimens which he is able to obtain. It would no doubt be a great gain to our collections of Coal-Measure plants if arrangements could be made in some collieries for the preservation of the finer specimens met with in the working of the seams, instead of breaking up the slabs of shale and consigning everything to the waste heaps. There is one more point which should be alluded to in connection with possible sources of error, and that is the essential importance of accuracy in the illustration of specimens, especially as regard

[1] Williamson has drawn attention to the occurrence of such borings and coprolites in Coal-Measure plant tissues. *E.g.* Williamson (80) Pl. 20, figs. 65 and 66.
[2] Renault (96) p. 437.   [3] Slide No. 1923 in the Williamson collection.

type-specimens. It is often impossible to inspect the original fossils which have served as types, and it is of the utmost importance that the published figures should be as faithful as possible. M. Crépin[1] of Brussels, in an article on the use of photography in illustrating, has given some examples of the confusion and mistakes caused by imperfect drawings. It does not require a long experience of palaeobotanical work to demonstrate the need of care in the execution of drawings for reproduction.

[1] Crépin (81).

CHAPTER VI.

NOMENCLATURE.

"I do not think more credit is due to a man for defining a species, than to a carpenter for making a box."

CHARLES DARWIN, *Life and Letters*, Vol. I., p. 371.

ANY attempt to discuss at length the difficult and thorny question of nomenclature would be entirely out of place in an elementary book on fossil plants, but there are certain important points to which it may be well to draw attention. When a student enters the field of independent research, he is usually but imperfectly acquainted with the principles of nomenclature which should be followed in palaeontological work. After losing himself in a maze of endless synonyms and confused terminology, he recognises the desirability of adopting some definite and consistent plan in his method of naming genera and species. It is extremely probable that whatever system is made use of, it will be called in question by some critics as not being in strict conformity with accepted rules. The opportunities for criticism in matters relating to nomenclature are particularly numerous, and the critic who may be but imperfectly familiar with the subject-matter of a scientific work is not slow to avail himself of some supposed eccentricity on the part of the author in the manner of terminology. The true value of work may be obscured by laying too much emphasis on the imperfections of a somewhat heterodox nomenclature. On the other hand good systematic work is often seriously spoilt by a want of attention to generally accepted rules in naming and defining species. It is essential that those who take up systematic research should pay attention to the necessary though secondary question of technical description.

In inventing a new generic or specific name, it is well to adhere to some definite plan as regards the form or termination of the words used. To deal with this subject in detail, or to recapitulate a series of rules as to the best method of constructing names whether descriptive or personal, would take us beyond the limits of a single chapter. The student should refer for guidance to such recognised rules as those drawn up by the late Mr Strickland and others at the instance of the British Association[1].

It is not infrequently the case that the same generic name has been applied to a fossil and to a recent species. Such a double use of the same term should always be avoided as likely to lead to confusion, and as tending to admit a divorce between botany and palaeobotany.

In the course of describing a collection of fossil species, various problems are bound to present themselves as regards the best method of dealing with certain generic or specific names. A few general suggestions may prove of use to those who are likely to be confronted with the intricacies of scientific and pseudoscientific terminology.

In writing the name of a species, it is important to append the name, often in an abbreviated form, of the author who first proposed the accepted specific designation. *Stigmaria ficoides* Brongn. written in this form records the fact that Brongniart was the author of the specific name *ficoides*. It means, moreover, that Brongniart not only suggested the name, but that he was the first to give either a figure or a diagnosis of this particular fossil. It is frequently the case that a specific name is proposed for a new species, without either figures or description; such a name is usually regarded as a *nomen nudum*, and must yield priority to the name which was first accompanied by some description or illustration sufficiently accurate to afford a means of recognition. A practice which may be recommended on the score of convenience is to write the name of the author of a

[1] Rules for Zoological Nomenclature, drawn up by the late H. E. Strickland, M.A., F.R.S., London, 1878.

species in brackets if he was not the first to use the generic as
well as the specific name. *Onychiopsis Mantelli* (Brongn.) tells
us that Brongniart founded the species, but made use of some
other generic name than that which is now accepted. This
leads us to another point of some importance. Brongniart
described this characteristic Wealden fern under the name
Sphenopteris Mantelli; *Sphenopteris* being one of those extremely useful provisional generic terms which are used
in cases where we have no satisfactory proof of precise
botanical affinity. *Sphenopteris* stands for fern fronds having
a certain habit, form of segment and venation, and in this wide
sense it necessarily includes representatives of various divisions
and genera of Filices. If an example of a sphenopteroid frond
is discovered with sori or spores sufficiently well preserved to
enable us to determine its botanical position within narrower
limits, we may with advantage employ another genus in place
of the purely artificial form-genus which was originally chosen
as a consequence of imperfect knowledge. Fronds of this
Wealden fern have recently been found with well defined fertile
segments having a form apparently identical with that which
characterises the polypodiaceous genus *Onychium*. For this
reason the name *Onychiopsis* has been adopted. It is safer and
more convenient to use a name which differs in its termination
from that of the recent plant with which we believe the fossil
to be closely related. A common custom is to slightly alter the
recent name by adding the termination *-opsis* or *-ites*. There
are several other provisional generic terms that are often used
in Fossil Botany, and which might be advantageously chosen
in many cases where the misleading resemblance of external
form has often given rise to the use of a name implying
affinities which cannot be satisfactorily demonstrated.

It was the custom of some of the earliest writers, in spite of
their habit of using the names of recent Flowering plants for
extinct Palaeozoic species of Vascular Cryptogams, to adopt
also general and comprehensive terms. We find such a name
as *Lithoxylon* employed by Lhwyd[1] in 1699 as a convenient
designation for fossil wood.

[1] Lhwyd (1699).

One of the most important and frequently disputed questions associated with the naming of species is that of priority. No name given to a plant in pre-Linnaean days need be considered, as our present system of nomenclature dates from the institution of the binominal system by Linnaeus As a general rule, which it is advisable to follow, the specific name which was first given to a plant, if accompanied by a figure or diagnosis, should take priority over a name of later date. If A in 1850 describes a species under a certain name, and in 1860 B proposes a new name for the same species, either in ignorance of the older name or from disapproval of A's choice of a specific term, the later name should not be allowed to supersede A's original designation. Such a rule is not only just to the original author, but is one which, if generally observed, would lead to less confusion and would diminish unnecessary multiplication of specific names. Some writers would have us conform in all cases to this rule of priority, which they consistently adhere to apart from all considerations of convenience or long-established custom. There are, however, cogent reasons for maintaining a certain amount of freedom. While accepting priority as a good rule in most cases, it is unwise to allow ourselves to be too servile in our conformity to a principle which was framed in the interests of convenience, if the strict application of the rule clearly makes for confusion and inconvenience. A name may have been in use for say eighty years, and has become perfectly familiar as the recognised designation of a particular fossil; it is discovered, however, that an older name was proposed for the same species ninety years ago, and therefore according to the priority rule, we must accustom ourselves to a new name in place of one which is thoroughly established by long usage. From a scientific point of view, the ideal of nomenclature is to be plain and intelligible. To prefer priority to established usage entails obscurity and confusion. If priority is to be the rule which we must invariably obey in the shadowy hope that by such means finality in nomenclature[1] may be reached, it becomes necessary for the student to devote no inconsiderable portion

---

[1] Knowlton (96) p. 82.

of his time to antiquarian research, with a view to discover
whether a particular name may be stamped with the hall-
mark of 'the very first.' While admitting the advisability of
retaining as a general principle the original generic or specific
name, the extreme subservience to the priority craze' without
regard to convenience, would seem to lead irresistibly to the
view that "botanists who waste their time over priority are
like boys who, when sent on an errand, spend their time in
playing by the roadside[1]."

There is another point which cannot be satisfactorily settled
in all cases by a rigid adherence to an arbitrary rule. How far
should we regard a generic name in the sense of a mere mark
or sign to denote a particular plant, or to what extent may we
accept the literal meaning of the generic term as an index of
the affinity or character of the plant? If we consider the
etymology of many generic names, we soon find that they are
entirely inappropriate as aids in recognizing the true taxonomic
position of the plants to which they are applied. The generic
name *Calamites* was first suggested by the supposed resemblance of this Palaeozoic plant to recent reeds. If considered
etymologically, it is merely a record of a past mistake, but it
would be absurd to discard such a well-known name on the
grounds that the genus is a Vascular Cryptogam and far
removed from reeds. On the other hand, there often arise
cases which present a real difficulty. The following example
conveniently illustrates two distinct points of view as regards
generic nomenclature. In 1875 Saporta described and figured
a fragment of a fossil plant from the Jurassic beds of France as
Cycadorachis armata[2]; the name being chosen in the belief that
the specimen was part of a cycadean petiole, and there were
good grounds for such a view. A few years ago Mr Rufford
discovered more perfect specimens, in the Wealden rocks of
Sussex, clearly belonging to Saporta's genus, and these afforded
definite evidence that Saporta had been deceived by the
imperfection of the specimens as to their true botanical
position. Owing to the obviously misleading name first given

to this plant, I ventured to substitute *Withamia*[1] for *Cycado-rachis*, and chose such a term in preference to one denoting affinity, on account of the difficulty of placing the plant in a definite class or family.　On the other hand, it has been objected that the original name, despite its meaningless meaning—if the expression may be used—should be retained. A friendly critic[2], in writing of the proposed change of *Cycado-rachis*, urges the importance of adhering to the name which was first applied to a genus.　The same author pertinently remarks that we can no more dispense with a nomenclature than we can dispense with language.　We may extend the comparison and point out that in language, as in scientific nomenclature, conciseness, clearness and convenience should be kept in view as guiding principles.

The student must judge for himself what course to follow in each case.　While adhering as far as possible to a consistent plan, he must take care that he does not allow his own judgment to be completely over-ridden by a blind obedience to fixed rules, which if pressed too far may defeat their own ends.

[1] Seward (95) p. 173.　　　　[2] Ward (96) p. 874.

PART II. SYSTEMATIC.

CHAPTER VII.

THALLOPHYTA.

THE divisions of the plant kingdom dealt with in the
following chapters of Volume I. are taken in their natural
sequence, beginning with the lowest and passing gradually to
the highest groups.   The list of the classes and families
included in Chapters VII.—XI. is given in the table of
contents preceding Chapter I.

Thallophytes are of the simplest type, but they exhibit a
very wide range as regards both the structure and differentiation
of the vegetative body and the methods of reproduction.   In
some cases the individual consists of a minute simple cell which
multiplies by cell-division; in others the body or thallus is
made up of a number of similar units, while in a great
number of forms there is a well-marked physiological division
of labour, as expressed both in the external division of the
thallus into distinct organs corresponding in function to the root,
stem, and leaves of the higher plants, and further in the high
degree of histological differentiation of the tissues.   In other
thallophytes, again, the thallus is a *coenocyte* either unseptate
or incompletely septate; that is, the individual consists of a
single cell differing from a true plant-cell, in the stricter sense
of the term, in possessing several nuclei, in other words, the
thallus is divided up into compartments by transverse septa,
but each division contains more than one nucleus.   Such

coenocytic plants may show well-marked external differentiation of the thallus into members or parts subserving different
functions.

A similar wide range is covered by the methods of reproduction among thallophytes.

I. PERIDINIALES.

The organisms included under this head are of little
importance from a palaeontological point of view, but a brief
reference may be made to them as a section of the Thallophyta.

The Peridiniales include very small single-celled organisms
which have often been described as occupying a position on the
borderland between animals and plants, lying on the "shadowy
boundary between animal and vegetable life." The individuals
are rarely naked, more frequently they are covered with a cellulose or mucilaginous investment which has frequently the form
of two or more minute armour-like plates of a limiting membrane. The chromatophores are green, yellow, brown or
colourless. Simple division is the usual method of reproduction, but spores have been described as occurring in some
species. The motile forms are provided with cilia. The
Peridiniaceae, a section of the Peridiniales, are regarded as
nearly related to the Diatoms.

The Peridiniales play an important rôle in the Plankton
flora of the sea and freshwater lakes, and have a world-wide
distribution. In the narrative of the Challenger cruise they
are described as occasionally filling the tow-nets with a yellow
coloured slime[1]. Some genera, such as *Ceratium*, are found in
enormous numbers off the British coast.

As an example of the occurrence of fossil representatives of
the Peridiniaceae reference may be made to one of two species
of *Peridinium* described by Ehrenberg in 1836. These were
found in a siliceous rock described as Cretaceous in age from
Delitszch in Saxony. A comparison of Ehrenberg's figures of
the fossil species *Peridinium pyrophorum* Ehrenb.[2], with those

---

[1] Challenger (85) p. 934.
[2] Ehrenberg (36) p. 117, Pl. i. figs. 1 and 4, and Ehrenberg (54) Pl. xxxvii.
fig. vii.

of the recent species *Peridinium divergens* Ehrenb., as given by Schütt[1] and other writers, brings out clearly the very close resemblance if not identity of the two forms. Bütschli[2] in his account of the Dinoflagellata in Bronn's *Thier-Reich* confirms Ehrenberg's determination of *Peridinium pyrophorum*, and points out its striking agreement with the recent species.

II. COCCOSPHERES AND RHABDOSPHERES.

(Organisms of doubtful affinity.)

Our knowledge of these minute calcareous organisms is derived from Huxley's description of coccoliths from the Atlantic in 1857, and from the accounts of Wallich, John Murray, and other writers. In the first volume of the narrative of the Challenger cruise[3] and in the volume on deep-sea deposits[4] these minute forms of life are figured and described. In the latter volume both genera are spoken of as extremely abundant in the surface waters of the tropical and temperate regions of the open ocean, and as forming an important constituent of the Globigerine ooze; they are said to occur entangled in the gelatinous substance of the Radiolarians, Diatoms, and Foraminifera, and are very common in the stomachs of Salps, Pteropods and other pelagic animals. Rhabdospheres are rare in regions where the temperature of the water sinks below 65° F.; the Coccospheres occur in tropical and temperate latitudes, and extend further north and south than the Rhabdospheres. As regards their botanical position, John Murray expresses the view that they are in all probability pelagic algae.

In the interesting memoir by Schütt on the *Pflanzenleben der Hochsee*[5] there occurs a short reference to the forms described in the Challenger Reports, but they were not obtained by the staff of the Hensen Plankton Expedition and Schütt's remarks are not based therefore on personal observations. While admitting the existence of such bodies, he points out that Zoologists have referred Coccospheres and Rhabdospheres

[1] Schütt (96) p. 22.   [2] Bütschli (83–87) p. 1028.
[3] Challenger Reports (85) p. 939.   [4] Challenger Reports (91) p. 257.
[5] Hensen (92), Schütt (93) p. 44.

to the algae as organisms which cannot be included in any group of animals, and Schütt is unable to recognise a sufficient reason for referring them to this class of plants. It is suggested indeed that they may be purely inorganic structures.

The most recent account of these two genera is by Messrs G. Murray and Blackman in a short notice in *Nature* for April 1, 1897[1]. Numerous examples of Coccospheres and Rhabdospheres were obtained by Capt. Milner of the R.M.S. Para during a voyage to Barbados by allowing the sea water to enter the feed-pipe of the boiler through a fine muslin net. All the forms described in the Challenger Reports were met with, and an examination of the material by means of extremely high

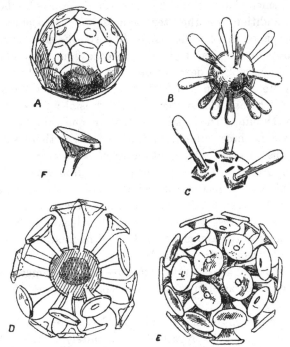

Fig. 25.   (From Murray and Blackman).

A, Coccosphere × 1300. *B*, Rhabdosphere × 900. *C*, Portion of the same × 1300. *D*, Rhabdosphere of another type, in optical section × 1900. *E*, The same in surface view × 1900. *F*, End of one of the trumpet-shaped appendages of *E*.

[1] Murray, G., and Blackman, V. H. (97).

objectives has confirmed the original account of the genera, and added some points to our previous knowledge.

Coccospheres (fig. 25 *A*). Spherical bodies of exceedingly small size, consisting of a central protoplasmic vesicle covered with overlapping circular calcareous scales, each of which is attached to the minute cell by a button-like projection. The scales are frequently found detached and are then spoken of as Coccoliths.

Rhabdospheres (fig. 25 *B—F*). Spherical bodies, extremely minute, consisting of a single cell, on the surface of which are embedded numerous calcareous plates bearing long blunt spines (fig. 25, *C*) or beautiful trumpet-like appendages (fig. 25, *D—F*). The detached plates of Rhabdospheres are known as Rhabdoliths.

In addition to the text-figures of Coccospheres and Rhabdospheres in the Challenger Reports, the same structures are shown in samples of globigerine ooze figured in Plate XI. of the Monograph on deep-sea deposits. In a recent number of *Nature* Messrs Dixon and Joly[1] have announced the discovery of Coccoliths and Coccospheres in the coastal waters off South County Dublin. They estimate that in one sample of water taken about three miles from the Irish coast there were 200 Coccoliths in each cubic centimetre of sea water.

The interest of these calcareous bodies from a palaeobotanical point of view lies in the fact that similar forms have been recognized in the Chalk and the Upper Lias. Sorby, in his memorable Address delivered before the Geological Society in 1879, refers to the abundance of Coccoliths in sections of chalk which he examined[2]. Rothpletz[3] has recently recorded the occurrence of numerous Coccoliths, 5—12 μ in diameter, associated with the skeleton of a horny sponge (*Phymatoderma*) of Liassic age.

The question of the nature of Coccospheres and Rhabdospheres cannot be regarded as definitely settled. It has been shown by J. Murray, and more recently by G. Murray and V. H. Blackman, that on the solution of the calcareous material by a weak acid there remains a small gelatinous body

---

[1] Dixon and Joly (97).   [2] Sorby (79) p. 78.
[3] Rothpletz (96), p. 909, Pl. xxiii. fig. 4.

apparently protoplasmic in nature. We may at least express the opinion that Schütt's suggestion as to their being inorganic must be ruled out of court. It would appear that they are extremely minute unicellular organisms characterised by a delicate calcareous armour consisting of numerous plates or scales. We know nothing as to their life-history, and cannot attempt to determine their affinities with any degree of certainty until further facts are before us. It is not improbable that they are algae of an extremely minute size, and the evidence so far obtained would lead us to regard them as complete individuals rather than the reproductive cells of some larger organism. Mr George Murray is of opinion that they are certainly algae, but he considers that they cannot be included in any existing family. It is conceivable that they may be minute eggs or reproductive cells of animals or plants, but on the whole the balance of probability would seem to be in favour of regarding them as autonomous organisms.

III. SCHIZOPHYTA.

I. SCHIZOPHYCEAE (CYANOPHYCEAE).

II. SCHIZOMYCETES.

In this group are included small single-celled plants of an extremely low type of organisation, in which reproduction takes the form of multiplication by simple cell-division, or the formation of spores The characteristic method of reproduction by division has given rise to the general term Fission-plants for this lowest sub-class in the vegetable kingdom. In many cases the members of this sub-class contain chlorophyll, and associated with it a blue-green colouring matter; such plants are classed together as the Blue-green algae, Cyanophyceae, or Schizophyceae. Others, again, are destitute of chlorophyll, and may be conveniently designated Schizomycetes or Fission-fungi. Seeing how close is the resemblance and relationship between the members of the sub-class, it has been the custom to include them as two parallel series under the general head, Schizophyta, rather than to incorporate them among the Algae and Fungi respectively.

I. SCHIZOPHYCEAE (CYANOPHYCEAE or Blue-green Algae).

Chroococcaceae. Thallus of a single cell, the cells may be either free, or more usually joined together in colonies enveloped by a common gelatinous matrix, formed by the mucilaginous degeneration of the outer portion of the cell-walls. Reproduction by means of simple division or resting cells.

Nostocaceae. Thallus consists of simple or branched rows of cells in which special cells known as *heterocysts* often occur. Reproduction by means of germ-plants or *hormogonia*, or by resting cells specially modified to resist unfavourable conditions.

In both families the individuals are surrounded by a gelatinous envelope, which in some genera assumes the form of a conspicuous and comparatively resistant sheath. Marine, freshwater, and aerial forms are represented among recent genera. Several species occur as endophytes, living in the tissues or mucilage-containing spaces in the bodies of higher plants. In addition to the frequent occurrence of blue-green algae in freshwater streams and on damp surfaces, certain forms are particularly abundant in the open sea[1], and in lakes or meres[2] where they are the cause of what is known in some parts of the country as "the breaking of the meres" ("Fleurs d'eau"). From the narrative of the cruise of the Challenger, we learn that the Oscillariaceae are especially abundant in the surface waters of the ocean. The "sea sawdust" so named by Cook's sailors[3], and the same floating scum collected by Darwin[4], affords an illustration of the abundance of some of these blue-green algae in the sea.

Another manner of occurrence of these plants has been recorded by different writers, which is of special importance from the point of view of fossil algae. On the shores of the Great Salt Lake, Utah, there are found numerous small oolitic calcareous bodies thrown up by the waves[5]. These are coated with the cells of *Glœocapsa* and *Glœotheca*, two genera of the Chroococcaceae. Sections of the grains reveal

---

[1] Challenger (85) *passim.* Schütt (93).   [2] Phillips W. (93).
[3] Kippis (78) p. 115.   [4] Darwin (90) p. 13.   [5] Rothpletz (92).

the presence of the same forms in the interior of the calcareous matrix, and it has been concluded on good evidence that the algae are responsible for the deposition of the carbonate of lime of the oolitic grains. By extracting the carbonic acid which they require as a source of food, from the waters of the lake, the solvent power of the water is decreased and carbonate of lime is thrown down. In similar white grains from the Red Sea[1] there is a central nucleus in the form of a grain of sand, and cells of Chroococcaceae occur in the surrounding carbonate of lime as in the Salt Lake oolite. Prof. Cohn of Breslau in 1862 demonstrated the importance of low forms of plant life in the deposition of the Carlsbad "Sprudelstein[2]." On the bottom of Lough Belvedere, near Mullingar in Ireland[3], there occur numerous spherical calcareous pebbles, of all sizes up to that of a filbert From a pond in Michigan (U.S.A.)[4] similar bodies have been obtained varying in diameter from one to three and a-half inches. In the former pebbles a species of *Schizothrix*, one of the Nostocaceae occurs in abundance, in the form of chains of small cells enclosed in the characteristic and comparatively hard tubular sheath, and associated with *Schizothrix fasciculata* there have been found *Nostoc* cells and the siliceous frustules of Diatoms. In the Michigan nodules the same *Schizothrix* occurs, associated with *Stigonema* and *Dichothrix*, other genera of the Nostocaceae. One of the Michigan pebbles is shown in section in fig. 32 *D*.

The connection between the well-known oolitic structure, characteristic of rocks of various ages in all parts of the world, and the presence of algal cells is of the greatest interest from a geological point of view. In recent years considerable attention has been paid to the structure of oolitic rocks, and in many instances there have been found in the calcareous grains tubular structures suggestive of simple cylindrical plants, which have probably been concerned in the deposition of the carbonate of lime of which the granules consist. In 1880 Messrs Nicholson and Etheridge[5] recorded the occurrence of such a

---

[1] Walther (88).   [2] Cohn (62).
[3] Murray, G. (95²).   [4] Thiselton-Dyer (91) p. 225.
[5] Nicholson and Etheridge (80) p. 23, Pl. IX. fig. 24.

tubular structure in calcareous nodules obtained from a rock of
Ordovician age in the Girvan district of Scotland. These
Authors considered the tubes to be those of some Rhizopod,
and proposed to designate the fossil *Girvanella.*

Girvanella (fig. 26).

Messrs Nicholson and Etheridge defined the genus as
follows:—

"Microscopic tubuli, with arenaceous or calcareous (?) walls, flexuous
or contorted, circular in section, forming loosely compacted masses. The
tubes, apparently simple cylinders, without perforations in their sides, and
destitute of internal partitions or other structures of a similar kind."

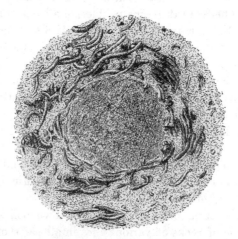

Fig. 26.   *Girvanella problematica*, Eth. and Nich.   Tubules of *Girvanella*
lying in various positions and surrounding an inorganic 'nucleus' or
centre.   From a section of Wenlock limestone, May Hill.   × 65

Since this diagnosis was published very many examples of
similar tubular fossils have been described by several writers
in rocks from widely separated geological horizons. The accompanying sketch (Fig. 26), drawn from a micro-photograph
kindly lent to me by Mr Wethered of Cheltenham, who has
made oolitic grains a special subject of careful investigation,
affords a good example of the occurrence of such tubular
structures in an oolitic grain of Silurian age from the Wenlock

limestone of May Hill, Gloucestershire[1]. In the centre is a crystalline core or nucleus round which the tubules have grown, and presumably they had an important share in the deposition of the calcareous substance. The nature of *Girvanella*, and still more its exact position in the organic world, is quite uncertain; it is mentioned rather as *à propos* of the association of recent Cyanophyceae with oolitic structure, than as a well-defined genus of fossil algae.

In the desciption of the calcareous nodules from Michigan, Murray speaks of the *Schizothrix* filaments at the surface of the pebbles as fairly intact, while nearer the centre only sheaths were met with. It is conceivable that in some of the tubular structures referred to *Girvanella* we have the mineralised sheaths of a fossil Cyanophyceous genus[2]. The organic nature of these tubules has been a matter of dispute, but we may probably assume with safety that in some at least of the fossil oolitic grains there are distinct traces of some simple organism which was in all likelihood a plant. Some authors have suggested that *Girvanella* is a calcareous alga which should be included in the family Siphoneae[3] As a matter of fact we must be content for the present to leave its precise nature as still *sub judice*, and while regarding it as probably an alga, we may venture to consider it more fittingly discussed under the Schizophyta than elsewhere.

Wethered[4] would go so far as to refer oolitic structure in general to an organic origin. While admitting that a Girvanella-like structure has been very frequently met with in oolitic rocks, it would be unwise to adopt so far-reaching a conclusion. It is at least premature to refer the formation of all oolitic structure to algal agency, and the evidence adduced is by no means convincing in every case. The discovery of *Girvanella* and allied forms in rocks from the Cambrian[5],

[1] Wethered (93) p. 237.

[2] For figures of the sheaths of Cyanophyceous algae, see Murray (95[2]), Pl. xix. fig. 5. Gomont (88) and (92); *etc.*

[3] Brown (94) p. 203.

[4] For references to the papers of Wethered and others, see Seward (94), p. 24.

[5] E. G. Bornemann (87), Pl. ii.

Ordovician, Silurian, Carboniferous, Jurassic and other systems
is a striking fact, and lends support to the view that oolitic
structure is in many cases intimately associated with the
presence of a simple tubular organism. Among recent algae
we find different genera, and representatives of different families,
growing in such a manner and under such circumstances as are
favourable to the formation of a ball-like mass of algal threads,
which may or may not be encrusted with carbonate of lime.
Similarly as regards oolitic grains of various sizes, and the
occurrence in rocks of calcareous nodules, the tubular structure
is not always of precisely the same type, and cannot always be
included under the genus *Girvanella.*

Several observers have recorded the occurrence of low forms
of plant-life in the waters of thermal springs. It has been
already mentioned that Cohn described the occurrence of
simple plants in the warm Carlsbad Springs, and fission-plants
of various types have been discovered in the thermal waters
of Iceland, the Azores[1], New Zealand, the Yellowstone Park,
Japan, India, and numerous other places.

A few years ago Mr Weed, of the geological survey of the
United States, published an interesting account of the formation of calcareous travertine and siliceous sinter in the Yellowstone Park district[2]. This author emphasizes the important
rôle of certain forms of plants in the building up of the calcareous and siliceous material. Among other forms of frequent
occurrence, *Calothrix gypsophila* and a species *Leptothrix* are
mentioned, the former being a member of the Nostocaceae, allied
to *Rivularia,* and the latter a genus of Schizomycetes. In many
of the springs there are found masses of algal jelly like those
previously described by Cohn in the Carlsbad waters. Sections
of such dried jelly showed a number of interlaced filaments
with glassy silica between them. Weed refers to the occurrence of small gritty particles in this mucilaginous material.
These are calcareous oolitic granules which are eventually
cemented together into a compact and firm mass of travertine
by the continued deposition of carbonate of lime. The presence
of the plant filaments is often difficult to recognise in the

[1] Moseley, H. N. (75), p. 321.   [2] Weed (87–88), *vide* also Tilden (97).

"leathery sheet of tough gelatinous material," or in "the skeins of delicate white filaments" which make up the travertine deposits.

Under the head of *Cyanophyceae*, mention should be made of the recent genus *Hyella*[1], which occurs as a perforating or boring alga in the calcareous shells of molluscs. On dissolving the carbonate of lime of shells perforated by this alga, the latter is isolated and appears to consist of rows of small cells, with possibly some sporangia containing spores. Other boring algae have been recorded among the Chlorophyceae, and recently a member of the Rhodophyceae[2] has been found living in the substance of calcareous shells. Such examples are worthy of note in view of the not infrequent occurrence of fossil corals, shells and fish-scales, which have evidently been bored by an organism resembling in form and manner of occurrence these recent algal borers.

The occurrence of small ramifying tubes in recent and fossil corals, fish-scales, and bones was long ago pointed out by Quekett[3], Kölliker[4], Rose[5] and other writers[6]. These narrow tubular cavities have generally been attributed to the boring action of some parasitic organism, either a fungus or an alga. In 1876 Duncan published two important papers[7] dealing with the occurrence of such tubes in recent corals, as well as in the calcareous skeleton of *Calceolina*, *Goniophyllum* and other Palaeozoic, Mesozoic and Tertiary species of corals. This writer attributed the formation of the cavities in the case of the fossil species to the action of a fungus which he named *Palaeachlya perforans*, and considered as very nearly related to *Achlya penetrans* found in the "dense sclerenchyma" of recent corals. In fig. 27 A. is reproduced one of the drawings given by Rose[8]

[1] Bornet and Flahault (89[2]) Pl. XI.   [2] Batters (92).

[3] Quekett (54), fig. 78.   [4] Kölliker (59) and (59[2]); good figures in the latter paper.

[5] Rose (55), Pl. I.

[6] For other references *vide* Bornet and Flahault (89[2]).

[7] Duncan (76) and (76[2]).

[8] Similar borings are figured by Kölliker (59[2]), Pl. XVI. 14, in a scale of *Beryx ornatus* from the Chalk.

in his paper published in 1855; it shows a section of a fish-scale from the Kimeridge clay which has been attacked by a boring organism.   Rose attributes the dichotomously branched canals to some "infusorial parasite."

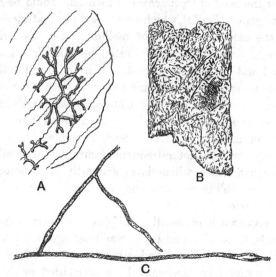

FIG. 27.   A, Section of a fish-scale from the Kimeridge Clay, showing branched canals, made by a boring organism, × 85.   B, Section of a Solen shell, penetrated in all directions by the boring thallus of *Ostracoblabe* (a fungus?), × 330.   C, Piece of the thallus of *Ostracoblabe* isolated by decalcification, × 745.   A, after Rose.   B and C, after Bornet and Flahault.

In the important paper by MM. Bornet and Flahault on perforating algae a full description is given of various boring forms belonging to the Chlorophyceae and the Cyanophyceae[1]. The canals which these algae produce in calcareous shells and other hard substances are of the same type as those previously described in fossil corals, fish-scales and bones.   In dealing with living perforating Thallophytes the colour and other cell-contents often enable us to distinguish between algae and fungi, but in fossil specimens such tests cannot be applied.   The fossil tubular borings may or may not show traces of the trans-verse septa and reproductive cells; it is often the case that no

[1] Bornet and Flahault (89[2]).

trace of the organism has been left, but only the canals by which it penetrated the calcareous or bony skeleton. In some of the examples of *Palaeachlya* figured by Duncan there appear to be numerous spores in some of the sections, but it is generally a very difficult and often an impossible task to discriminate between the borings of fungi and algae in fossil material.

Fig. 27 B, which is copied from one of Bornet and Flahault's drawings, represents a piece of Solen shell riddled with small canals made by the organism which has been named by the French authors *Ostracoblabe implexa*, and regarded by them as a fungus. Fig. 27 C represents a small piece of the vegetative body of *Ostracoblabe* obtained from a decalcified shell. In endeavouring to determine the organism which has produced borings in fossil corals or shells, it must be borne in mind that some forms of canals or passages may have been the work of perforating sponges, but these are larger in diameter than those made by algae or fungi. By some writers[1] the tubular cavities in shells have been referred to true algae, but others consider them to be of fungal origin.

As an example of a fossil alga referred to the Cyanophyceae, the genus *Zonatrichites*[2] may be quoted. Bornemann, who first described the specimens, points out the close resemblance in habit to some members of the recent Rivulariaceae.

Zonatrichites.

The author of the genus defines it as follows:—

"A calcareous alga, with radially arranged filaments, forming hemispherical or kidney-shaped layers, growing on or enclosing other bodies. Parallel or concentric zones are seen in cross-section, formed by the periodic growth of the alga, the older and dead layers serving as a foundation on which the young filaments grow in radially arranged groups."

The nodules which are apparently formed by species of this genus occur in various sizes and shapes; Bornemann describes one hemispherical mass 8 cm. broad and 4 cm. thick. In some

[1] E. G. Wedl (59). Good figures are given in this paper.
[2] Bornemann (86), p. 126, Pls. v. and vi.

cases the organism has given rise to oolitic spherules, which in
radial section exhibit the branched tubular cells spreading
in fan-shaped groups from the centre of the oolitic grain.   The
section parallel to the surface of a nodule presents the appearance
of a number of circular or elliptical tubes cut across transversely
or more or less obliquely.   The resemblance between the fossil
and a specimen of the recent species *Zonatrichia calcivora*
Braun, is certainly very close, but it is very difficult, in the
absence of material exhibiting more detailed structure than is
shown in the specimens described by Bornemann, to decide with
any certainty the true position of the fossil.   The figures do not
enable us to recognise any trace of cells in the radiating tubes.
It is possible that we have in *Zonatrichites* an example of a
Cyanophyceous genus in which only the sheaths of the filaments have been preserved.   In any case it is probable that this
Mesozoic species affords another instance of a fossil alga which
has been responsible for certain oolitic or other structures in
limestone rocks.

The species described by Bornemann was obtained from a
Breccia near Lissau in Silesia, of Keuper age.

M. Renault has recently described certain minute structures
in a Palaeozoic coprolite to which he gives the name *Gloioconis
Borneti*[1], and which he regards as a Permian gelatinous alga
similar to the well-known recent genus *Gloeocapsa*.   The appearances revealed in a section of the coprolite are interpreted by this
author as a collection of small colonies of a unicellular gelatinous
alga in various stages of development.   Renault's figure shows
a spherical group of faintly outlined and cloudy bodies, most
of which include one or two small dark spots.   The latter are
regarded as the cells of the alga, and the surrounding cloudy
substance is described as the gelatinous sheath.   The absence
of a nucleus in these extremely minute fossil cells ($8-10 \mu$ in
diameter) is referred to as an argument in favour of referring
the organism to the Cyanophyceae rather than to the Chlorophyceae.   It is possible that the ill-defined structure described
by Renault may be a petrified alga, but there is not sufficient
evidence to warrant a decided opinion; the absence of nuclei

[1] Renault (96[1]) p. 446.

can hardly be taken seriously in such a case as this as an argument in favour of the Cyanophyceae.

Although our exact knowledge of fossil Cyanophyceae is extremely small, it is probable that such simple forms of plants existed in abundance during the past ages in the earth's history. Several writers have expressed the opinion that the blue-green algae may be taken as the modern representatives of those earliest plants which first existed on an archaean land-surface. The living species possess the power of resisting un-favourable conditions in a marked degree, and are able to adapt themselves to very different surroundings. Their occurrence in hot springs proves them capable of living under conditions which are fatal to most plants, and suggests the possibility of their occurrence in the heated waters which probably constituted the medium in which vegetable life began. An interesting example of the growth of blue-green algae under unfavourable conditions was recorded in 1886 by Dr Treub[1] of the Buitenzorg Gardens, Java. In 1883 a considerable part of the island Krakatoa, situated in the Straits of Sunda, between Sumatra and Java, was entirely destroyed by a terrific volcanic ex-plosion. What remained had been reduced to a lifeless mass of hot volcanic ashes. Three years later, Treub visited the island, and found that several plants had already established themselves on the volcanic rocks. Various ferns and flowering plants were recorded in Treub's description of this newly established flora. It seemed that the barren rocky surface had been prepared for the more highly organised plants by the action of certain forms of Cyanophyceae, which were able to live under conditions which would be fatal to more complex types.

In the petrified tissues of fossil plants there are occasionally found small spherical vesicles, with delicate limiting membranes, in the cavities of parenchymatous cells or in the elements of vascular tissue. Some of these spherical inclusions have been described as possibly simple forms of endophytic algae[2], such as we are now familiar with in species of the Cyanophyceae and other algae. So far, however, no recorded instance of such fossil endophytic algae is entirely satisfactory. Some of the

[1] Treub (88).　　　　　　　[2] Williamson (88).

cells figured by Williamson as possibly algae, endophytic in the tissues of Coal-Measure plants, are no doubt thin-walled vesicles which formed part of a highly vacuolated cell-contents. Examples of such vesicles in living and fossil cells are shown in fig. 42. The fact that the contents of living plant tissues have been erroneously described as endophytic organisms, should serve as a warning against describing fossil endophytes without the test of good evidence to support them.

The description of a fossil *Nostoc* by the late Prof. Heer[1] from the Tertiary rocks of Switzerland cannot be accepted as a trustworthy example of a fossil plant, much less of a genus of recent algae. The application of recent generic names to fossils which are possibly not even organic must do more harm than good.

II. SCHIZOMYCETES (Bacteria).

It is impossible to draw a sharp line between the two sub-divisions of the Schizophyta. The so-called Fission-Fungi or Bacteria differ from the Schizophyceae or Fission-Algae in the cell-contents being either colourless, blood-red or green, but never blue-green. We may regard the Bacteria, generally, as the lowest forms of plants; they are extremely simple organisms which have been derived from some primitive types which possessed the power of independent existence and contained chlorophyll—that important substance which enables a plant to obtain its carbon first-hand from the carbon dioxide of the atmosphere.

Bacteria may be briefly described as single-celled plants, and as de Bary suggested comparable in shape to a billiard ball, a lead pencil or a corkscrew[2]. A single spherical or cylindrical cell measures about 1μ in diameter[3]. They occur either singly or in filaments, or as masses of various shapes consisting of numberless bacterial cells. The nature and manner of life of

[1] Heer (55) vol. I. p. 21, Pl. IV. fig. 2.

[2] de Bary (87) p. 9. A good account of the Schizomycetes has lately been written by Migula in Engler and Prantl's *Pflanzenfamilien*, Leipzig, 1896.

[3] $1\mu = 0.001$ millimetres.

Bacteria, and their extraordinary power of successfully resisting
the most unfavourable conditions, render it probable that they
constitute an extremely ancient group of organisms.

The wonderful perfection of preservation of many fossil
plants enables us to investigate the contents of petrified cells
and to examine in minutest detail the histology of extinct
plants. To those who are familiar with the possibilities of
microscopical research as applied to silicified and calcified
fossil tissues, it is by no means incredible that evidence has been
detected of the existence of Bacteria as far back in the history
of the earth as the Carboniferous and Devonian periods.

Were there no trustworthy records of the occurrence of Bacteria in Palaeozoic times, it would still be a natural supposition
that these ubiquitous organisms must have been abundantly
represented. It has been suggested as a probable conclusion that
some forms of Bacteria, which produced chemical changes in the
soil necessary for the nutrition of plants, must have existed
contemporaneously with the oldest vegetation[1].

The paper-coal of Toula, which in some places reaches a
thickness of 20 cm., is a plant-bed of exceptional interest. It
differs from ordinary coal in being made up of numberless thin
brown-papery sheets associated with a darker coloured substance
largely composed of ulmic acid. Prof. Zeiller[2], in an interesting
account of the papery layers, has shown that they consist of the
cuticles of a Lepidodendroid plant, *Bothrodendron.* An examination of a piece of one of the sheets at once reveals the existence
of a regular network of which the walls of the meshes are the
outlines of the epidermal cells, the meshes being bridged across
by a thin light brown membrane which represents the layer
of cuticularised cell-wall of each epidermal cell. At regular
intervals and disposed in a spiral arrangement, we find small
gaps in the papery cuticle which mark the position of the
Bothrodendron leaves. These Palaeozoic cuticles are not petrified; they are only slightly altered, and have retained the power
of swelling in water, being able to take up stains like recent

[1] James (93[2]), translation of a paper by M. Ferry in the *Revue Mycologique,*
1893.
[2] Zeiller (82).

tissues. It may reasonably be assumed that the persistent cuticles owe their preservation to a greater power of resistance to destructive agents than was possessed by the other tissues of the plant. It is by no means unlikely, as Renault[1] has recently suggested, that as the *Bothrodendron* stem-fragments lay in the swamps or marshes the tissues were gradually eaten away by Bacteria, but the cuticles successfully resisted the attacks of the bacterial saprophytes. The same observer has described what he regards as the actual organism which effected this wholesale destruction, under the name *Micrococcus Zeilleri*. He finds, after treating the cuticles with ammonia to remove the ulmic acid, that there occur numerous minute spherical bodies, each surrounded by a thin envelope, either singly or in groups on the surface of the cuticular membrane. These vary in size from 5μ to 1μ in diameter. I have not been able to detect any satisfactory proof of such *Micrococci* in specimens of the paper-coal which were treated according to Renault's method, but it is extremely probable that this unusual method of preservation of stem-cuticles is the result of selective bacterial action.

Renault believes that some of the minute spherulitic structures which are seen in sections of decayed tissues of Palaeozoic plants owe their origin, in part, to the ravages of bacteria. The disorganisation of parenchymatous cells gives rise to a gelatinous substance in which needle-like crystals of silica may be deposited, from a siliceous solution, in a matrix which has resulted from bacterial activity. In some of the sections of tissues figured by Renault[2] the outlines of a few cells are still indicated by fragments of the partially decayed wall, while in other cells the walls have been completely destroyed by Bacteria of which some are preserved in the centre of the cell-area, forming a kind of nucleus to the siliceous spherulites.

In addition to the *Micrococcus* described by Renault from the Toula paper-coal, there are a host of other forms which

---

[1] Renault (95[1]), (96[1]) p. 478, (96[2]) p. 106. (Several figures of the cuticles are given in these publications.)

[2] Renault (96[1]) p. 492.

have been minutely diagnosed and figured by Profs. Renault and
Bertrand[1]. These authors have discovered what they believe
to be well-defined species of *Micrococcus* and *Bacillus* ranging
in age from Devonian to Jurassic. The material which has
afforded the somewhat startling results of their investigations
consists partly of the coprolites of reptiles and fishes, and of
silicified and calcified plant tissues.

Bacillus Permicus. Ren. and Bert.[2] (Fig. 28 B.)

This *Bacillus,* which was discovered in sections of a Permian
coprolite from Central France, has the form of cylindrical rods
$12—14\mu$ in length, and $1\cdot3—1\cdot5\mu$ broad, rounded at each end.
The rods occur either singly or occasionally, two or three individuals are joined end to end. Fig. 28 B represents a piece of one
of Renault and Bertrand's sections; the small rods are clearly
seen lying in various directions in the homogeneous matrix

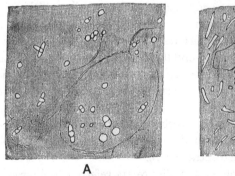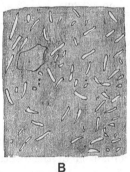

A　　　　　　　　　　　　　B

Fɪɢ. 28.　A, *Bacillus Tieghemi* Ren. and *Micrococcus Guignardi* Ren.
B, *Bacillus Permicus* Ren. (After Renault.)

of the coprolite. Each individual is said to be surrounded
by an extremely minute empty space $\cdot4\mu$ in width, originally
occupied by the Bacillus membrane, the central rod representing
the mineralized cell-contents. In this example the petrifying
substance was probably derived from the phosphate of calcium

[1] Renault (95[2]), (96[1]), (96[2]).
[2] Renault and Bertrand (94). See also Renault (95[2]) p. 3, (96[1]) p. 449,
Pl. ʟxxxɪx. (96[2]) p. 94, and (96[3]) p. 280, fig. 3.

of bones which were attacked by Bacteria. I am indebted to Prof Renault for an opportunity of examining specimens of this and other fossil Bacteria, and in this particular case there is undoubtedly strong evidence in favour of the author's determination.

Bacillus Tieghemi Ren.[1] and *Micrococcus Guignardi* Ren.[2] (Fig. 28 A.)

Renault has given the name *Bacillus Tieghemi* to certain minute rods 6—10μ in length, and 2·2—3·8μ broad, often containing a dark coloured spherical spore-like body 2μ in diameter, which have been found in the tissues of a Coal-Measure plant.

The name *Micrococcus Guignardi* has been applied to more or less spherical bodies 2·2μ in diameter, also met with in silicified plants.

A portion of one of Renault's figures is reproduced in Fig. 28 A. The faint and broken lines mark the position of the middle lamellae of parenchymatous cells from the pith of a Calamite. The tissue has been almost completely destroyed, but the more resistant middle lamellae have been partially preserved. The short and broad rods represent what Renault terms *Bacillus Tieghemi*; the small circle in the middle of some of these being referred to as a spore, and in one specimen shown in the figure, the second rod at right angles to the first is described as a small daughter-Bacillus formed by the germination of the central spore.

The isolated circles in the figure are referred to *Micrococcus*.

It is unnecessary to give an account of the numerous examples of *Micrococci* and *Bacilli* described by Renault from Devonian, Carboniferous, Permian and Jurassic rocks. We may, however, in a few words consider the general question of the existence and possible determination of fossil Bacteria.

In 1877 Prof. Van Tieghem[3] of Paris drew attention to the method of operation and plan of attack of *Bacillus amylobacter*

[1] Renault (95²) p. 17, fig. 9, (96¹) p. 460, fig. 102, and (96³) p. 292, fig. 10.
[2] Renault (96³) p. 297, fig. 14.   [3] Van Tieghem (77).

as a destructive agent in the decay of plant débris in water. He was able to follow the gradual disorganisation of the tissues and the various steps in the 'butyric fermentation' effected by this *Bacterium*. Similarly the same author[1] was able to detect the action of an allied organism in some silicified tissues from the Carboniferous nodules of Grand-Croix, a well-known locality for petrified plants near Saint-Étienne. He recognised also the traces of the *Bacillus* itself in the partially destroyed plant tissues. The Palaeozoic Bacteria made use of some cellulose-dissolving ferment of which the action is clearly demonstrated in sections of silicified tissues. Many of the phenomena described by Renault and Bertrand as due to similar Bacterial action, afford additional evidence that the gradual disorganisation of vegetable tissues was effected in precisely the same manner as at the present day.

In some cases we have I believe trustworthy examples of the Bacteria themselves, both in coprolites and plant-tissues, but it is more than probable that some of the recorded examples are not of any scientific value. The examination of petrified tissues under the higher powers of a microscope often reveals the existence of numerous spherical particles and rod-like bodies which agree in shape with *Micrococci* or *Bacilli*. Minute crystals of mineral substances may occur in the siliceous or calcareous matrix of a petrified plant which simulate minute organic forms. Vogelsang[2] in his important work *die Krystalliten* has thrown considerable light on the ontogeny of crystals, and the minute globulites and other forms of incipient crystallisation might well be mistaken for Bacterial cells. Granting, however, that we have satisfactory evidence, both direct and indirect, that some forms of Bacteria lived in the decaying tissues of Palaeozoic plants, and in the intestines of reptiles and other animals, we cannot safely proceed to specific diagnoses and determinations[3].

---

[1] Van Tieghem (79).   [2] Vogelsang (74). *Vide* also Rutley (92).

[3] I am indebted to Prof. Kanthack for calling my attention to an interesting account of Bacilli in small stones found in gall-bladders; a manner of occurrence comparable to that of the fossil forms in petrified tissues. *Vide* Naunyn (96) p. 51.

Renault has pointed out that fossil Bacteria may often be more readily detected than living forms owing to the presence of a brown ulmic substance which results from the carbonisation of the protoplasm. He is forced to admit, however, that such diagnostic characters as are obtained by Bacteriologists by means of cultures cannot be utilised when we are dealing with fossil examples! We are told that "Partout où nous avons cherché des Bacteriaceés, nous en avons rencontré."[1] This indeed is the danger; an extended examination of fossil sections under an immersion-lens must almost inevitably lead to the discovery of minute bodies of a more or less spherical form which *might* be *Micrococci*. To measure, and name such bodies as definite species of *Micrococci* is, I believe, but wasted energy and an attempt to compass the impossible.

Specialists tell us that the accurate determination of species of recent Bacteria is practically hopeless: may we not reasonably conclude that the attempt to specifically diagnose fossil forms is absolutely hopeless? "The imagination of man is naturally sublime, delighted with whatever is remote and extraordinary—", but it is to be deplored if the fascination of fossil bacteriology is allowed to warp sound scientific sense.

IV. ALGAE.

A. DIATOMACEAE. (Diatoms.)
B. CHLOROPHYCEAE. (Green algae.)
C. RHODOPHYCEAE. (Red algae.)
D. PHAEOPHYCEAE. (Brown algae.)

The presence of chlorophyll is one common characteristic of the numerous plants included in the Algae. The generally adopted classification rests in part on an artificial distinction, namely the prevailing colour of the plant.

It must be definitely admitted, at the outset, that palaeo-botany has so far afforded extremely little trustworthy information as to the past history of algae. Were we to measure

---

[1] Renault (96³) p. 277.

the importance of the geological history of these plants by the number of recorded fossil species, we should arrive at a totally wrong and misleading estimate. By far the greater number of the supposed fossil algae have no claim to be regarded as authentic records of this class of Thallophytes. It has been justly said that palaeontologists have been in the habit of referring to algae such impressions or markings on rocks as cannot well be included in any other group. " A fossil alga," has often been the *dernier ressort* of the doubtful student.

Before discussing our knowledge, or rather lack of knowledge, of fossil algae at greater length, it will be well to briefly consider the manner of occurrence and botanical nature of existing forms. In the sea and in fresh water, as well as in damp places and even in situations subject to periods of drought, algae occur in abundance in all parts of the world. We find them attaining full development and reproducing themselves at a temperature of − 1° C. in the Arctic Seas, and again living in enormous numbers in the waters of thermal springs. Around the coast-line of land areas, and on the floor of shallow seas algae exhibit a remarkable wealth of form and luxuriance of growth. As regards habit and structure, there is every gradation from algae in which the whole individual consists of a thin-walled unseptate vesicle, to those in which the thallus attains a length unsurpassed by any other plant, and of which the anatomical features clearly express a well-marked physiological division of labour such as occurs in the highest plants.

The large and leathery seaweeds which flourish in the extreme northern and southern seas are plants which it is reasonable to suppose might well have left traces of their existence in ancient sediments. Sir Joseph Hooker, in his account of the Antarctic flora[1], investigated during Sir James Ross's voyage in H.M. ships Erebus and Terror, has given an exceedingly interesting description of the gigantic brown seaweeds of southern latitudes. The trunks are described as usually 5—10 feet long, and as thick as a human thigh, dividing towards the summit into numerous pendulous branches which are again broken up into sprays with linear 'leaves.' Hooker

---

[1] Hooker, J. D. (44) p. 457. Pls. CLXVII. CLXVIII. and CLXXI. D.

records how a captain of a brig employed his crew for two
bitterly cold days in collecting *Lessonia* stems which had been
washed up on the beach, thinking they were trunks of trees fit
for burning. On our own coasts we are familiar with the
common *Laminaria*, the large brown seaweed with long and
strap-shaped or digitate fronds which grows on the rocks below
low-tide level. The frond passes downwards into a thick and
tough stipe firmly attached to the ground by special holdfasts.
A transverse section of the stalk of a fairly old plant presents
an appearance not unlike that of a section of a woody plant.
In the centre there is a well-defined axial region or pith
consisting of thick walled, long and narrow tubes pursuing a
generally vertical though irregular course, and embedded in a
matrix of gelatinous substance derived from the mucilaginous
degeneration of the outer portions of the cell-walls. The greater
part of such a section consists, however, of regularly disposed

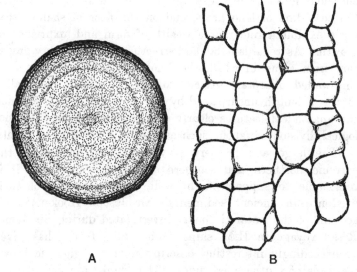

A   B

FIG. 29. A, Transverse section of the stipe of a *Laminaria*, slightly enlarged.
B, A small piece of the tissue between the central 'pith' and 'cortex'
showing the radially disposed secondary elements more highly magnified.

rows of cells which have obviously been formed by the

activity of a zone of dividing or meristematic elements. The occurrence of distinct concentric rings in this secondary tissue clearly points to some periodicity of growth which is expressed by the alternation of narrow and broader cells. In the Antarctic genus *Lessonia*, the stem reaches a girth equal to that of a man's thigh, and in structure it agrees closely with the smaller stem of *Laminaria*. In these large algal stems, the cells are not lignified as in woody plants, and in longitudinal section they have for the most part the form of somewhat elongated parenchyma, differing widely in appearance from the tracheids or vessels of woody plants. At the periphery of the *Laminaria* stem, represented in fig. 29, there occur numerous and comparatively large mucilage ducts.

In certain algae of different families the thallus is encrusted with carbonate of lime, and is thus rendered much more resistant. The Diatoms, on the other hand, possess still more durable siliceous tests which are particularly well adapted to resist the solvent action of water and other agents of destruction. It is these calcareous and siliceous forms which supply the greater part of the trustworthy data furnished by fossil algae.

It remains to consider some of the causes to which we may attribute the scarcity of fossil algae, and the possible sources of error which beset any attempt to describe or assign names to impressions and casts simulating algal forms.

In the first place, the delicate nature of algal cells is a serious obstacle to fossilisation. Even in plants in which the woody stems have been preserved by a siliceous or calcareous solution, we frequently find the more delicate cells represented by a mass of crystalline matter without any trace of the cell-walls being preserved. In such plants as algae, where the cell-walls are not lignified, but consist of cellulose or some special form of cellulose, which readily breaks down into a mucilaginous product, the tissues have but a small chance of withstanding the wear and tear of fossilisation.

The danger of relying on external form as a means of recognition is especially patent in the case of those numerous markings or impressions frequently met with on rocks, and which resemble in outline the thallus of recent algae. Among

animals, such as certain Polyzoa, the flat branching body of
various algae is closely simulated, and in other plants, such as
the frondose liverworts, the same thalloid and branched form
of body is again met with. Some of the much dissected
Aphlebia leaves of ferns (e.g. *Rhacophyllum* species) bear a
striking resemblance to fossil algae; and numerous other
examples might be quoted. In palaeobotanical literature we
find a host of names, such as *Chondrites*, *Fucoides*[1], *Caulerpites*
and others applied to indefinite and indistinct surface markings
which happen to resemble in shape certain of the better
known genera of recent seaweeds.

The close parallelism in outward form displayed by different
genera and families of algae is in itself sufficient argument
against the use of recent generic names for fossils of which
the algal nature is often more than doubtful. Were external
form to be accepted as a trustworthy guide, in the absence
of internal structure and reproductive organs, such a genus
as *Caulerpa*[2] would afford material for numerous generic
designations. A comparison of the different species of this
Siphoneous green alga brings out very clearly the exceedingly
protean nature of this interesting genus, and serves as one
instance among many of the small taxonomic value which
can be attached to external configuration. *Caulerpa pusilla*
Mart. and Her., *C. taxifolia* (Vahl.), *C. plumaris* Forsk.,
C. abies-marina J. Ag., *C. ericifolia* (Turn.), *C. hypnoides*
(R. Br.), *C. cactoides* (Turn.), *C. scalpelliformis* (R. Br.), and
others clearly illustrate the almost endless variety of form exhibited by the species of a single genus of algae. We constantly
find in the several classes of plants a repetition of the same
form either in the whole or in the separate members of the
vegetative body, and but a slight acquaintance with plant
types should lead us to use the test of external resemblance
with the greatest possible caution. To emphasize this danger
may seem merely the needless reiteration of a self-evident fact,

[1] An American writer has recently discussed the literature and history of
Fucoides; he gives a list of 85 species. It is very doubtful if such work as this
is worth the labour. (James [93].)

[2] Wille (97) p. 136, also Murray, G. (95) p. 121.

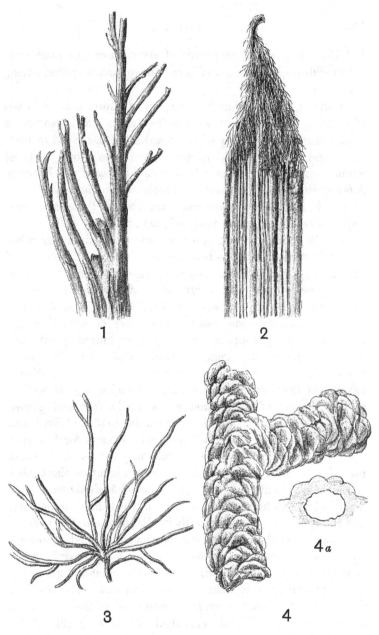

Fig. 30. 1. Rill-mark (after Williamson). 2. Trail made by a seaweed dragged along a soft plaster of Paris surface (after Nathorst). 3. Tracks made by *Goniada maculata*, a Polychaet (after Nathorst). 4. Burrow of an insect. 4 *a*. Section of the gallery (after Zeiller).

but there is, perhaps, no source of error which has been more responsible for the creation of numerous worthless species among fossil plants.

There is, however, another category of impressions and casts of common occurrence in sedimentary rocks which requires a brief notice. Very many of the fossil algae described in text-books and palaeobotanical memoirs have been shown to be of animal origin, and to be merely the casts of tracks and burrows. A few examples will best serve to illustrate the identity of many of the fossils referred to algae with animal trails and with impressions produced by inorganic agency.

Dr Nathorst of Stockholm has done more than any other worker to demonstrate the true nature of many of the species of *Chondrites, Cruziana, Spirophyton, Eophyton,* and numerous other genera. In 1867 there were discovered in certain Cambrian beds of Vestrogothia, long convex and furrowed structures in sandstone rocks which were described as the remains of some comparatively highly organised plant, and described under the generic name *Eophyton*[1]. By many authors these fossils have been referred to algae, but Nathorst has shown that the frond of an alga trailed along the surface of soft plaster of Paris produces a finely furrowed groove (fig. 30, 2) which would afford a cast similar to that of *Eophyton*. The same author has also adduced good reasons for believing that the Eophytons of Cambrian rocks may represent the trails made by the tentacles of a *Medusa* having a habit similar to that of *Polydonia frondosa* Ag. Impressions of *Medusae* have been described by Nathorst from the beds in which *Eophyton* occurs; and the specimens in the Stockholm Museum afford a remark-able instance of the rare preservation of a soft-bodied organism[2]. By allowing various animals to crawl over a soft-prepared surface it is possible to obtain moulds and casts which suggest in a striking manner the branched thallus of an alga. The tracks of the Polychaet, *Goniada maculata* Örstd.[3], one of the Glyceridae, are always branched and very algal-like in form (fig. 30, 3).

---

[1] Linnarsson (69) Pl. xi. fig. 3. There are many good specimens of this fossil in the Geological Survey Museum, Stockholm.

[2] Nathorst (81[2]), and (96).　　　　　　　　[3] Nathorst (81) p. 14.

Many of the so-called fossil algae are undoubtedly mere tracks or trails of this type. In the fossil-plant gallery of the British Museum there are several specimens of small branched casts, clearly marked as whitish fossils on a dark grey rock of Upper Greensand age from Bognor; these were described by Mantell and Brongniart[1] as an alga, but there is little doubt of their being of the same category as the track shown in fig. 30, 3.

The well-known half-relief casts met with in Cambrian, Silurian and Carboniferous rocks, and known as *Cruziana* or Bilobites, are probably casts of the tracks of Crustaceans. The impression left by a King-Crab (*Limulus*) as it walks over a soft surface affords an example of this form of cast. It has been suggested that some of the Bilobites may be the casts of an organism like *Balanoglossus*[2], a worm-like animal supposed by some to have vertebrate affinities. The resemblance between some of the lower Palaeozoic Bilobites and the external features of a *Balanoglossus* is very striking, and such a comparison is worth considering in view of the fact that soft-bodied animals have occasionally left distinct impressions on ancient sediments.

The literature on the subject of fossil algae *versus* inorganic and animal markings is too extensive and too wearisome to consider in a short summary; the student will find a sufficient amount of such controversial writing—with references to more—in the works quoted below[3].

In the Stockholm Museum of Palaeobotany there is an exceedingly interesting collection of plaster casts obtained by Dr Nathorst in his experiments on the manufacture of fossil 'algae,' which afford convincing proof of the value and correctness of his general conclusions.

The pressure of the hand on a soft moist surface produces a raised pattern like a branched and delicate thallus. The well-known *Oldhamia antiqua* Forbes and *Oldhamia radiata* Forbes[4], from the Cambrian rocks of Ireland may, in part at least, owe their origin to mechanical causes, and we have no sufficient

---

[1] Mantell (33) p. 166. *Vide* also Morris (54) p. 6.   [2] Bateson (88).
[3] Nathorst (81), (86) &c. Dawson (88) p. 26 *et seq*. Dawson (90) Delgado (86) Williamson (85) Hughes (84) Zeiller (84) Saporta (81) (82) (84) (86) Fuchs (95) Rothpletz (96).   [4] Kinahan (58).

evidence for including them among the select class of true fossil algae. Sollas[1] has shown that the structure known as *Oldhamia radiata* is not merely superficial but that it extends across the cleavage-planes. *Oldhamia* is recorded from Lower Palaeozoic rocks in the Pyrenees[2] by Barrois, who agrees with Salter, Göppert and others in classing the fossil among the algae. The photograph accompanying Barrois' description does not, however, add further evidence in favour of accepting *Oldhamia* as a genus of fossil algae.

The burrows made by *Gryllotalpa vulgaris* Latr., the Mole-cricket, have been shown by Zeiller to bear a close resemblance to a branch of a conifer in half-relief (fig. 30, 4), or to such a supposed algal genus as *Phymatoderma*[3].

In fig. 30, 1, we have what might well be described as a fossil alga. This is merely a cast of a miniature river-system such as one frequently sees cut out by the small rills of water flowing over a gently-sloping sandy beach. A cast figured and described by Newberry as an alga, *Dendrophycus triassicus*[4], from the Trias of the Connecticut Valley, is practically identical with the rill-marks shown in fig. 30, 1. The cracks produced in drying

FIG. 31. *Chondrites verisimilis* Salt. Wenlock limestone, Dudley. From a specimen in the British Museum (V. 2550). Slightly reduced.

and contracting sediment may form moulds in which casts are subsequently produced by the deposition of an overlying layer of sand, and such casts have been erroneously referred to algal

[1] Sollas (86).

[2] Barrois (88). References to other records of this genus may be found in Barrois' paper.

[3] Zeiller (84). *Phymatoderma* is probably a horny sponge (*vide* p. 154).

[4] Newberry (88) p. 82, Pl. xxi. There are some large specimens of this supposed alga in the National Museum, Washington; they are undoubtedly of the nature of rill-marks.

impressions[1]. Dawson[2] has figured two good examples of
Carboniferous rill-marks from Nova Scotia in his paper on
Palaeozoic burrows and tracks of invertebrate animals.

The specimen represented in fig. 31 affords an example of
a fairly well-known fossil from the Wenlock limestone, originally
described by Salter as *Chondrites verisimilis* Salt, from Dudley[3].
He regarded it as an alga, and the graphitic impression agrees
closely in form with the thallus of some small seaweeds. A
closer examination of the fossil reveals a curious and characteristic irregular wrinkling on the graphite surface, which suggests
an organism of more chitinous and firmer material than that of
an alga.

A similar and probably an identical fossil is described and
figured by Lapworth[4] in an appendix to a paper by Walter
Keeping on the geology of Central Wales, under the name
of *Odontocaulis Keepingi* Lap. and regarded as a dendroid
graptolite. In any case we have no satisfactory grounds for
including these fossils in the plant-kingdom.

How then are we to recognise the traces of ancient
algae? There is no golden rule, and we must admit the
difficulty of separating real fossil algae from markings made by
animal or mechanical agency. The presence of a carbonaceous
film is occasionally a help, but its occurrence is no sure test of
plant origin, nor is its absence a fatal objection to an organic
origin. While being fully alive to the small value of external
resemblance, and to the numerous agents which have been
shown to be capable of producing appearances indistinguishable
from plant impressions, we must not go too far in a purely
negative direction.

An important contribution to the subject of fossil algae has
lately appeared by Prof. Rothpletz[5]. He deals more particularly
with the much discussed Flysch[6] Fucoids of Tertiary age, and
while refusing to accept certain examples as fossil algae, he
brings forward weighty arguments in favour of including several
other forms among the algae. He is of opinion that most of the

---

[1] *Vide* Williamson (85).   [2] Dawson (90) p. 615.   [3] Salter (73) p. 99.
[4] Lapworth (81) p. 176, Pl. vii. fig. 7.   [5] Rothpletz (96).
[6] A term applied to a certain facies of Eocene and Oligocene rocks in
Central Europe.

main divisions of the algae are represented among the Flysch
Fucoids, but considers that the Phaeophyceae are the most
numerous.

Rothpletz's work is chiefly interesting as illustrating the
application of microscopic examination and chemical analysis to
the determination of fossil algae. Although he makes out a
good case in favour of restoring many of the Tertiary fossils to
the plant kingdom, the material at his disposal does not admit
of satisfactory botanical diagnosis.

No doubt some of the fossils from the Silurian and Cambrian
rocks are true algae, and Nathorst has pointed out that such a
species as Hall's *Sphenothallus angustifolius*[1] may well be an alga.
Additional examples might be quoted from Bornemann and
other writers, but in view of the attempts which are sometimes
made to trace the development of more recent plants to more
than doubtful Lower Palaeozoic Algae, one must agree with
Nathorst's opinion,—" Je crois que l'on rend un bien mauvais
service à la théorie de l'évolution, en essayant de baser l'arbre
généalogique des algues fossiles sur des corps aussi douteux
que les Bilobites, Crossochorda, Eophyton, etc.[2]"

There are many carbonaceous impressions on rocks of
different ages which it is reasonable to refer to algal origin,
and although such are of little or no botanical value, it may be
a convenience to refer to them under a definite term. The
comprehensive generic name *Algites*[3] has been suggested as
a convenient designation for impressions or casts which are
probably those of algae.

Some of the fossils described by Mr Kidston from British
Carboniferous rocks as probably algae present an undoubted
algal appearance, and might be placed in the genus *Algites*;
but in some cases—e.g. *Chondrites plumosa*[4] Kidst. from the
Calciferous Sandstone of Eskdale, one feels much more doubtful;
in this particular instance the impressions suggest the fine
roots of a water-plant.

[1] Hall (47) Pl. LXVIII. 1 and 2, p. 261.
[2] Nathorst (83) p. 453.
[3] Seward (94²) p. 4.
[4] Kidston (83) Pl. XXXII. fig. 2. Specimens of this form may be seen in the
British Museum collection.

The statement is occasionally made that the numerous fossil algae and the absence of higher plants in the older strata justify the description of the oldest rocks as belonging to the 'age of algae.' Such an assertion rests on an unsound basis, and is rather the expression of what might be expected than what has been proved to be the case. The oldest plants with which we are at all closely acquainted are of such a type as to forcibly suggest that in the lowest fossiliferous rocks we are still very far from the sediments of that age which witnessed the dawn of plant life.

Many of the obscure markings on rock surfaces which have been referred to existing genera of algae or described as new genera, are much too doubtful to be included even under such a comprehensive name as *Algites*. Space does not admit of further reference to determinations of this type which abound in palaeontological literature.

It would be very difficult to produce satisfactory evidence for the algal nature of many of the supposed fossil algae from Cambrian rocks[1]; there has been a special tendency to recognise algal remains in the oldest fossiliferous strata, due in part no doubt to the fallacy that in that period nothing higher than Thallophytes is likely to have existed. The so-called *Phycodes* referred to by Credner[2] as characteristic of the Cambrian rocks of the Fichtelgebirge (" Phycoden-Schiefer ") is probably of inorganic origin, and comparable to the genus *Vexillum* of Saporta[3] and other writers, which Solms-Laubach has described as being formed every day in the soft mud of our ponds where local currents are checked by branches and other obstacles[4]. There are several good specimens of *Phycodes* in the Berg-akademie of Berlin and in the Leipzig Museum which, I believe, clearly demonstrate the absence of all satisfactory evidence of an algal origin.

We may next pass to a short description of a few representative types of algae, which may reasonably be classed under

---

[1] *Cf.* Matthew, G. F. (89). Hall called attention in 1852 to the prevalent habit of describing 'algae' from the older strata, without any evidence for a vegetable origin. (Hall [52] p. 18.)

[2] Credner (87) p. 431.   [3] Saporta (84) p. 45, Pl. VII.

[4] Solms-Laubach (91) p. 51.

definite families, and accepted as evidence possessing some botanical value.

A. DIATOMACEAE (BACILLARIACEAE).

This family occupies a somewhat isolated position among the algae, and is best considered as a distinct subdivision rather than as a family of the Phaeophyceae or Brown algae, with which it possesses as a common characteristic a brown-colouring matter.

Single-celled plants consisting of a simple protoplasmic body containing a nucleus and brown colouring matter (diatomin) associated with the chlorophyll. The cell-wall is in the form of two halves, known as *valves*, which fit into one another like the two portions of a pill-box. The cell-wall contains a large amount of silica, and the siliceous cases of the diatoms are commonly spoken of as the valves of the individual, or the *frustules*. Diatoms exhibit a characteristic creeping movement, and are reproduced by division, also by the development of spores in various forms[1].

The recent members of the family have an exceedingly wide distribution, occurring both in freshwater and in the sea. Owing to the lightness of the frustules, they are frequently carried along in the air, and atmospheric dust falling on ships at sea has been found to contain large numbers of diatoms[2]. The siliceous valves are abundant in guano deposits, and they have been found also in association with volcanic material. Diatomaceous deposits are now being formed in the Yellowstone Park district; "they cover many square miles in the vicinity of active or extinct hot spring vents of the park, and are often three feet, four feet, and sometimes five to six feet thick[3]." The gradual accumulation of the siliceous tests on the floor of a fresh-water lake results in the formation of a sediment consisting in part of pure silica. Such deposits, often spoken of as *kieselguhr*

---

[1] A monograph on the Diatomaceae has recently been written by Schütt for Engler and Prantl's systematic work. See also Murray, G. (97) and Pfitzer (71).

[2] Darwin (90) p. 5.　　　　　　　　　　　[3] Weed (87).

or *diatomite*, and used as a polishing material, occur in many parts of Britain, marking the sites of dried-up pools or lakes. At the northern end of the island of Skye there occurs an unusually pure deposit of diatomite overlain by peat and turf, and extending over an area of fifty-eight square miles. Many of the individuals in this deposit were in all probability carried into the lake by running water, while others lived in the lake and after death their tests contributed to the siliceous deposit[1]. The late Dr Ehrenberg published numerous papers on diatomaceous deposits in different parts of the world, and in his great work, *Zur Mikrogeologie*[2], he gave numerous and beautifully executed illustrations of such siliceous accumulations. In many of the samples he figures one sees fragments of plant tissues, spores of conifers and ferns, associated with the diatom tests. The occurrence of the pollen grains of coniferous trees in lacustrine and marine deposits is not surprising in view of their abundance in Lake Constance and other lakes. It is stated that the pollen of conifers in the Norwegian fiords plays an important part in the nourishment of the Rhizopod *Saccamina*[3].

In the waters of the ocean diatoms are of frequent occurrence, and very widely distributed. Sir Joseph Hooker records the existence of masses of diatomaceous ooze over a wide area in Antarctic regions[4]. Along the shores of the Victoria Barrier, a perpendicular wall of ice, between one and two hundred feet above sea-level, the soundings were found to be invariably charged with diatom remains, and from the base of the ice-wall there appeared to be in process of formation a bank of these tests stretching north for a distance of 200 miles. The more extended researches conducted during the cruise of the Challenger have clearly proved the enormous accumulations of diatoms now being formed on the ocean-bed[5]. South of latitude 45° S. there is now being built up a vast deposit which may be eventually upraised as a fairly pure siliceous rock. From extreme northern latitudes Nansen has recently

---

[1] Wilson (87).   [2] Ehrenberg (54).
[3] Noll (95) p. 248.   [4] Hooker, J. D. (44) vol. i. p. 503.
[5] Murray, J. and Renard (91) p. 208.

recorded the occurrence of these lowly organised plants. He writes,—"I found a whole world of diatoms and other micro-scopical organisms, both vegetable and animal, living in the fresh-water pools on the Polar drift-ice, and constantly travelling from Siberia to the east coast of Greenland[1]." In warmer latitudes diatoms abound in the surface waters, but there they are associated with numerous other forms of the Plankton vegetation. The waters of the Amazon carry with them into the sea large numbers of fresh-water forms, which are floated out to sea and finally added to the rock-building material which is constantly accumulating on the ocean floor[2]. No definite results have so far been obtained as to the geographical and bathymetrical distribution of marine diatoms.

The enormous number of recent species precludes any attempt to give a description of the better-known forms. It is more important for us to realize how common and widely distributed are the living genera. The hard and almost indestructible valves have been frequently found in a fossil condition, often forming thick and extensive masses of siliceous rock. From diatom-beds now forming in lakes and on the ocean-bed we pass to deposits such as those in Skye and elsewhere, which mark the site of recently dried-up sheets of water, and so to older rocks of Tertiary age formed under similar conditions. Among the many examples of diatomaceous deposits of Tertiary and Cretaceous age mention should be made of those of Berlin, Königsberg, Bilin in Bohemia, and Richmond in Virginia. The diatoms in the beds of Berlin are regarded as fresh-water, and those of Richmond as marine. It has been pointed out by Pfitzer that it is a comparatively easy matter to distinguish between fresh-water and marine forms of diatoms. The diatomaceous rocks of Bilin are known as polishing slates; they attain a thickness of 50 feet. In these, as in many other cases, the deposit has become cemented together as a hard flinty or glassy rock, in which the cementing material was formed by the solution of some of the diatom tests[3]. In many cases in which calcareous and siliceous rocks

---

[1] Nansen, Daily Chronicle, Nov. 2, 1896.   [2] Schütt (93) p. 10.
[3] Ehrenberg (36) p. 77.

reveal no direct evidence of organic origin it is probable that they were originally formed by the accumulation of plants of which the structure has been completely obliterated by secondary causes. The genus *Gallionella* plays an important part in the composition of the Bilin beds. Occasionally impressions of leaves and other organic remains are found associated with the diatoms in the siliceous rocks. In the British Museum (Botanical department) a large block of white powdery rock is exhibited as an example of a diatomaceous deposit of Tertiary age from Australia. It is described as being largely made up of the tests of fresh-water diatoms, such as *Navicula*, *Gomphonema*, *Cymbella*, *Synedra*, and others.

The abundance of Diatoms in Cretaceous rocks of the Paris basin has recently been recorded by Cayeux[1]; it would seem that these algae had already assumed an important rôle as rock-builders in pre-Tertiary times. Cayeux points out that the silica of these Cretaceous diatomaceous frustules has often been replaced by carbonate of calcium.

In addition to the occurrence of Diatoms in the various diatomaceous deposits, their siliceous tests may occasionally be recognised in argillaceous or other sediments. Shrubsole and Kitton[2] have described several species of Diatoms from the London Clay of Lower Eocene age. In many localities in the London basin the clay obtained from well-sinkings presented the appearance of being dusted with sulphur-like particles of a dark bronze or golden colour which glistened in the sunlight. These yellow bodies have been found to be diatomaceous frustules in which the silica has been replaced by iron pyrites. The genus *Coscinodiscus* is one of the commonest forms recorded from the London Clay[3].

Without further considering individual examples of diatomaceous rocks we may briefly notice the general facts of the geological history of the family. As Ehrenberg pointed out several years ago, the Tertiary and Cretaceous species of diatoms show a very marked resemblance to living forms. In

[1] Cayeux (92), (97).　　　　　　[2] Shrubsole and Kitton (81).

[3] I am indebted to Mr Murton Holmes for specimens of these London Clay Diatoms.

many cases the species are identical, and the fossil deposits as a whole seem to differ in no special respect from those now being built up.

With the exception of two species of Liassic Diatoms, no trustworthy examples of the Diatomaceae have been found below the Cretaceous series. The oldest known Diatoms were discovered by Rothpletz[1] among the fibres of an Upper Lias sponge from Boll in Würtemberg. They occur as small thimble-shaped siliceous tests with coccoliths and foraminifera in the horny skeleton of *Phymatoderma*, a genus formerly regarded as an alga. Rothpletz describes two species which he includes in the genus *Pyxidicula*, *P. bollensis* and *P. liasica*. This generic name of Ehrenberg is used by Schütt[2] as a sub-genus of *Stephanopyxis*.

Seeing how great a resemblance there is between the recent and Cretaceous species, and how many examples there are of Tertiary diatom deposits, it is not a little surprising that the past history of these plants has not been traced to earlier periods. In 1876 Castracane[3], an Italian diatomist, gave an account of certain species of diatoms said to have been found in a block of coal from Liverpool obtained from the English Coal-Measures. The species were found to be identical with recent forms. It is generally agreed that these specimens cannot have been from the coal itself, but that they must have been living forms which had come to be associated with the coal. The late Prof. Williamson spent many years examining thin sections and other preparations of coal from various parts of the world, but he never found a trace of any fossil diatom. There is no apparent reason why diatoms should not be found in Pre-Cretaceous rocks, and the microscopic investigation of old sediments may well lead to their discovery. Prof. Bertrand of Lille, who has devoted himself for some time past to a detailed microscopical examination of coal, informs me that he has so far failed to discover any trace of Palaeozoic diatomaceous tests.

The genus *Bactryllium* is often quoted in text-books as a probable example of a Triassic diatom. It was first described

[1] Rothpletz (96) p. 910, fig. 3, Pl. xxiii. fig. 203.
[2] Schütt (96) p. 62.   [3] Castracane (76).

by Heer[1] from the Trias of Switzerland and North Italy, also from the neighbourhood of Heidelberg, and regarded as an extinct member of the Diatomaceae. Heer defined the genus as follows:

"Small bodies, with parallel sides, rounded at either end, the surface traversed by one or two longitudinal grooves."

(fig. 32, C.) Several species have been figured by Heer from beds of Muschelkalk, Keuper and Rhaetic age. He describes the wall as thick and firm (fig. 32, C. ii.) and probably com-

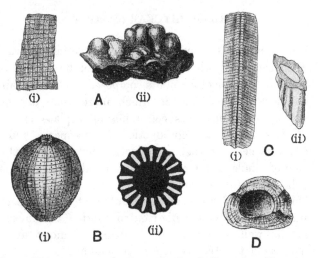

FIG. 32. *A, Lithothamnion mamillosum* Gümb. (i) In section, (ii) surface view [after Gümbel. (i) × 320, (ii) nat. size]. *B, Sycidium melo* Sandb. (i) Surface view, (ii) transverse section (after Deecke). *C, Bactryllium deplanatum* Heer. (i) Surface view, (ii) section, showing the thick wall and hollow interior (after Heer). *D*, Calcareous pebble from a lake in Michigan. Rather less than nat. size (after Murray).

posed of silica, with a hollow interior. The specimen shown in fig. 32, C. was found in the Rhaetic beds, and named by Heer *Bactryllium deplanatum*; it has a length of 4·5 mm.; the surface is transversely striated and traversed by a single longitudinal groove. Stefani[2] has given reasons in favour of removing

---

[1] Heer (76) p. 66, Pl. xxiii. and (53) p. 117, Pl. vi.
[2] Stefani (82) p. 103.

Bactryllium from the plant to the animal kingdom; he points
out that the specimens are too large for diatoms, and moreover
that they are asymmetrical in form and possessed a calcareous
and not a siliceous shell. He would place the fossil among
the Pteropods, comparing it with such genera as *Cuvierina*
and *Hyalaea*. In view of Stefani's opinion we cannot attach
any importance to this supposed diatom, especially as it has
generally been regarded as at best but an unsatisfactory genus.

B. CHLOROPHYCEAE (Green Algae).

Thallus unseptate, having the form of a vesicle or a variously
branched coenocyte, which may or may not be encrusted with
carbonate of lime, or of filaments composed of cells containing a
single nucleus, or of cells in which more than one nucleus
occurs; in other instances consisting of a plate of cells or
a cell-mass. Asexual reproduction by zoospores and other
reproductive cells; sexual reproduction by means of the conjugation of similar gametes or by the fertilisation of a typical
egg-cell by a motile spermatozoid.

This family of algae is represented at the present day
by numerous and widely distributed marine and fresh-water
genera, as well as by genera growing in moist air or as
endophytes in the tissues of higher plants[1].

Seeing how very few fossil forms have been described which
have any claim to inclusion in this subdivision of the Algae,
it is unnecessary to enumerate or define the various families
of the Chlorophyceae. It is true that many species have been
figured as examples of different genera of green algae, but few
of these possess any scientific value. There is, however, one
division of the Chlorophyceae, the Siphoneae, which must be
treated at some length on account of its importance from a
palaeobotanical and geological point of view.

---

[1] The Chlorophyceae have recently been exhaustively dealt with by Wille
(97) in Engler and Prantl's *Pflanzenfamilien*.

α. SIPHONEAE.

Thallus consisting of simple or branched cells very rarely divided by septa, and containing many nuclei. In certain genera the branches form a pseudoparenchymatous tissue by their repeated branching, and as a result of the intimate felting together of the branched cells. Reproduction is effected either by the conjugation of similar gametes or by the fertilisation of an egg-cell.

Vaucheria and *Botrydium* are two well-known British genera of this order, but most of the recent representatives live in tropical and sub-tropical seas. The most striking characteristic feature of this division of the Chlorophyceae is the fact that the thallus of a siphoneous alga consists of an unseptate coenocyte; the plant may be extremely small and simple, or it may reach a length of several inches, but in all cases the body does not consist of more than one cell or coenocyte.

From a palaeontological standpoint the Siphoneae are of exceptional interest. It is impossible to do more than refer to a few of the living and fossil genera. There are numerous fossil representatives already known, and there can be little doubt that further research would be productive of valuable results.

As examples of the order, a few genera may be described belonging to the three families Caulerpaceae, Codiaceae, and Dasycladaceae.

α. **Caulerpaceae.**

Thallus unseptate, showing an extraordinary variation in the external differentiation of the plant-body. Reproduction is effected by means of detached portions of the parent plant.

The genus *Caulerpa*, represented by a few species in the Mediterranean and by many tropical forms, has already been alluded to as a striking example of a plant which appears under a great many different forms[1]. As a recent writer has said, "Nature seems to have shown in this genus the utmost possibilities of the siphoneous thallus[2]." Fragments of

---

[1] *Vide* p. 142.   [2] Murray G. (95) p. 123.

coniferous twigs, the tracks and burrows of various animals and other objects have been described by several authors as fossil species of *Caulerpa*. As an illustration of the identification of a very doubtful fossil as a species of *Caulerpites*, reference may be made to such a form as *C. cactoides* Göpp.[1] from Silurian and Cambrian rocks. There are several examples of this fossil in the Brussels Museum which probably owe their origin to some burrowing animal, and may be compared with Zeiller's figures of the tunnels made by the mole-cricket (fig. 30, 4)[2].

Mr Murray, of the British Museum, has recently described what he regards as a trustworthy example of a fossil *Caulerpa* from the Kimeridge Clay near Weymouth[3]. Specimens of the fossil were first figured in a book on the geology of the Dorset coast as casts of an equisetaceous plant[4].

To this fossil Murray has assigned the name *Caulerpa Carruthersi*, and given to it a scientific diagnosis. The best specimens have the form of a slender central axis, giving off at fairly regular intervals whorls of short and somewhat clavate branches; they bear a superficial resemblance to such a recent species as *Caulerpa cactoides* Ag. An examination of several examples of this fossil leads me to express the opinion that there is not sufficient reason for assigning to them the name of a recent genus of algae[5]. To use the generic name of a recent plant without following the common custom of adding on the termination "ites" (i.e. *Caulerpites*) is as a general rule to be avoided in dealing with fossil forms; and there are, I believe, no satisfactory grounds for referring to these fossils as trustworthy examples of a Mesozoic alga.

In the present case I am disposed to regard the *Caulerpa*-like casts as of animal rather than plant origin. The clavate branches have the form of very deep moulds in the hard brown rock which have been filled in with blue mud. It is hardly conceivable that the branches of a soft watery plant such as *Caulerpa* could leave more than a faint impression on an old sea-floor. The specimens occur in different positions in the matrix

[1] Göppert (60) p. 439. Pl. xxxiv. fig. 8.
[2] Zeiller (84).   [3] Murray G. (92) p. 11; also (95) p. 127.
[4] Damon (88) Pl. xix. fig. 12.   [5] *Vide* also Rothpletz (96) p. 894.

of the rock and they are not confined to the lines of bedding; in none of the examples is there any trace of carbonaceous matter in association with the deep moulds. On the whole, then, this Kimeridge fossil cannot, I believe, be accepted as an authentic example of a Mesozoic *Caulerpa*.

It is not improbable that some of the supposed fossil algae may be casts of egg-cases or spawn-clusters of animals. In Ellis' Natural History of the Corallines[1] there is a drawing representing a number of disc-like ovaries attached to a tough ligament, and referred to the mollusc *Buccinum*, which bears a certain resemblance to the Weymouth fossil. A similar body is figured by Fuchs[2] in an important memoir on supposed fossil algae.

It is not suggested that the *Caulerpa Carruthersi* of Murray should be regarded as the cast of some molluscan egg-case attached to a slender axis, but it is important to bear in mind the possibility of matching such extremely doubtful fossils with other organic bodies than the thallus of a *Caulerpa*. In an example of an egg-case in the Cambridge Zoological Museum, referred to a species of *Pyrula,* there is a hard, long and slender axis, bearing a series of semicircular chambers divided into radial compartments. The whole is hard and horny and might well be preserved as a fossil.

β. **Codiaceae.**

The members of this Order present a considerable diversity of form as regards the shape of the plant-body; the thallus of some species is encrusted with carbonate of lime. The order is widely distributed in tropical and temperate seas.

Among the recent genera *Penicillus* and *Codium* may be chosen as important types from the point of view of fossil representatives.

Codium.

The thallus of *Codium* consists of a spongy mass of tubular cell-branches which are differentiated into two fairly distinct regions, an outer peripheral layer in which the branches have

---

[1] Ellis (1755) Pl. xxxiii. *a* p. 86.   [2] Fuchs (95) Pl. viii. fig. 3.

long club-shaped terminations, and an inner region consisting
of loosely interwoven filaments.

Codium Bursa L. and *C. tomentosum* Huds. are two well-known
British species, the former presents the appearance of a spongy
ball of cells, and in the latter the thallus is divided up into
dichotomously forked branches[1]. In this genus the thallus
is not encrusted with carbonate of lime, at least in recent
species.

Sphaerocodium. Fig. 37, D.

Rothpletz[2] instituted this genus for certain small spherical
or oval bodies varying from 1 mm. to 2 cm. in diameter, which
have been found on crinoid stems or shell fragments of Triassic
age. Each spherical body consists of dichotomously branched
single-celled filaments, between 50 and 100μ in breadth, and
from $300-500\mu$ in height. The tubular cavities occasionally
swell out into spherical spaces which are regarded by Rothpletz
as sporangia.

There is not sufficient evidence that *Sphaerocodium Bornemanni* Roth. has been correctly referred to the Codiaceae.
The sporangia-like swellings described by the author of the
species are not by any means conclusive as characters of
important taxonomic value. Figure 37, D, illustrates the
general structure of the fossil as seen in a transverse section
of one of the calcareous grains.

Like *Girvanella*, which has been referred by some writers
to the Siphoneae, *Sphaerocodium* occurs in the form of oolitic
grains. In the Triassic Raibler and St Cassian beds of the
Tyrol, as well as in rocks of Rhaetic age in the Eastern Alps, it
makes up large masses of limestone. Rothpletz compares the
structure of this genus with that of the recent alga *Codium
adhaerens* Ag., but it is wiser to regard such tubular structures
as *Girvanella*, *Siphonema*[3] and *Sphaerocodium* as closely allied
organisms, which are probably algae, but too imperfectly known
to be referred to any particular family.

[1] Murray G. (95) Pl. III. figs. 1 and 2.
[2] Rothpletz (90), and (91) Pls. XV. and XVI.
[3] Bornemann (87) p. 17, Pl. II. pp. 1-4.

Penicillus.

The recent genus *Penicillus* is one of those algae formerly included among animals. Fig 33, O, has been copied from a drawing of a species of *Penicillus* given by Lamouroux[1] under the generic name of *Nesea* in his treatise on the genera of Polyps published in 1821. He describes the genus as a brush-like Polyp with a simple stem.

The thallus consists of a stout stem terminating in a brush-like tuft of fine dichotomously-branched filaments. The apical branches are divided by regular constrictions into short oval or rod-like segments which may be encrusted with carbonate of lime. A few of the segments from the terminal tuft of a recent *Penicillus* are shown in fig. 35, E. Each of these calcareous segments has the form of an oval shell perforated at each end, and the wall is pierced by numerous fine canals. *Penicillus* is represented by about 10 recent species, which with one exception live in tropical seas.

The recognition of *Penicillus*, or a very similar type, in a fossil condition is due to Munier-Chalmas[2]. This keen observer has rendered great service to palaeobotany by directing attention to the calcareous algae in the Paris basin beds, and by proving that many of the fossils from these Tertiary deposits have been erroneously included by previous writers among the Foraminifera[3]. It is greatly to be desired that Prof. Munier-Chalmas may soon publish a monograph on the fossil Siphoneous forms of which he possesses a unique collection.

Ovulites. Figs. 33, K, L. and 35, F.

In his Natural History of Invertebrate Animals, Lamarck described some small oval bodies from the Calcaire Grossier (Eocene) of the Paris basin under the name of *Ovulites*. He

---

[1] Lamouroux (21) Pl. xxv. fig. 5, p. 23.

[2] Munier-Chalmas (79).

[3] For references to genera of calcareous algae previously referred to Foraminifera, *vide* Sherborn (93).

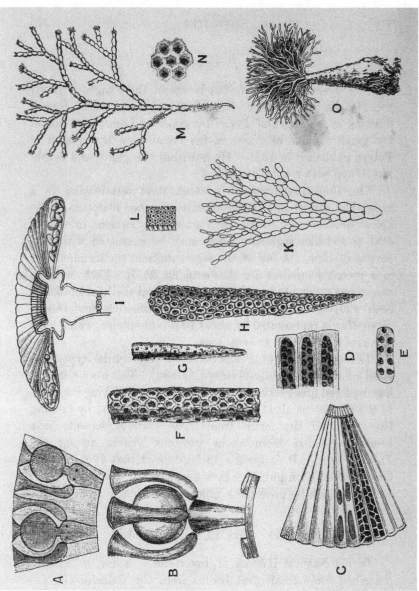

Fig. 33. A and B, *Cymopolia barbata* (L.); A, transverse section of the calcareous cylinder. B, verticillate branches and sporangium after removal of the calcareous matrix (A and B after Munier-Chalmas). C and D, *Acicularia Andrussowi* Solms (C, after Andrussowi; D, after Solms). E, *Acicularia Miocenica* Reuss; section of a spicula (after Reuss). F and G, *Acicularia sp.* (after Carpenter), F×40; G×20. H, *Acicularia Schencki* (Möb.) (after Solms). I, *Acetabularia Mediterranea* Lamx.; section of the cap (after Falkenberg). K and L, *Ovulites margaritula* (Lam.) (after Munier-Chalmas); K slightly enlarged; L, a piece of the thallus more highly magnified. M, *Corallina barbata* (L.) (after Ellis, nat. size). N, *C. barbata* (L.); the surface of the thallus; magnified. O, *Penicillus pyramidalis* (Lamx.) (after Lamouroux, nat. size).

defined them as follows :—" Polypier pierreux, libre, ovuliforme ou cylindracé, creux intérieurement, souvent percé aux deux bouts. Pores très petits, régulièrement disposés à la surface[1]."

The specimens are referred to two species, *Ovulites margaritula* and *O. elongata.*

By some subsequent writers[2] these calcareous fossils, like miniature birds' eggs with a hole at either end, were included among the Zoophytes. Carpenter and others afterwards referred *Ovulites* to the Foraminifera, and compared the genus with *Lagena*[3]. The single specimens of *Ovulites* have a length of 2—6 mm. At each end there is usually a fairly large and somewhat irregular hole (fig. 35, F), and in some rarer cases there may be two apertures at the broader end of an Ovulite. A good example of *Ovulites margaritula* with two pores at the broader end is figured by Michelin[4]. The surface of the shell when seen under a low magnifying power appears to be covered over with regularly arranged circular pores, which are the external openings of fine canals (fig. 33, L).

In 1878 Munier-Chalmas expressed the opinion, which was supported by strong evidence, that *Ovulites* should be referred to the siphoneous algae[5]. · He regarded it as generically identical with *Penicillus* (*Coralliodendron*, Kützing). It has already been pointed out that in *Penicillus* the apical tuft of filaments is partially calcareous (fig. 33, O)[6]. The individual calcareous segments agree almost exactly with the fossil *Ovulites*. As a rule the Ovulites occur as separate egg- or rod-like bodies, but Munier-Chalmas informs me that occasionally two or three have been found joined end to end in their natural position. The terminal holes in the fossil specimens represent the apertures left after the detachment of the calcareous segments from the uncalcified filaments of the alga. The segments with two holes at the broader end were no doubt situated at the base of dichotomising branches as shown in fig. 33, K.

[1] Lamarck (16) p. 193.

[2] Defrance (26) Pl. xlviii. fig. 2, and Pl. l. fig. 6.

[3] Carpenter (62) p. 179, Pl. xii. figs. 9 and 10.

[4] Michelin (40–47) Pl. xlvi. fig. 24.

[5] Munier-Chalmas (79).

[6] Lamouroux (21) Pl. xxv. fig. 5, p. 23.

The restoration of *Ovulites*, shown in fig. 33, K, bears a striking resemblance to the figure of an Australian *Penicillus* given by Harvey in his *Phycologia Australica*[1].

It is probable that these Eocene forms agreed closely in habit with the recent species of *Penicillus*. The portions preserved as fossils are segments of the filaments which probably formed a terminal brush of fine branches supported on a stem. The retention of the original generic name *Ovulites* is on the whole better than the inclusion of the fossil species in the recent genus. The Tertiary species lived in warm seas of the Lower and Middle Eocene of England, Belgium, France and Italy.

Halimeda.

An example of an Eocene species of *Halimeda* has been recorded by Fuchs from Greifenstein under the name of *Halimeda Saportae*[2]. The impression has the form of a branched plant consisting of wedge-shaped or oval segments, and there is a close resemblance to the thallus of a recent *Halimeda*, e.g. *H. gracilis* Harv. It is not improbable that Fuchs' determination is correct, but without more definite evidence than is afforded by a mere impression it is a little rash to make use of the recent generic name.

γ. **Dasycladaceae.**

In this family of Siphoneae are included a number of genera represented by species living in tropical and subtropical seas.

The thallus consists of an elongated axial cell fixed to the substratum by basal rhizoids, and bearing whorls of lateral appendages of limited growth which may be either simple or branched. Many of the lateral branches bear sporangia or spores. The thallus is in many species encrusted with carbonate of lime.

---

[1] Harvey (58) Vol. I. Pl. xxii. fig. 3.
[2] Fuchs (94).

The two genera *Acetabularia* and *Cymopolia* may be briefly described as recent types which are represented by trustworthy fossil forms.

Fig. 34. *Acetabularia mediterranea* Lamx. From a specimen in the Cambridge Botanical Museum (nat. size).

Acetabularia. Figs. 33, I, and 34.

With the exception of *A. mediterranea* Lamx. (fig. 34) the few living species of this genus are confined to tropical seas.

The habit of *Acetabularia* is well illustrated by the photograph of a cluster of plants of *A. mediterranea* Lamx.[1] reproduced

[1] Lamouroux gives a figure of *Acetabularia*, and includes this genus with several other algae in the animal kingdom (Lamouroux [21] p. 19, Pl. lxix.).

in fig. 34. The thallus consists of a delicate stalk attached to the
substratum by a tuft of basal holdfasts, and expanded distally into
a small circular disc 10—12 mm. in diameter and more or less
concave above. This terminal cap is made up of a number of
laterally fused appendages given off from the upper part of the
stalk in the form of a crowded whorl. The whole thallus resembles a small and long-stalked calcareous fungus. In each radially
elongated compartment of the fertile cap (fig. 33, I) there are
several sporangia (*gametangia*) developed; these eventually
open and produce numerous ciliated gametes which give rise to
zygospores by conjugation. Fig. 33, I, represents the cap of an
Acetabularia in radial section and surface-view; the two radial
compartments seen in section contain the elliptical gametangia;
the circular markings at the base of the figure are scars of
sterile deciduous branches.

The whole plant is unicellular, each chamber in the disc
being in open communication with the stem of the plant.

Acicularia. Fig. 33, C—H.

In a recent monograph on the Acetabularieae, Solms-Laubach[1]
has described a new type of these algae which is of special
importance from the point of view of the past history of the
family. Möbius described an example of *Acetabularia* in 1889
under the name *A. Schencki*; this species has since been
placed in D'Archiac's genus *Acicularia*[2]. *Acicularia Schencki*[3]
bears a close resemblance as regards external form to *Acetabularia mediterranea*. In the latter species the walls of the
terminal disc compartments are calcified, and the cavity of each
of the laterally fused members contains numerous free spores;
in *Acicularia*, the cavity of each disc-ray is occupied by a calcareous substance in the form of a spicule containing numerous
cavities in each of which is a single sporangium. A single
spicule is seen in fig. 33, H, showing the spherical pockets in
which the sporangia were originally situated. This species,

---

[1] Solms-Laubach (95³).   [2] D'Archiac (43) p. 386, Pl. xxv. fig. 8.
[3] Solms-Laubach *loc. cit.* p. 33, Pl. iii.

Acicularia Schencki, has been recorded from Martinique, Guadeloupe, Brazil, and a few other places.

The genus *Acicularia* was founded by D'Archiac for certain minute calcareous spicules found in the Eocene sands (Calcaire Grossier) of the Paris basin. D'Archiac describes one species, *Acicularia pavantina*, which he defines as follows:—" Polypier aciculaire, élargi, et légèrement comprimé à sa partie supérieure, qui est échancrée au milieu. Surface couverte de petits pores simples, nombreux, disposés irrégulièrement[1]." The same species is figured also in Michelin's *Iconographie Zoophytologique*, and described as an organism of which the exact zoological position is uncertain[2]. After these fossils had been placed in various divisions of the animal kingdom, Carpenter[3] described several specimens as portions of foraminifera. Finally, Munier-Chalmas removed *Acicularia* to the plant kingdom, and "with rare divination" placed the genus among the Acetabularieae. The history of our knowledge of the true nature of *Acicularia* is of unusual interest. Some of the specimens of this genus figured in Carpenter's monograph have the form of imperfect long and narrow bodies tapering to a point at one end and broad at the other (fig. 33, F and G); they are joined together laterally and pitted with numerous small cavities. From the resemblance of such specimens to a fragment of the terminal fertile disc of the recent Acetabularias, Munier-Chalmas referred the fossils to this type of algae. In the living species which were then known the radiating chambers of the disc contained loose sporangia, without any calcareous matrix filling the cavity of the chambers. In the fossil Acicularias, on the other hand, the manner of preservation of the pitted calcareous spicules pointed to the occurrence of sporangia embedded in cavities in a calcareous matrix. Subsequent to Munier-Chalmas' somewhat daring conclusions as to the relation of *Acicularia* to *Acetabularia*, Solms-Laubach found that the species originally described by Möbius as *Acetabularia Schencki* from Guadeloupe presented exactly those characters in which the fossil specimens differ

---

[1] D'Archiac (43) p. 386, Pl. xxv. fig. 8.
[2] Michelin (40) p. 176, Pl. xlvi. fig. 14.
[3] Carpenter (62) p. 137, Pl. xi. figs. 27–32.

from *Acetabularia*. The genus *Acicularia* formerly restricted to fossil species is now applied also to this single living species *Acicularia Schencki*.

The genus is thus defined by Solms-Laubach :—

"Discus fertilis terminalis e radiis inter se conjunctis formatus, coronis et inferiore et superiore praeditis, sporae massa mucosa calce incrustata coalitae, pro radio spiculam solidam cuneatam formantes[1]."

As Solms-Laubach points out in his recent monograph, Munier-Chalmas' conjecture, "which had little to support it in the fossil material, has been more recently proved true in the most brilliant fashion by the discovery of a living species of this genus."

1. *Acicularia Andrussowi* Solms[2]. Fig. 33, C and D. This species was first described by Andrussow[3] as *Acetabularia miocenica* from the Crimea. It occurs in Miocene rocks south of Sevastopol, and, with *Ostrea* and *Pecten*, forms masses of white limestone.

In each sporangial ray of the disc the cavity contains a calcareous spicula bearing spore cavities in four rows. "Round each spore-cavity there is a circular zone which stands out, when viewed in reflected light, through its white colour against the central mass of the spicule, though a sharp contour is not visible[4]." Fig. 33, C, is taken from a somewhat diagrammatic sketch by Andrussow; it shows ten of the fertile rays of the disc. The thick walls of the chambers are seen in the two lowest rays, and in the next two rays the spore-cavities are represented. A more accurate drawing, from Solms-Laubach's memoir, is reproduced in fig. 33, D. The calcareous spicule with numerous spore-cavities shown in fig. 33, H, is from a fertile ray of the recent species *Acicularia Schencki*. This corresponds to the spore-containing calcareous matrix in each ray of the disc of *Acicularia Andrussowi* Solms. The spicule copied in fig. 33, F from one of Carpenter's drawings[5] of an

---

[1] Solms-Laubach *loc. cit.* p. 32.   [2] *Ibid.* p. 34, Pl. III. fig. 13.
[3] Andrussow (87).   [4] Solms-Laubach (95³) p. 11.
[5] Carpenter (62) Pl. XI. fig. 32.

Eocene specimen bears the closest resemblance to the recent spicule of fig. 33, H, and emphasizes the very close relationship between the fossil forms and the single rare tropical species.

2. *Acicularia miocenica* Reuss. Another Tertiary species has been described under this name by Reuss[1] from the Miocene of the Vienna district, from the Leithakalk of Moravia and elsewhere. It agrees very closely with the recent species *A. Schenckii.* A section of one of the spicules of this species is shown in fig. 33, E; the dark patches represent the pockets in the calcareous spicule which were originally occupied by sporangia and spores.

Cymopolia. Fig. 33, A, B, M and N.

The genus *Cymopolia* is at present represented by two species, *C. barbata* (L.) and *C. mexicana*, Ag., living in the Gulf of Mexico and off the Canary Islands.

Cymopolia and *Acetabularia*, with several other calcareous algae, are figured by Ellis and other writers as members of the animal kingdom. Ellis speaks of the species of *Cymopolia* which he figures as the Rosary Bead-Coralline of Jamaica.

Fig. 33, M, has been drawn from a figure published by Ellis in his *Natural History of the Corallines* published in 1755[2]. The thallus has the form of a repeatedly forked body, of which the branches are divided into cylindrical joints thickly encrusted with carbonate of lime, but constricted and uncalcified at the limits of each segment. A tuft of hairs is given off from the terminal segment of each branch. The axis of each branch of the thallus is occupied by a cylindrical and unseptate cell which gives off crowded whorls of lateral branches. In the lower part of fig. 33, M, the calcareous investment has been removed, and the branches are seen as fine hair-like appendages of the central cell. The branches given off from the constricted portions of the axis are unbranched simple appendages, but the others terminate in bladder-like swellings, each of which bears an apical sporangium. The sporangia are surrounded and enclosed by the swollen tips of four to six branches which spring from the summit of the sporangial branch. Fig. 33, A, represents part

---

[1] Reuss (61) p. 8, figs. 5–8.   [2] Ellis (1755) Pl. xxv. C.

of a transverse section through the calcareous outer portion
of a branch of *Cymopolia*; the darker portions or cavities in
the calcareous matrix were originally occupied by the lateral
branches and sporangia[1].

In Fig. 33, B, the sporangial branch with the terminal
sporangium and three of the investing branches are more
clearly shown, the surrounding calcareous investment and the
thallus having been removed by the action of an acid.

In a transverse section of a branch from which the organic
matter had been removed, and only the calcareous matrix left,
one would see a central circular cavity surrounded by a thick
calcareous wall perforated by radially disposed canals and containing globular cavities; the canals and cavities being occupied
in the living plant by branches and sporangia respectively.

The two circular cavities shown in the figure mark the
position of the sporangia which are borne on branches with
somewhat swollen tips. From the summit the left-hand
sporangial branch shown in fig. 33, A, three of the secondary
branches are represented by channels in the calcareous matrix;
the two black dots on the face of the sporangiaphore being the
scars of the remaining two secondary branches.

By the lateral contact of the swollen ends of the ultimate
branches enclosing the sporangia the whole surface of the
thallus, when examined with a lens, presents a pitted appearance.
Each pit or circular depression (fig. 33, N) marks the position
of the swollen tip of a branch.

This form of thallus represents a type which is met with in
several members of the Dasycladaceae. It would carry us
beyond the limits of a short account to describe additional
recent genera which throw light on the numerous fossil species.
For further information as to the recent members of the family,
the student should refer to Murray's *Seaweeds*[2], and for a more
detailed memoir on the group to Wille's recent contribution to
the *Pflanzenfamilien*[3] of Engler and Prantl. Among the various
special contributions to our knowledge of the Dasycladaceae,

---

[1] Solms-Laubach (91) p. 38 gives a detailed description with two figures
of a recent species of *Cymopolia*.

[2] Murray G. (95).   [3] Wille (97).

those by Munier-Chalmas[1], Cramer[2], Solms-Laubach[3], and Church[4], may be mentioned.

The publication of a short preliminary note by Prof. Munier-Chalmas in the *Comptes Rendus* for 1877 was the means of calling attention to the exceptional importance of the calcareous Siphoneae as algae possessing an interesting past history, of which satisfactory records had been preserved in rocks of various ages. Decaisne had pointed out in 1842 that certain marine organisms previously regarded as animals should be transferred to the plant kingdom. Such seaweeds as *Halimeda, Udotea, Penicillus* and others were thus assigned to their correct position. Many fossil algae belonging to this group continued to be dealt with as Foraminifera until Munier-Chalmas demonstrated their true affinities. In Gümbel's monograph on the so-called Nullipores found in limestone rocks, published in 1871[5], several examples of siphoneous algae are included among the fossil Protozoa.

In recent years there have been several additions to an already long list of fossil Siphoneae. In addition to the numerous and well-preserved specimens, representing a large number of generic and specific forms, which have been collected from the Eocene of the Paris basin, there is plenty of evidence of the abundance of the members of the Dasycladaceae in the Triassic seas. In the Triassic limestones of the Tyrol. as well as in other regions, the calcareous bodies of siphoneous algae have played no inconsiderable part as agents of rock-building[6]. Genera have been recorded from Silurian and other Palaeozoic horizons, and there is no doubt that the Verticillate Siphoneae of to-day are the remnants of an extremely ancient family, which in former periods was represented by a much more widely distributed and more varied assemblage of species. There is probably no more promising field of work in the domain of fossil algae than the further investigation of the numerous forms included in Munier-Chalmas' class of Siphoneae

[1] Munier-Chalmas (77).   [2] Cramer (87) (90).
[3] Solms-Laubach (91) (93) (95³).   [4] Church (95).
[5] Gümbel (71).   [6] Benecke (76).

Verticillatae. A brief description of a few genera from different
geological horizons must suffice to draw attention to the
character of the data for a phylogenetic history of this group.

The fossil examples of the genus *Cymopolia* (*Polytrypa*)
were originally described by Defrance[1] in the *Dictionnaire des
Sciences Naturelles* as small polyps under the generic name
Polytrypa.

In the Eocene sands of the Paris basin there have been
found numerous specimens of short, calcareous tubes which
Munier-Chalmas has shewn are no doubt the isolated segments
of an alga practically identical with the recent *Cymopolia*. A
section[2] through one of the fossil segments presents precisely
the same features as those which are represented in fig. 33, A.
The habit of the Eocene alga and its minute structure were
apparently almost identical with those of the recent species,
Cymopolia barbata. The two drawings of *Cymopolia* reproduced
in fig. 33, A and B, have been copied from Munier-Chalmas'
note in the *Comptes Rendus*[3]; the corresponding figures given
by this author of the Eocene species (*Cymopolia elongata* Deb.)
are practically identical with figs. A and B, and show no
points of real difference. The segments of the thallus of the
fossil species, as figured by Defrance[4], appear to be rather
longer than those of the recent species. The calcareous
investment of the axial cell of the thallus was traversed by
regular verticils of branches or 'leaves'; the central branch
of each whorl terminates in an oval sporangial cavity, exactly
as in fig. 33, A and B; and from the top of this branch there
is given off a ring of slender prolongations which terminate
on the surface of the calcareous tube as regularly disposed
depressions, which were no doubt originally occupied by their
swollen distal ends as in the recent species.

Vermiporella.

This generic name was proposed by Stolley for certain
branched and curved tubes found in Silurian boulders from the

---

[1] Defrance (26) p. 453.   [2] Munier-Chalmas (77) p. 815.
[3] Munier-Chalmas *ibid.*   [4] Defrance (26) Pl. XLVIII. fig. 1.

North German drift¹. The tubes have a diameter of ·5—1 mm.,
and are perforated by radial canals which probably mark the
position of verticils of branches given off at right angles to the
central axis. The surface of the tubes is divided into regular
hexagonal areas.

The resemblance of these Silurian fossils to *Diplopora*
and other genera favours their inclusion in the Verticillate
Siphoneae.

<h2 style="text-align:center">*Sycidium.* Fig. 32, B.</h2>

The fossils included in this genus were first described by
Sandberger from the middle Devonian rocks of the Eifel, and
referred by him to the animal kingdom. More recently Deecke
has suggested the removal of the genus to the calcareous
Siphoneae, and such a view appears perfectly reasonable,
although without more data it is not possible to speak with
absolute certainty.

Sycidium melo. (Sandb.) Fig. 32, *B*. The specimen represented in fig. 32, B (i), (ii), drawn from Deecke's figures², has
the form of a small oval calcareous body, 1 mm. in transverse
diameter and 1—1·3 mm. in longitudinal diameter. It is pointed
at one end and flattened at the other. At the flatter end there
is a circular depression, continued into a funnel-shaped cavity,
and on the walls of this cavity there are 18—20 radially disposed
ribs, which extend over the surface of the whole body. A series
of transverse ribs intersects the vertical ribs at right angles.
The calcareous wall is perforated by numerous whorls of circular
pores, and the internal cavity is a simple undivided space. Each
of these oval bodies (fig. 33, B) is probably the segment of a
thallus, and the perforations in the wall may have been originally
occupied by lateral prolongations from the unseptate axial cell
of the thallus. *Sycidium* bears a fairly close resemblance to the
Tertiary *Ovulites*.

¹ Stolley (93).　　　² Deecke (83).

Diplopora. Fig. 35, A and B.

This genus of algae is characteristic of Triassic rocks, and is especially abundant in Muschelkalk and Lower Keuper limestones of the Alps, Silesia, and elsewhere. The thallus, or rather the calcareous portion of the thallus, has the form of a thick-walled tube, with a diameter of about 4 mm., and occasionally reaching a length of 50 mm. At one end the tube has a rounded and closed termination, and the wall is pierced throughout its whole length by regular whorls of fine canals. *Diplopora* agrees with *Cymopolia* in its main features.

Fig. 35, A affords a diagrammatic view of a *Diplopora* tube, and shews the arrangement of the numerous whorls of canals. In fig. 35, B, a piece of limestone is represented containing several Diploporas cut across transversely and more or less

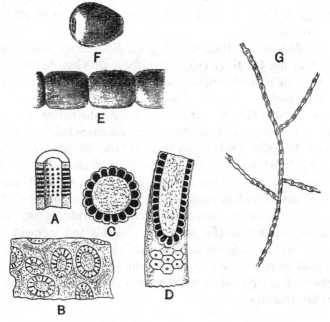

Fig. 35. A, B, *Diplopora*. × 2. C, D, *Gyroporella* (after Benecke. × 4). E, Calcareous segments of *Penicillus*, from a specimen in the British Museum. × 5. F, a single segment of *Ovulites margaritula* Lam. × 4. G, *Confervites chantransioides* Born. (after Bornemann. × 150).

obliquely. In an obliquely transverse section of a tube perforated by horizontal canals the cavities of the canals necessarily appear as holes or discontinuous canals in the substance of the calcareous wall. The manner of occurrence of the specimens points to the abundance of this genus in the Triassic seas, and suggests that the calcareous tubes of *Diplopora* may have been important factors in the building up of limestone sediments[1]. In many instances no doubt the carbonate of lime of the thallus has been dissolved and recrystallised, and the original form completely obliterated. As in the rocks built up largely of calcareous Florideae (p. 185) which have lost their structure, it is a legitimate inference that some of the limestone rocks which shew no trace of organic structure may have been in part derived from the calcareous incrustation of various algal genera.

Gyroporella. Fig. 35, C and D.

In this genus from the Alpine Trias the structure of the calcareous tube is very similar to that in *Diplopora*, but in *Gyroporella* the canals form less distinct whorls and are closed externally by a small plate, as seen in figs. 35, C and D.

As Solms-Laubach has pointed out, the branch-systems of *Diplopora*, *Gyroporella* and other older genera are much simpler than in the Tertiary genera *Dactylopora* and others[2].

A species of *Gyroporella*, *G. bellerophontis*, has recently been described by Rothpletz[3] from Permian rocks in the Southern Tyrol. The thallus is tubular in form and has a diameter of ·5—1 mm.

Dactylopora.

The genus *Dactylopora* was founded by Lamarck[4] on some fossil specimens from the Calcaire Grossier and included among the Zoophytes. D'Orbigny afterwards included it among the Foraminifera, and the structure of the calcareous body has been described by Carpenter[5] and other writers on the Foraminifera.

[1] Benecke (76) Pl. xxiii.   [2] Solms-Laubach (91) p. 42.
[3] Rothpletz (94) p. 24.   [4] Lamarck (16) p. 188.
[5] Carpenter (62) Pl. x.

In a specimen of *Dactylopora cylindracea* Lam. from the Paris basin, for which I am indebted to Munier-Chalmas, the tubular thallus measures 4 mm. in diameter; at the complete end it is closed and bluntly rounded. The wall of the tube is perforated by numerous canals, and contains oval cavities which were no doubt originally occupied by sporangia. The shape of the specimens is similar to that of *Diplopora*, but the canals and cavities present a characteristic and more complex appearance, when seen in a transverse section of the wall, than in the older genus *Diplopora*. Gümbel has given a detailed account of this Tertiary genus in his memoir on *Die sogenannten Nulliporen*[1]; he distinguishes between *Dactyloporella* and *Gyroporella* by the existence of cavities in the calcareous wall of the tube in the former genus, and by their absence in the latter. The oval cavities in a *Dactyloporella* were originally occupied by sporangia; in *Diplopora* and *Gyroporella* the sporangia were probably borne externally and on an uncalcified portion of the thallus.

In addition to the few examples of fossil species described above there are numerous others of considerable interest, which illustrate the great wealth of form among the Tertiary and other representatives of the Verticillate Siphoneae.

Reference has already been made to *Vermiporella* as an example of a Silurian genus. Other genera have been described by Stolley from Silurian boulders in the North-German drift under the names *Palaeoporella, Dasyporella* and *Rhabdoporella*[2]; the latter genus is compared with the Triassic *Diplopora*, and the two preceding with the recent *Bornetella*.

Schlüter has transferred a supposed Devonian Foraminiferal genus, *Coelotrochium*[3], to the list of Palaeozoic Siphoneae. Munier-Chalmas regards some of the fossils described by Saporta under the name of *Goniolina*[4], and classed among the inflorescences of pro-angiospermous plants, as examples of Jurassic Siphoneae. The shape and surface-features of some of the

---

[1] Gümbel (71). *Vide* also Solms-Laubach (91) p. 39.
[2] Stolley (93).   [3] Schlüter (79).   [4] Saporta (91) Pl. xxxii. &c.

examples of *Goniolina* suggest a comparison with Echinoid spines, but the resemblance which many of the forms in the Sorbonne collection present to large calcareous Siphoneae is still more striking. A comparison of Saporta's fig. 5, Pl. xxxiii. and fig. 4, Pl. xxxii. in volume iv. of the *Flore Jurassique*, with the figures given by Solms-Laubach[1] and Cramer[2] of species of *Bornetella* brings out a close similarity between *Goniolina* and recent algae; the chief difference being the greater size of the fossil forms. The possibility of confounding Echinoid spines with calcareous Siphoneae is illustrated by Rothpletz[3], who has expressed the opinion that Gümbel's *Haploporella fasciculata* is not an alga but the spine of a sea-urchin.

Among Cretaceous forms, in addition to *Goniolina*, which passes upwards from Jurassic rocks, *Triploporella*[4] and other genera have been recorded.

Uteria[5] is an interesting type of Tertiary genera; it occurs in the form of barrel-shaped rings, which are probably the detached segments of a form in which the central axial cell was encrusted with carbonate of lime, but the sporangia and the whorls of branches differed from those of *Cymopolia* in being without a calcareous investment.

b. CONFERVOIDEAE.

Without attempting to describe at length the fossil forms referred to this division of the Chlorophyceae, there is one fossil which deserves a passing notice. Brongniart in 1828[6] instituted the generic term *Confervites* for filamentous fossils resembling recent species of confervoid algae. Numerous fossils have been referred to this genus by different authors, but they are for the most part valueless and need not be further considered. In 1887 Bornemann described some new forms which he referred to this genus from the Cambrian rocks of Sardinia. He describes the red marble of San Pietra, near

---

[1] Solms-Laubach (93), Pl. IX. figs. 1, 8.   [2] Cramer (90).
[3] Rothpletz (92²) p. 235.   [4] Steinmann (80).
[5] Solms-Laubach (91), p. 40. fig. 3. *Vide* also Deecke (83) Pl. I. fig. 12.
[6] Brongniart (28) p. 211.

Masne, as being in places full of the delicate remains of algae having the form of branched filaments, and appearing in sections of the rock as white lines on a dark crystalline matrix. In fig. 35, G, one of these Sardinian specimens is represented. This form is named *Confervites Chantransioides*[1]; the thallus consists of branched cell-filaments, having a breadth of 6—7μ, and composed of ovate cells. It is possible that this is a fragment of a Cambrian alga, but the figures and descriptions do not afford by any means convincing evidence. From post-Tertiary beds various genera, such as *Vaucheria* and others, have been recorded, but they possess but little botanical value.

INCERTAE SEDIS.

Fossils in Boghead 'Coal' referred by some authors to the Chlorophyceae.

During the last few years much has been written by two French authors, Dr Renault and Prof. Bertrand, on the subject of the so-called Boghead of France, Scotland, and other countries. They hold the view that the formation of the extensive beds of this carbonaceous material was due to the accumulation and preservation of enormous numbers of minute algae which lived in Permo-Carboniferous lakes.

In an article contributed to *Science-Progress* in 1895 I ventured to express doubts as to the correctness of the conclusions of MM. Renault and Bertrand[2]. Since then Prof. Bertrand has very kindly demonstrated to me many of his microscopic preparations of various Bogheads, and I am indebted to Prof. Bayley Balfour of Edinburgh for an opportunity of examining a series of sections of the Scotch Boghead. The examination of these specimens has convinced me of the difficulties of the problems which many investigators have tried to solve, but it has by no means led me to entirely adopt the views expressed by MM. Bertrand and Renault.

The Boghead or Torbanite of Scotland was rendered famous by a protracted lawsuit tried in Edinburgh from July 29th to

---

[1] Bornemann (91) p. 485. Pls. 42 and 43.   [2] Seward (95²) p. 367.

August 4th, 1853. A lease had been granted by Mr and Mrs Gillespie, of Torbanehill, in Fifeshire, to Messrs James Russell and Son, coal-masters of Falkirk, of "the whole *coal*, ironstone, iron-ore, limestone and fire-clay (but not to comprehend copper, or any other minerals whatsoever, except those specified) with lands of Torbanehill[1]." After the Boghead had been worked for two years the Gillespies challenged the right of Messrs Russell, and argued that the valuable *mineral* Torbanite was not included among the substances named in the agreement. The defendants maintained that it was a *coal*, known as gas-, cannel- or parrot-coal. A verdict was given for the defendants. Some of the scientific experts who gave evidence at the trial considered that the Boghead afforded indications of organic structure, while others regarded it as essentially mineral in origin.

The Torbanite or Boghead is a close-grained brown rock, of peculiar toughness and having a subconchoidal fracture. It contains about $65^o/_o$ carbon, with some hydrogen, oxygen, sulphur, and mineral substances. A thin section examined under the microscope presents the appearance of a dark and amorphous matrix, containing numerous oval, spherical and irregularly shaped bright orange-yellow patches. Fig. 36, 1 shows the manner of occurrence of the yellow bodies in a piece of Scotch Boghead, as seen in a slightly magnified horizontal section. Under a higher power the light patches in the figure reveal traces of a faint radial striation, which in some cases suggests the occurrence of a number of oval or polygonal cells.

The Autun Boghead possesses practically the same structure. The yellow bodies are often sufficiently abundant to impart a bright yellow colour to a thin section. If the section is vertical the coloured bodies are seen to be arranged in more or less regular layers parallel to the plane of bedding.

The Kerosene shale of New South Wales agrees closely with the Scotch and French Boghead; it is approximately of the same geological age, and is largely made up of orange or yellow bodies similar to those of the European Boghead, but much more clearly preserved.

[1] Report of the Trial (62).

The nature and manner of formation of the various forms of coal should be dealt with in a later chapter devoted to the subject of plants as rock-builders, but in view of the recent statements as to the algal nature of these bituminous deposits it may not be out of place to state briefly the main conclusions of the French authors.

MM. Renault and Bertrand regard each of the yellow bodies in the European and Australian Boghead as the thallus of an alga. To the form which is most abundant in the Kerosene shale they have given the generic name of *Reinschia*, while that in the Scotch and French Boghead is named *Pila*.

Reinschia. Fig. 36, 3.

A section of a piece of Kerosene shale at right angles to the bedding appears to be made up of fairly regular layers of flattened elliptical sacs of an orange or yellow colour. Each sac or thallus is about 300μ in length and 150μ broad (fig. 36, 3). A single row of cells constitutes the wall surrounding the central globular cavity. The cells are more or less pyriform in shape, and the cell-cavities are filled with a dark substance, described by Renault and Bertrand as protoplasm, and the cell-

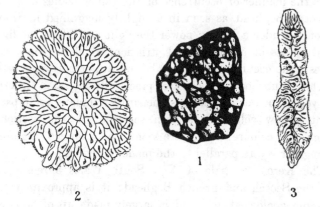

2   3

Fig. 36, 1. Section of a piece of Scotch Torbanite. Slightly enlarged. 2. *Pila bibractensis* from the Autun Boghead, × 282 (after Bertrand). 3. *Reinschia Australis*, from the Kerosene shale of New South Wales, × 592 (after Bertrand).

walls are fairly thick. In some of the larger specimens there
are often found a few smaller sacs enclosed in the cavity of the
partially disorganised mother-thallus. In the larger specimens
the wall is usually invaginated in several places, giving the
whole thallus a lobed or brain-like appearance. The supposed
alga, which makes up $\frac{9}{10}$ths of the contents of a block of
Kerosene shale, is named *Reinschia Australis*; it is regarded
by the authors of the species as nearly related to the Hydro-
dictyaceae or Volvocineae.

In the Kerosene shale from certain localities in New South
Wales Bertrand recognises a second form of thallus, which he
refers to the genus *Pila*, characteristic of the European Bogheads.

Pila. Fig. 36, 2.

The "thallus" characteristic of the Scotch Boghead has been
named *Pila scotica*, and that of the Autun Boghead, *Pila
bibractensis*.

In the latter form, which has been studied in more detail
by MM. Renault and Bertrand, the thallus consists of about
6—700 cells, and is irregularly ellipsoidal in form, from ·189—
·225 mm. in length; and 136—·160 mm. broad. The surface-
cells are radially disposed and pyramidal in shape, the internal
cells are polygonal in outline and less regularly arranged
(fig. 36, 2). The Pila thalli make up $\frac{3}{4}$ths of the mass in an
average sample of the Autun Boghead. The Autun Boghead
often contains siliceous nodules, and sections of these occasionally
include cells of a *Pila* in which the protoplasmic contents and
nuclei have been described by the French authors. The evidence
for the existence of these supposed nuclei is, however, not entirely
satisfactory ; sections of silicified thalli which were shown to
me by Prof. Bertrand did not satisfy me as to the minute
histological details recognised by Bertrand and Renault.

The species of *Pila* are compared with the recent genus
Celastrum, and regarded as most nearly allied to the Chroococcaceae or Pleurococcaceae among recent algae. Prof. Bornet[1]
has suggested *Gomphosphaeria* as a genus which presents a
resemblance to the Autun *Pila*.

[1] Bertrand and Renault (92) p. 29.

In addition to the Bogheads of Autun, Torbanehill, and
New South Wales, there are similar Palaeozoic deposits in
Russia, America, and various other parts of the world. Full
details of the structure of Boghead and the supposed algae
referred to *Reinschia, Pila,* and other genera will be found in
the writings of Bertrand and Renault[1]

The Kerosene shale of New South Wales affords the most
striking and well-preserved examples of the cellular orange and
yellow bodies referred to as the globular thalli of algae. It is
almost impossible to conceive a purely inorganic material
assuming such forms as those which occur in the Australian
Boghead. On the other hand, it is hardly less easy to
understand the possibility of such explanations as have been
suggested of the organic origin of these characteristic bodies.

The ground-mass or matrix of the Boghead is referred to a
brown ulmic precipitate thrown down on the floor of a Permian
or Carboniferous lake, probably under the action of calcareous
water. In this material there accumulated countless thalli of
minute gelatinous algae, which probably at certain seasons
completely covered the surface of the waters, as the *fleurs d'eau*
in many of our fresh-water lakes. In addition to the thalli of
Reinschia and *Pila* the Bogheads contain a few remains of
various plant fragments, pollen-grains, and pieces of wood.
Fish-scales and the coprolites of reptiles and fishes occur in some
of the beds. On a piece of Kerosene shale in the Woodwardian
Museum, Cambridge, there are two well-preserved graphitic
impressions of the tongue-shaped fronds of *Glossopteris Browni-
ana,* Brongn. There can be little doubt that the beds of
Boghead were deposited under water as members of a regular
sequence of sedimentary strata. The yellow bodies which form
so great a part of the beds are practically all of the same type.
Reinschia and *Pila* cannot always be distinguished, and it
would seem that there are no adequate grounds for instituting
two distinct genera and referring them to different families of
recent algae.

Stated briefly, my conclusion is that the algae of the

[1] Bertrand (93), Bertrand and Renault (92) (94), Bertrand (96), Renault (96).
Additional references may be found in these memoirs.

French authors may be definite organic bodies, but it is unwise to attempt to determine their affinities within such narrow limits as have been referred to in the above *résumé*. The structure of the bituminous deposits is worthy of careful study, and it is by no means impossible that further research might lead us to accept the view of the earlier investigators, that the brightly coloured organic-like bodies may be inorganic in origin.

C. RHODOPHYCEAE. (FLORIDEAE. RED ALGAE.)

The thallus of the members of this group assumes various forms, and consists of branched cell-filaments of a more or less complex structure. Cells of the thallus contain a red colouring matter in addition to the green chlorophyll. The reproduction is asexual and sexual; the formation of asexual reproductive cells (*tetraspores*) in groups of four in sporangia is a characteristic method of reproduction. Sexual reproduction is effected by means of distinct male and female cells.

With the exception of a few fresh-water genera all the red algae are marine. The Rhodophyceae, like the Cyanophyceae and Chlorophyceae, include a shell-boring form which has been found in the common razor-shell[1]. Several genera live as endophytes in the tissues of other algae. The recent species of this section of algae are characteristic of temperate and tropical seas. One subdivision of the red algae, the Corallinaceae, is extremely important from a geological point of view and must be dealt with in some detail.

CORALLINACEAE.

The thallus is usually encrusted with carbonate of lime; it is of a branched cylindrical form in the well-known *Corallina officinalis*, Linn. of the British coasts, of an encrusting and foliaceous type, in the genus *Lithophyllum*, and of a more coral-like form in the genus *Lithothamnion*. The reproductive organs occur in conceptacles, having the form of small depressed

---

[1] Batters (92). *Vide* also Schmitz (97) p. 315.

cavities in the thallus, or projecting as warty swellings above the surface of the plant. Asexual reproduction is by means of tetraspores formed in conceptacles resembling those containing the sexual cells. The Corallinaceae may be subdivided into the two families Melobesieae and Corallineae[1].

Melobesieae. Thallus encrusting, leaf- or coral-like; unsegmented.

(*Melobesia, Lithophyllum, Lithothamnion.*)

Corallineae. Cylindrical filamentous and segmented thallus. (*Amphiroa* and *Corallina.*)

The genus *Corallina* is the best known British representative of the Corallinaceae. With other members of the group it was long regarded as a coralline animal, and it is only comparatively recently that the plant-nature of these forms has been generally admitted. *Lithophyllum, Lithothamnion, Melobesia,* and other genera of the Corallinaceae and some of the Siphoneae play a very important part in the building and cementing of coral-reefs. The pink or rose-coloured calcareous thallus of some of these calcareous algae or Nullipores imparts to coral-reefs a characteristic appearance. In some cases, indeed, the coral-reefs are very largely composed of algae. Saville Kent[2] describes the Corallines or Nullipores of the Australian Barrier-reef as furnishing a considerable quota towards the composition of the coral rock. Mr Stanley Gardiner, who accompanied the coral-boring expedition to the island of Funafuti, has kindly allowed me to quote the following extract from his notes, which affords an interesting example of the importance of calcareous algae as reef-building organisms. "It is quite a misnomer to speak of the outer edge of a reef like this (Rotuma Island) as being formed of coral. It would be far better to call it a Nullipore reef, as it is completely encrusted by these algae, while outside in the perfectly clear water, 10 to 15 fathoms in depth, the bottom has a most brilliant appearance from masses of red, white and pink Nullipores, with only a stray coral here and there."

[1] Hauck (85) in Rabenhorst's *Kryptogamen Flora,* vol. ii.
[2] Kent (93) p. 140.

Agassiz[1] has given an account of the occurrence of immense masses of Nullipores (*Udotea, Halimeda* etc.) in the Florida reefs; his description is illustrated by good figures of these algae.

In the Mediterranean there are true Nullipore reefs, which are interesting geologically as well as botanically. Walther[2] has described one of these limestone-banks in the Gulf of Naples which occurs about 1 kilometre from the coast and 30 metres below the surface of the water. Every dredging, he says, brings up numberless masses of *Lithothamnion fasciculatum* (Lamarck), and *L. crassum* (Phil.). Between the branches of the algae, gasteropods and other animals become completely enclosed by the growing plants, while diatoms, foraminifera, and other forms of life are abundant. Water percolating through the mass gradually destroys the structure of the algal thalli, and in places reduces the whole bank to a compact structureless limestone.

The same author[3] has also called attention to the importance of *Lithophyllum* as a constructive element in the coral-reefs off the Sinai peninsula.

Lithothamnion a typical genus of the Corallinaceae may be briefly described.

Lithothamnion. Fig. 37.

Philippi[4] was the first writer to describe this and other genera as plants. He gave the following definition of *Lithothamnion*:

"Stirps calcarea rigida, e ramis cylindricis vel compressiusculis dichotoma ramosis constans."

The thallus of *Lithothamnion* grows attached to the face of a rock or other foundation, and forms a hard, stony mass, assuming various coralline shapes. The exposed face may have the form of numerous short branches or of an irregular warty surface.

[1] Agassiz (88) vol. I. p. 82.　　　[2] Walther (85).
[3] *Ibid.* (88) p. 478.　　　[4] Philippi (37) p. 387.

In section (fig. 37, A.) the lower part of the thallus is seen to
be made up of rows of cells radiating out from a central point,

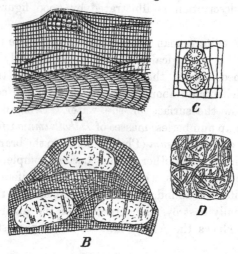

FIG. 37, *A*.   Section of a recent *Lithothamnion* (after Rosanoff[1], × 200).
   B. Section of *Lithothamnion suganum*, Roth (after Rothpletz[2], × 100).
   C. A conceptacle with tetraspores from a Tertiary *Lithothamnion* (after
   Früh[3], × 300).   *D*.   *Sphaerocodium Bornemanni* Roth. (after Rothpletz,
   × 150).

and the upper portion consists of vertical and horizontal rows
of cells.   The whole body is divided up into a large number of
small cells by anticlinal and periclinal walls, and possesses an
evident cellular as distinct from a tubular structure.   Conceptacles containing reproductive organs are either sunk in the
thallus or project above the surface.   The two types of structure
in a single thallus are shown in fig. 37, A, also a conceptacle
containing tetraspores.

In the closely allied *Lithophyllum* the thallus is encrusting,
and in section it presents the same appearance as the lower
part of a Lithothamnion thallus.

Species of *Lithothamnion* occur in the Mediterranean Sea,
and are abundant in the arctic regions[4], while on the British
coasts the genus is represented by four species[5].   Some large

[1] Rosanoff (66) Pl. VI. fig. 10.   [2] Rothpletz (91) Pl. XVII. fig. 4.
[3] Früh (90) fig. 12.   [4] Kjellman (83).
[5] Holmes and Batters (90) p. 102.

specimens of *Lithothamnion* and *Lithophyllum* are exhibited in
one of the show-cases in the botanical department of the British
Museum. For the best figures and descriptions of recent species
reference should be made to the works of Hauck, Rosanoff,
Rosenvinge, Kjellman and Solms-Laubach[1].

It is to be expected that such calcareous algae as *Lithothamnion* should be widely represented by fossil forms. In
addition to the botanical importance of the data furnished by
the fossil species as to the past history of the Corallinaceae,
there is much of geological interest to be learnt from a study of
the manner of occurrence of both the fossil and recent representatives. As agents of rock-building the coralline algae are
especially important. The late Prof. Unger[2] in 1858 gave an
account of the so-called Leithakalk of the Tertiary Vienna
basin, and recognised the importance of fossil algae as rock-
forming organisms. The Miocene Leithakalk, which is widely
used in Vienna as a building stone[3], consists in part of limestone
rocks consisting to a large extent of *Lithothamnion*.

Since the publication of Unger's work several writers have
described numerous fossil species of *Lithothamnion* from various
geological horizons. A few examples will suffice to illustrate
the range and structure of this and other genera of the
Corallinaceae. In dealing with the fossil species it is often
impossible to make use of those characters which are of primary
importance in the recognition of recent species. The fossil
thallus is usually too intimately associated with the surrounding
rock to admit of any use being made of external form as a
diagnostic feature. The size and form of the cells must be
taken as the chief basis on which to determine specific differences. In the absence of conceptacles or reproductive organs it
is not always easy to distinguish calcareous algae from fossil
Hydrozoa or Bryozoa. In many instances, however, apart from
the nature and size of the elements composing the thallus, the
conceptacles afford a valuable aid to identification. An example

---

[1] Hauck (85). Rosanoff (66). Rosenvinge (93) p. 779. Kjellman (83) p. 88.
Solms-Laubach (81). 　　　　　　　　　　　　　[2] Unger (58).

[3] A microscopic section of the Vienna Leithakalk is figured in Nicholson and
Lydekker's *Manual of Palæontology* (89) vol. II. p. 1497.

of a fossil conceptacle containing tetraspores is shown in fig. 37, C; it is from a Tertiary species of *Lithothamnion*, described by Früh from Montévraz in Switzerland.

1. *Lithothamnion mamillosum* Gümb. Fig. 32, A (i) and (ii). (p. 155.) This species was first recorded by Gümbel[1] from the Upper Cretaceous (Danian) rocks of Petersbergs, near Maëstricht, on the Belgian frontier. It was originally described as a Bryozoan. The thallus has the form of an encrusting calcareous structure bearing on its upper surface thick nodular branches, as shown in fig. 32, A (ii); in section, A (i), the thallus consists of a regular series of rectangular cells.

The specific name *mamillosum* has also been given to a recent species by Hauck[2], but probably in ignorance of the existence of Gümbel's Cretaceous species.

2. *Lithothamnion suganum* Roth. Fig. 37, B. The section of this form given in fig. 37, B shows three oval conceptacles filled with crystalline material. The two lower conceptacles originally communicated with the surface of the thallus, but as in recent species the deeper portions of the algal body became covered over by additions to the surface, forming merely dead foundations for new and overlying living tissues.

The cells of the thallus have a breadth of $7-9\mu$, and a length of $9-12\mu$.

The specimen was obtained from a Lithothamnion bank, probably of Upper Oligocene age, in Val Sugana[3], in the Austrian Tyrol.

Numerous other species of Jurassic, Cretaceous and Tertiary age might be quoted, but the above may suffice to illustrate the general characters and mode of occurrence of the genus. It is important that the student should become familiar with the *Lithothamnion* and *Lithophyllum* types of thallus, in view of their frequent occurrence in crystalline limestone rocks and in such comparatively recent deposits as those of upraised coral-reefs. The coral-rock of Barbadoes and other West-Indian islands

---

[1] Gümbel (71) Pl. ii. fig. 7, p. 41.　　　[2] Hauck (85) p. 272.
[3] Rothpletz (91) Pl. xvii. fig. 4.

affords a good illustration of the manner of occurrence of fossil coralline algae in association with corals and other organisms[1].

In the fossil species of *Lithothamnion* hitherto recorded there do not appear to be any important features in which they differ from recent forms; the geological history of the genus so far as it is known, favours the view that the generic characters are of considerable antiquity.

Solenopora. Fig. 38.

Mr A. Brown[2], of Aberdeen, has recently brought forward good evidence for including various calcareous fossils, described by several authors under different names and referred to various genera of fossil animals, in the genus *Solenopora*, which he places among the coralline algae.

Species of this genus have been described from England, Scotland, Esthonia, Russia, and other countries. The geological range of *Solenopora* appears to be from Ordovician to Jurassic rocks; in some cases it is an important constituent of beds of limestone.

Solenopora compacta (Billings). Fig. 38. This species was originally described by Billings as *Stromatopora compacta*,

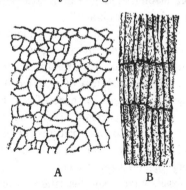

A   B

FIG. 38.   *Solenopora compacta* (Billings).   A.   Tangential section. × 100.
B.   Vertical section.   × 50.   (After Brown.)

[1] *Vide* Walther (88) p. 499; also Jukes-Browne and Harrison (91) *passim*. I am indebted to Mr G. F. Franks, who has studied the Barbadian reefs, for the opportunity of examining sections of West-Indian coral-rock.
[2] Brown A. (94).

and afterwards defined by Nicholson and Etheridge. The thallus forms sub-spheroidal masses, from the size of a hemp-seed to that of an orange. The external surface is lobulate; the fractured surface has a porcellanous and sometimes a fibrous appearance, and is usually white or light brown in colour. In vertical section (fig. 38, B) the cells are elongated and arranged in a radiating and parallel fashion; they often occur in concentric layers. The cells have a diameter of about $\frac{1}{17}$ mm. and possess distinctly undulating walls, as seen in a tangential section (fig. 38, A). Brown describes certain larger cells in the thallus (fig. 38, A) as sporangia[1], but it is difficult to recognise any distinct sporangial cavities in the drawing. The example figured is from the Trenton limestone of Canada; a variety of the same species has been recorded from the Ordovician rocks of Girvan in Ayrshire. There appear to be good reasons for accepting Brown's conclusion that *Solenopora* belongs to the Corallinaceae rather than to the Hydrozoa, among which it was originally included. After comparing *Solenopora* with recent genera of Florideae, Brown concludes that "the forms of the cells and cell-walls, the method of increase, and the arrangement of the tissue cells in the various species of *Solenopora* bear strong evidence of relationship between that genus and the calcareous algae[2]."

The importance of the calcareous Rhodophyceae has been frequently emphasised by recent researches, and our knowledge of the rock-building forms is already fairly extensive. We possess evidence of the existence of species of different genera in Ordovician seas, as well as in those of the Silurian, Triassic, Jurassic, and more recent periods. It is reasonable to prophesy that further researches into the structure of ancient limestones will considerably extend our knowledge of the geological and botanical history of the Corallinaceae.

Numerous fossils have been described as examples of other genera[3] of Rhodophyceae than those included in the Corallinaceae, but these possess little or no scientific value and need not be considered.

---

[1] Brown A. (94) p. 147.   [2] *ibid.* p. 200.   [3] *e.g.* Saporta (82) p. 12.

D. PHAEOPHYCEAE (BROWN ALGAE).

Olive-brown algae, thallus often leathery in texture, composed of cell-filaments or parenchymatous tissue, in some cases exhibiting a considerable degree of internal differentiation. The sexual reproductive organs may be either in the form of passive egg-cells and motile antherozoids or of motile cells showing no external sexual difference.

With one or two exceptions all the genera are marine. They have a wide distribution at the present day, and are especially characteristic of far northern and extreme southern latitudes. The gigantic forms *Lessonia*, *Macrocystis* and others already alluded to, belong to this group; also the genus *Sargassum*, of which the numberless floating plants constitute the characteristic vegetation of the Sargasso Sea.

Palaeobotanical literature is full of descriptions of supposed fossil representatives of the brown algae, but only a few of the recorded species possess more than a very doubtful value; most of them are worthless as trustworthy botanical records. Many of the numerous impressions referred to as species of *Fucoides* and other genera present a superficial resemblance to the thallus of the common Bladder-wrack and other brown seaweeds. Such similarity of form, however, in the case of flat and branched algal-like fossils is of no scientific value. In many instances the impressions are probably those of an alga, but they are of no botanical interest. The flat and forked type of thallus of *Fucus*, *Chondrus crispus* (L.) and other members of the Phaeophyceae is met with also among the red and green algae, to say nothing of its occurrence in the group of thalloid Liverworts, or of the almost identical form of various members of the animal kingdom. The variety of form of the thallus in one species is well illustrated by the common *Chondrus crispus* (L.). This alga was described by Turner[1] in his classic work on the *Fuci* under the name of *Fucus crispus* as "a marine Proteus." It affords an interesting example of the different appearance presented by the same species under different conditions, and at the same time it furnishes another proof of the

[1] Turner (11) vol. II. p. 51.

futility of relying on imperfectly preserved external features as
taxonomic characters of primary importance.

An example of a supposed Jurassic *Fucus* is shown in fig. 49,
and briefly described in the Chapter dealing with fossil Bryophytes.

Several species of Flysch Algae have recently been referred
by Rothpletz[1] to the Phaeophyceae under the provisional
generic name *Phycopsis*, but they are of no special botanical
interest.

The extremely interesting genus *Nematophycus* has lately
been assigned by a Canadian author[2] to a position in the
Phaeophyceae. Although the particular points on which he
chiefly relies are not perhaps thoroughly established, there
are certain considerations which lead us to include *Nematophycus* as a doubtful member of the present group of algae.

Nematophycus.

The stem attains a diameter of between 2 and 3 feet in the
largest specimens; it is made up either of comparatively wide
and loosely arranged tubes pursuing a slightly irregular vertical
course accompanied by a plexus of much narrower tubes, or of
tubes varying in diameter but not divisible into two distinct
types. Rings of growth occur in some forms but not in others.
Radially elongated or isodiametric spaces occur in the stem
tissues in which the tubes are less abundant.

Reproductive organs unknown, with the possible exception
of some very doubtful bodies described as spores.

In 1856 Sir William Dawson proposed the generic name
Prototaxites for some large silicified trunks discovered in the
Lower and Middle Devonian rocks of Canada. A few years
later the same writer[3] published a detailed account of the new
fossils and arrived at the conclusion that the Devonian stem
showed definite points of affinity with the recent genus *Taxus*,
and the generic name suggests that he regarded it as the type
of Coniferous trees belonging to the sub-family Taxineae. The

[1] Rothpletz (96).   [2] Penhallow (96) p. 45.   [3] Dawson (59).

reasons for this determination were afterwards shown by Carruthers to be erroneous. Dawson thought he recognised pits and spiral thickenings in the walls of the tubular elements, as well as pointed ends in some of the latter. The spiral markings were in reality small hyphal tubes passing obliquely across the face of the wider tubes, and the apparent ends of the supposed tracheids were deceptive appearances due to the fact that the tubes had in some cases been cut through in an oblique direction. In 1870 Carruthers[1] expressed the opinion that Dawson's *Prototaxites* was a "colossal fossil seaweed" and not a coniferous plant. The same author[2] in 1872 published a full and able account of the genus, and conclusively proved that *Prototaxites* could not be accepted as a Phanerogam; he brought forward almost convincing evidence in favour of including the genus among the algae. The name *Prototaxites* was now changed for that of *Nematophycus*. Carruthers compares the rings of growth in the fossil stems with those in the large Antarctic *Lessonia* stems, but he regards the histological characters as pointing to the Siphoneae as the most likely group of recent algae in which to include the Palaeozoic genus.

We may pass over various notes and additional contributions by Dawson, who did not admit the corrections to his original descriptions which Carruthers' work supplied. In 1889 an important memoir appeared by Penhallow[3] of Montreal in which he confirmed Carruthers' decision as to the algal nature of *Prototaxites*; he contributed some new facts to the previous account by Carruthers, and expressed himself in favour of regarding the fossil plant as a near ally of the recent Laminariae. The next addition to our botanical knowledge of this genus was made by Barber[4] who described a new specific type of *Nematophycus*—*N. Storriei*—found by Storrie in beds of Wenlock limestone age near Cardiff. Solms-Laubach[5], in a recent memoir on Devonian plants, recorded the occurrence of another species of this genus in Middle Devonian rocks near Gräfrath on the Lower Rhine. Lastly Penhallow[6], in describing a new species,

---

[1] *Vide* 'Academy' 1870, p. 16.　　　[2] Carruthers (72).

[3] Penhallow (89).　　　　　　　　　 [4] Barber (92).

[5] Solms-Laubach (95²).　　　　　　　[6] Penhallow (96).

lays stress on the resemblance of some of the tubular elements in the stem to the sieve-hyphae of the recent seaweeds *Macrocystis* and *Laminaria*. He concludes that the new facts he records make it clear that *Nematophycus* "is an alga, and of an alliance with the Laminarias." The recent evidence brought forward by Penhallow is not entirely satisfactory; the drawings and descriptions of the supposed trumpet-shaped sieve-hyphae are not conclusive. On the whole it is probably the better course to speak of *Nematophycus* as a possible ally of the brown algae rather than as an extinct type of the Siphoneae, but until our knowledge is more complete it is practically impossible to decide the exact position of this Siluro-Devonian genus.

Solms-Laubach[1] has suggested that the generic name *Nematophyton*, used by Penhallow in preference to Carruthers' term *Nematophycus*, is the more suitable as being a neutral designation and not one which assumes a definite botanical position. In view of the nature of the evidence in favour of the algal affinities of the fossil, the reasons for discarding Carruthers' original name are hardly sufficient.

Before discussing more fully the distribution and botanical position of *Nematophycus* we may describe at length one of the best known species, and give a short account of some other forms.

1. *Nematophycus Logani* (Daws.). Fig. 39, A—E. The stem possesses well marked concentric rings of growth due to a periodic difference in size of the large tubular elements. The tissues consist of two distinct kinds of tubular elements, the larger tubes loosely arranged and pursuing a fairly regular longitudinal course, and having a diameter of 13–35 μ; the smaller tubes, with a diameter of 5–6 μ, ramify in different directions and form a loose plexus among the larger and more regularly disposed elements. Branching occurs in both kinds of tubes; septa have been recognised only in the smaller tubes. Irregular and discontinuous radial spaces traverse the stem tissues, having a superficial resemblance in their manner of occurrence to the medullary rays of the higher plants.

[1] *loc. cit.* p. 83.

The best specimens of this species were obtained by Sir
William Dawson from the Devonian Sandstones of Gaspé in
New Brunswick. The largest stems had a diameter of 3 feet
and reached a length of several feet[1]; in some examples
Dawson found lateral appendages attached to the stem which
he described as "spreading roots." Externally the specimens
were occasionally covered with a layer of friable coal, and
internally the tissues were found to be more or less perfectly
preserved by the infiltration of a siliceous solution. Most of
the examples of *Nematophycus* from Britain and Germany are
much smaller and less perfectly preserved than those from
Canada. The Peter Redpath Museum, Montreal, contains
several very large blocks of *Nematophycus*, in many of which
one sees the concentric rings of growth clearly etched out by
weathering agents in a cross section of a large stem.

In fig. 39, A, a sketch is given of a thin transverse section of
a stem, drawn natural size. The lines of growth are clearly
seen, and as in coniferous stems the breadth of the concentric
zones varies considerably. The short lines traversing the
tissues in a radial direction represent the medullary-ray-like
spaces referred to in the specific diagnosis. A transverse section
examined under a low-power objective presents the appearance
of a number of thick-walled and comparatively wide tubes
loosely arranged; they may be in contact or separated from
one another. If the microscope be carefully focussed through the
thickness of the section the transversely-cut tubes appear to move
laterally, producing a curiously dazzling effect if the objective
is raised or lowered rapidly. This lateral movement is due
to the undulating vertical course of the tubes. Under a
higher power the lighter-coloured matrix in which the tubes
are embedded shows a number of very much smaller and
thinner-walled hyphal elements; some of these are cut across
transversely, others more or less obliquely and others again
longitudinally. These smaller tubes constitute an irregular
plexus surrounding and ramifying between the larger elements.
The diameter of the larger tubes decreases for a certain
distance in a radial direction as seen in a transverse section,

[1] Dawson (59), also (71) p. 17.

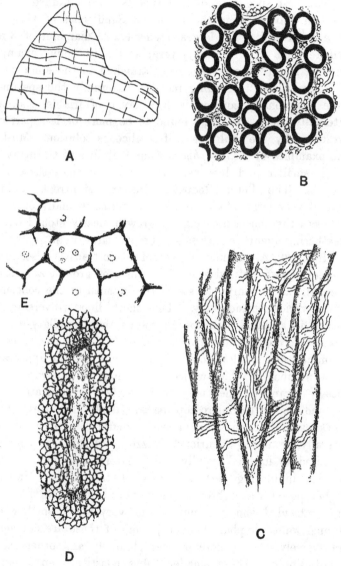

FIG. 39. *Nematophycus Logani* (Daws.). A. Part of a transverse section from a specimen in the British Museum. (Nat. size.) B. Transverse section from specimens in Mr Barber's possession. C. Longitudinal section. (B and C × 160.) D. Transverse section showing a radial space. E. Transverse section; a few 'cells' more highly magnified. D and E from a specimen in the British Museum.

and this change in size gives rise to the appearance of concentric lines indicating periodic changes in growth.

The radial spaces are characterised by the partial absence of
the larger tubes, and as seen in longitudinal sections these spaces
constitute regions in which the smaller tubes branch very freely.
Fig. 39, B, represents a small piece of a transverse section seen
under a fairly high power. In fig. 39, C, the tubes are seen in
longitudinal section. The larger elements are unseptate and
not very regular in their vertical course through the stem; the
smaller elements are seen as fine tubes lying between and across
the larger tubes. In the sections I have examined no undoubted transverse septa were detected in any of the tubular
elements.

The question as to the possible connection between the
larger and smaller elements is one which is not as yet satisfactorily disposed of. Penhallow[1] regards the finer hyphal elements
as branches of the larger tubes, but Barber[2], who has carefully
examined good material of *Nematophycus Logani*, was unable to
detect any organic connection between the two. My own
observations are in accord with those of Barber. Further
details and numerous figures of this species of *Nematophycus*
will be found in the memoirs of Carruthers, Penhallow and
Barber

Some specimens of silicified *Nematophycus* stems afford particularly instructive examples of the state of preservation or
method of mineralisation as a source of error in histological
work. The sketches reproduced in fig. 39, D and E, were made
from a section of a large specimen of *Nematophycus* in the
British Museum. In fig. D we have one of the radial spaces
containing some indistinct small elements, the tissue surrounding the space appears to consist of polygonal cells
suggesting ordinary parenchymatous tissue. In fig. E a few of
these 'cells' are seen more clearly, they have black and ragged
walls, and often contain very small and faint circles of which
the precise nature is uncertain. The true interpretation of

---

[1] Penhallow (89) and (96) p. 46.   [2] Barber (92) p. 336.

this form of structure was first supplied by Penhallow[1]. The
black network simulating parenchymatous tissue consists of
the substance of *Nematophycus* tubes which has been completely redistributed during fossilisation and collected along
fairly regular lines, as seen in figs. D and E. The original
structure has been almost completely destroyed, and the
material composing the walls of the large tubes has finally been
rearranged as a network, interrupted here and there by the
characteristic radial spaces which remain as evidence of the
original *Nematophycus* characters. It is possible in some cases
to trace every gradation from sections exhibiting the normal
structure through those having the appearance shown in
figs. D and E to others in which the structure is completely
lost. Penhallow describes this method of fossilisation in *N.
crassus* (Daws.); an examination of several specimens in the
National Collection leads me to entirely confirm his general
conclusions, and also to the opinion that *N. Logani* shows
exactly the same manner of mineralisation as *N. crassus*. The
chief point of interest as regards this method of preservation
lies in the fact that a fossil described by Dawson[2] as *Cellulo-
xylon primaevum*, and referred to as a probable conifer, is
undoubtedly a badly preserved *Nematophycus*. Penhallow
examined Dawson's specimens and obtained convincing evidence of their identity with certain forms of highly altered
Nematophycus stems.

2. *Nematophycus Storriei* Barber. Fig. 40. The specimens
on which Barber[3] founded this species were obtained by
Mr Storrie from the Tymawr quarry near Cardiff, in beds of
Wenlock age. The fragmentary nature of the material is
largely compensated for by the excellence of the preservation.
We may briefly define the species as follows:

The stem consists of separate interlacing undivided and
usually unbranched tubes of varying diameter. Spaces more
or less isodiametric in dimensions are scattered through the
tissue. The spaces constitute regions in which the tubular
elements branch freely.

[1] Penhallow (89) and (93).   [2] Dawson (81) p. 302.   [3] Barber (92).

The main distinguishing features of this British species are (i) the absence of two distinct and well-defined forms of tubular elements. The main part of the stem consists of thick walled

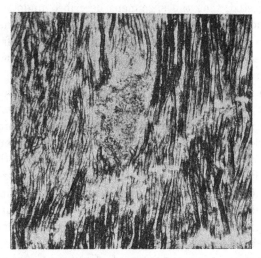

Fig. 40. *Nematophycus Storriei* Barb. Longitudinal section, from a photograph by Mr C. A. Barber. × 45.

tubes similar to those of *N. Logani,* but the spaces between them are occupied by thinner-walled and smaller tubes varying considerably in diameter; (ii) the form of the spaces which are not radially elongated as in *N. Logani.*

Fig. 40 shows the undulating course of the tubes as seen in a longitudinal section; the black colour of some of the elements is due to the fact that the surface of the wall is seen, while in the lighter-coloured portions of the tubes the wall has been cut through. The lighter patch about the middle of the figure shows the form of one of the spaces in which the tubes are freely branched.

In addition to the two species already described six others have been recorded, but with these we need not concern ourselves in detail. One of these species, *N. Hicksi,* was found by Dr Hicks[1] in the Denbighshire grits quarry of Pen-y-Glog near Corwen in North Wales. The position of these beds has

[1] Hicks (81) p. 490.

recently been determined by Mr Lake[1] as corresponding to that
of the Wenlock limestone. This species and *N. Storriei* are
both Silurian examples of the genus. It is possible, as Barber
has suggested, that the specimens described under these two
names should be referred to one species. The specimens found
by Hicks were small and imperfectly preserved fragments
Etheridge has given a full description of their structure, and
Barber has subsequently examined the material. The preservation is not such as will admit of any very precise specific
diagnosis; the fragments are correctly referred to *Nematophycus*,
but their specific characters cannot be clearly determined.

Solms-Laubach[2] has described some fragments of another
species of *Nematophycus* from the Devonian rocks of the Lower
Rhine. His specimens are chiefly interesting as extending the
geographical range of the genus, and as affording examples of a
curious method of preservation. The specimens obtained were
small fragments, flattened and very dark brown in colour. The
tubular elements consisted of an external membrane of black
coal, enclosing a central core of dark red iron-oxide. On
burning the fragment on a piece of platinum foil the coal
composing the wall of the tubes was removed and the deep-red
casts of the tube-cavities remained[3]. The investigation of the
structural characters of this imperfect material was conducted
by reflected light. Under certain conditions, when it is impossible to obtain thin sections for examination by transmitted
light, it is possible to accomplish much, as shown by Solms-
Laubach's work, by means of observation with direct light.

The last species to be noticed is *Nematophycus Ortoni*
recently described by Penhallow. There are no concentric
rings of growth, no radial spaces and no smaller hyphae in the
tissues of this type of stem. In longitudinal section, the tubes
show occasional local expansions of the lumen which Penhallow
compares with the 'trumpet-hyphae' of some recent brown
algae. No actual sieve-plates or transverse walls have been
detected, but the general appearance of the tubes is considered

---

[1] Lake (95) p. 22.   [2] Solms-Laubach (95[2]).

[3] A similar method of fossilisation has been noted by Rothpletz in the case
of the Lower Devonian alga *Hostinella*. [Rothpletz (96) p. 896.]

to afford distinct evidence of the original existence of such walls. The figures accompanying the description do not carry conviction as to the correctness of the reference of the tubes to imperfectly preserved sieve-hyphae.

The following list, taken, with a few alterations, from Penhallow's memoir[1], shows the geographical and geological range of the species of *Nematophycus* hitherto recorded.

Nematophycus Logani (Daws.)	⎧ Lower Devonian of Gaspé. ⎨ Silurian [Wenlock] of England. ⎩ Silurian of New Brunswick.
N. Hicksi (Eth.)	Silurian (Wenlock) of N. Wales.
N. crassus (Daws.)[2]	Middle Devonian of Gaspé and New York.
N. laxus (Daws.)	Lower Devonian of Gaspé.
N. tenuis (Daws.)	Lower Devonian of Gaspé.
N. Storriei (Barb.)	Silurian (Wenlock) of Wales (Cardiff).
N. dechenianus (Pied.)	Upper Devonian of Germany (Gräf-rath).
N. Ortoni (Pen.)	Upper Erian of Ohio.

In summing up our information as to the structure of *Nematophycus* we find there are certain points not definitely settled, and which are of considerable importance. The few recorded instances of spore-like bodies by Penhallow and Barber are not satisfactory; we are still ignorant of the nature of the reproductive organs. Such instances of lateral appendages as have been referred to do not throw much light on the habit of the plant. So far as we know at present the stem of *Nematophycus* was not differentiated internally into a cortical and central region. It may be that the specimens have been only partially preserved, and the coaly layer which occasionally surrounds a stem may represent a carbonised cortex which has never been petrified. The large and loosely arranged tubes constitute the chief characteristic feature of the genus; in some cases (*N. Logani*) there is an accompanying plexus of smaller hyphae, in others (*N. Storriei*) there is no definite division of the tissue into two sets of tubes of uniform size, and in *N. Ortoni* the tubular elements are all of the large type.

---

[1] Penhallow (96) p. 47.

[2] Carruthers (72) p. 162 regards this species as identical with *N. Logani*.

Penhallow has recognised the branching of large tubes in
N. Logani and *N. crassus* giving rise to the small hyphal
elements. In most specimens, however, no such mode of origin
of the smaller tubes can be detected. The spaces which
interrupt the homogeneity of the tissues in some forms have
been described as branching depots, on account of the frequent
occurrence in these areas of much branched hyphae. The
function of these spaces (fig 39, D, and fig. 40) may be connected
with aeration of the stem-tissues.

As Carruthers first pointed out the unseptate nature of the
elements and the occurrence of large and small tubes forming
a comparatively lax tissue suggested affinities with such recent
genera as *Penicillus, Halmeda, Udotea* and other members of
the Siphoneae. In those fossil stems which possess tubes of
two distinct sizes, we cannot as a rule trace any organic
connection between the two sets of tubular elements. Transverse septa have been detected in the tubes of some specimens
of *N. Logani.* These considerations and the large size and habit
of growth of the stem leave one sceptical as to the wisdom
of assigning the fossil genus to the Siphoneae. On the other
hand, apart from the doubtful sieve-hyphae of Penhallow, the
manner of growth of the plant, the concentric rings, marked by a
decrease in the diameter of the tubes, the lax arrangement and
irregular course of the elements, afford points of agreement with
some recent Phaeophyceae. The stem of a *Laminaria* (fig. 29)
or of a *Lessonia* are the most obvious structures with which
to compare *Nematophycus.* The medullary region of a *Laminaria* or *Fucus* and of other genera presents a certain resemblance
to the tissues of the fossil stems. On the whole we may be
content to leave *Nematophycus* for the present as probably an
extinct type of alga, more closely allied to the large members
of the Phaeophyceae than to any other recent seaweeds.

Pachytheca.
(A fossil of uncertain affinity.)

There is another fossil occasionally associated with *Nematophycus* and referred by many writers to the Algae, which calls

for a brief notice. *Pachytheca* is too doubtful a genus to justify a detailed treatment in the present work. Although, as I have elsewhere suggested[1], we are hardly in a position to speak with any degree of certainty as to its affinity, it is not improbable that it may eventually be shown to be an alga.

Without attempting a full diagnosis of the genus, we may briefly refer to its most striking characters.

Pachytheca usually occurs in the form of small spherical bodies, about 5 cm. in diameter, in Old Red Sandstone or Silurian rocks. In section a single sphere is found to consist of two well marked regions; in the centre, of a number of ramifying and irregularly placed narrow tubes, and in the peripheral or cortical region, of numerous regular and radially disposed simple or forked septate tubes. The tubular elements of the two regions are in organic connection.

The name was proposed by Sir Joseph Hooker for some specimens found by Dr Strickland[2] in the Ludlow bone-bed (Silurian) of Woolhope and May-Hill. Examples were subsequently recorded from the Wenlock limestone of Malvern and from Silurian and Old Red Sandstone rocks of other districts. Hicks[3] found *Pachytheca* in the Pen-y-Glog grits of Corwen in association with *Nematophycus*, and the two fossils have been found together elsewhere. This association led to the suggestion that *Pachytheca* might be the sporangium of *Nematophycus*, and Dawson[4], in conformity with his belief in the coniferous character of the latter plant, referred to *Pachytheca* as a true seed.

The best sections of this fossil have been prepared with remarkable skill by Mr Storrie of Cardiff; they were carefully examined and described by Barber in two memoirs[5] published in the *Annals of Botany*, the account being illustrated by several well executed drawings and microphotographs.

Among other difficulties to contend against in the interpretation of *Pachytheca* there is that of mineralisation. The preservation is such as to render the discrimination of original structure as distinct from structural features of secondary origin,

[1] Seward (95[3]).   [2] Strickland and Hooker (53).   [3] Hicks (81) p. 484.
[4] Dawson (82) p. 104.   [5] Barber (89) and (90).

consequent on a particular manner of crystallisation of the siliceous material, a matter of considerable difficulty.

Suggestions as to the nature of *Pachytheca* have been particularly numerous; it has been referred to most classes of plants and relegated by some writers to the animal kingdom. The most recent addition to our knowledge of this problematic fossil was the discovery of a specimen by Mr Storrie in which the *Pachytheca* sphere rested in a small cup, like an acorn fruit in its cupule. This specimen was figured and described by Mr George Murray[1] in 1895; he expresses the opinion that the discovery makes the taxonomic position of the genus still more obscure. Solms-Laubach briefly refers to *Pachytheca* in connection with *Nematophycus*, and regards its precise nature almost as much an unsolved riddle now as it was when first discovered. For a fuller account of this fossil reference must be made to the contributions of Hooker[2], Barber[3] and others. The literature is quoted by Barber and more recently by Solms-Laubach[4]. There are several specimens and microscopic sections of *Pachytheca* in the geological and botanical departments of the British Museum. The genus has been recorded from Shropshire, North Wales, Malvern, Herefordshire, Perthshire and other British localities, as well as from Canada; it occurs in both Silurian and Old Red Sandstone rocks.

Algites.

A generic name for those fossils which in all probability belong to the class Algae, but which, by reason of the absence of reproductive organs, internal structure, or characters of a trustworthy nature in the determination of affinity, cannot be referred with any degree of certainty to a particular recent genus or family.

This term was suggested in 1894[5] as a provisional and comprehensive designation under which might be included such impressions or casts as might reasonably be referred to Algae. The practice of applying to alga-like fossils names suggestive of a definite alliance with recent genera is as a rule unsound.

---

[1] Murray G. (95[3]).   [2] Hooker J. D. (89).   [3] *loc. cit.*
[4] Solms-Laubach (95[2]) p. 81.   [5] Seward (94[2]) p. 4.

It would simplify nomenclature, and avoid the multiplication of generic names, if the term *Algites* were applied to such algal fossils from rocks of various ages as afford no trustworthy data by which their family or generic affinity can be established.

V. MYXOMYCETES (MYCETOZOA).

This class of organisms affords an interesting example of the impossibility of maintaining a hard and fast line between the animal and plant kingdom.   Zoologists and Botanists usually include the Myxomycetes[1] in the text-books of their respective subjects, and the name Animal-fungi which has been applied to these organisms expresses their dual relationship.   They constitute one of three groups which we may include in that intermediate zone or buffer-state' between the two kingdoms. From a palaeobotanical point of view the Myxomycetes are of little interest, but a very brief reference may be made to them rather for the sake of avoiding unnecessary incompleteness in our classification than from their importance as possible fossils.

They are organisms without chlorophyll, consisting of a naked mass of protoplasm, known as the *plasmodium*, which may attain a size of several inches.   Such plasmodia creep over the surface of decaying organic substrata, and in forming their asexual reproductive cells they are converted into somewhat complex fruits containing spores.   The spores produce motile swarm-cells, which eventually coalesce together to form a new plasmodium.

A few examples of fossil Myxomycetes have been recorded from the Palaeozoic and more recent formations, but none of them are entirely beyond suspicion.   We may mention three examples of fossils referred to this group, but only one of these is entitled to serious consideration.

Myxomycetes Mangini Ren.[2]   It is not uncommon to find

---

[1] An excellent monograph on the Mycetozoa has lately been issued by the Trustees of the British Museum under the authorship of Mr A. Lister (94).   *Vide* also Schröter (89) in Engler and Prantl's *Natürlichen Pflanzenfamilien.*

[2] Renault (96) p. 422, figs. 75 and 76.

distinct traces of original or secondary cell-contents in well
preserved petrified plant-tissues. There is often a difficulty,
however, in distinguishing between the true cell-contents and
the cells of some parasitic or saprophytic intruder. In some
petrified corky tissue in a silicified nodule from the Permo-
Carboniferous beds of Autun, Renault has recently discovered
what he believes to be traces of a Myxomycetous plasmodium.
The cork-cells would be without protoplasmic contents of their
own, and their cavities contain a number of fine strands
stretching from the cell-walls in different directions and uniting
in places as irregular or more or less spherical masses. The
drawings given by Renault of these irregular reticulated structures with scattered patches of what may possibly be petrified
plasmodial protoplasm bear a striking resemblance to the plasmodium of a Myxomycete. A figure of the capillitium of a species
of *Leocarpus* figured by Schröter[1] in his account of the Myxomycetes in Engler and Prantl's work is very similar to that of
Renault's plasmodium.'

It is by no means inconceivable that the *Myxomycetes
Mangini* may be correctly referred to this group, but the wisdom
of assigning a name to such structures may well be questioned.

The other two examples call for little notice. Messrs Cash
and Hick[2] in a paper on fossil fungi from the Coal-Measures
refer to some small spherical bodies as possibly the spores of a
Myxomycete. They might be referred equally well to numerous
other organisms.

Göppert and Menge[3] in their monograph on plants in the
Baltic Tertiary Amber, express the opinion that an ill-defined
tangle of threads which they figure may be a Myxomycete.

It would serve no useful purpose to quote other instances
of possible representatives of fossil Mycetozoa; but the consideration of the above examples may serve to emphasize the
desirability of refraining from converting a possibility into an
apparently recognised fact by the application of definite generic
and specific names.

---

[1] Schröter (89) p. 32, fig. 18 B.
[2] Cash and Hick (78²) Pl. vi. fig. 3.
[3] Göppert and Menge (83) Pl. xiii. fig. 106.

VI. FUNGI.

The most striking difference between the fungi and algae is the absence of chlorophyll in the former, and the consequent inability of fungi to manufacture their organic compounds from inorganic material. Fungi live therefore either as parasites or saprophytes, and as the same species may pass part of its life in a living host to occur at another stage of its development as a saprophyte, it is impossible to distinguish definitely between parasitic and saprophytic forms. The vegetative body of a fungus, that is the portion which is concerned with providing nourishment and preparing the plastic food-substance for the reproductive organs, is known as the *mycelium*. It consists either of a single and branched tubular cell known as a *hypha*, or of several hyphae or thread-like elements (filamentous fungi). The hyphal filaments may be closely packed together and form a felted mass of compact tissue, which in cross section closely simulates the parenchyma of the higher plants. This pseudo-parenchymatous form of thallus is particularly well illustrated by the so-called *sclerotia*; these are sharply defined and often tuberous masses of hyphal tissue covered by a firm rind and containing supplies of food in the inner hyphae. They are able to remain in a quiescent state for some time, and to resist unfavourable conditions until germination and the formation of a new individual take place. The reproductive structures assume various forms; in some of the simpler fungi (Phycomycetes) sexual organs occur, as in the parallel group of Siphoneae among the algae, but in the higher fungi the reproduction is usually entirely asexual. An interesting case has recently been recorded among the more highly differentiated fungi in which distinct sexuality has been established[1]. In addition to the reproductive organs, such as oogonia and antheridia, the asexual cells or spores are borne either in special sporangia, or they occur as exposed *conidia* on supporting hyphae or *conidiophores*. Thick-walled and resistant resting-spores of various forms are also met with.

[1] Harper (95).

Without going into further details we may very briefly refer
to the larger subdivisions of this group of Thallophytes.

PHYCOMYCETES. Mycelium usually consisting of a single cell.
ZYGOMYCETES, Reproduction by means of conidia, and in many
OOMYCETES, cases also by the conjugation of two similar
including hyphae or by the fertilisation of an egg-cell con-
Chytridiaceae, &c. tained in an oogonium.

MESOMYCETES,
including the Intermediate between the Phycomycetes and
Sub-classes the higher fungi. Multicellular hyphae. No
HEMIASCI and sexual organs.
HEMIBASIDII.

MYCOMYCETES. Septate vegetative mycelium. No sexual re-
including the production—as a general rule. Asexual conidia
Sub-classes and other forms of spores. In the Ascomycetes the
ASCOMYCETES and spores are found in characteristic club-shaped cases
BASIDIOMYCETES. or asci ; in the Basidiomycetes the spores are borne
on special branches from swollen cells known as
basidia. The sporophore or spore-bearing body
in this group may attain a considerable size
(e.g. *Agaricus, Polyporus*, &c.) and exhibit a
distinct internal differentiation.

Before describing a few examples of fossil fungi, it is
important to consider the general question of their manner of
occurrence and determination. Considering the small size and
delicate nature of most fungi, it is not surprising that we have
but few satisfactory records of well-defined fossil forms. The
large leathery sporophores of *Polyporus* and other genera of
the Basidiomycetes, which are familiar objects as yellow or brown
brackets projecting from the trunks of diseased forest trees, have
been found in a fairly perfect condition in the Cambridgeshire
peat-beds, and examples have been described also by continental
writers[1]. As a general rule, however, we have to depend on
the chance mineralisation or petrifaction of the hyphae of a
fungus-mycelium which has invaded the living or dead tissues
of some higher plant. In the literature on fossil plants there
are numerous recorded species of fungi founded on dark coloured
spots and blotches on the impression of a leaf. Most of such
records are worthless; the external features being usually too

[1] *e.g.* Ludwig (57) Pl. xvi. fig. 1.

imperfect to allow of accurate identification. The occurrence of recent fungi as discolourations on leaves is exceedingly common, and the characteristic *perithecia* or compact and more or less spherical cases enclosing a group of sporangia in certain Ascomycetous species, might be readily preserved in a fossil condition.

Some examples of possible Ascomycetous fungi have been recently recorded by Potonié from leaves and other portions of plants of Permian age. There is a distinct superficial resemblance between the specimens he figures and the fructifications of recent Ascomycetes, but in the absence of internal structure, it would be rash to do more than suggest the probable nature of the ·markings he describes. For one of the fungus-like impressions Potonié proposes the generic name *Rosellinites*; he compares certain irregularly shaped projections on a piece of Permian wood with the perithecia of *Rosellinia*, a member of the Sphaeriaceae, and describes them as *Rosellinites Beyshlagii* Pot.[1] Various other records of similar Ascomycetes-like fossils may be found in palaeobotanical literature[2], but it is unnecessary to examine these in detail. Unless we are able to determine the nature of the supposed fungus by microscopical methods our identifications cannot in most cases be of any great value.

An example of the perithecia of a fungus (*Rosellinia congregata* [Beck])[3] has been recorded from the Oligocene of Saxony, which would appear to rest on a more satisfactory basis than is often the case. In this particular instance the small projections on a piece of fossil coniferous stem present a form which naturally suggests a fungus perithecium. In cases where the black spots on a fossil stem or leaf possess a definite form and structure, it is perfectly legitimate to refer them to a group of fungi; but in very many instances the forms referred to such genera as *Sphaerites* and others are of little or no value.

---

[1] Potonié (93) p. 27, Pl. ɪ. fig. 8.

[2] References are given by Potonié to illustrations by Zeiller (92²) Pl. xv. fig. 6, Grand' Eury (77) Pl. xxxɪɪɪ. fig. 7, and others in which possible fungi are represented.

[3] Engelhardt (87).

Many forms of scale-insects and galls on leaves present an obvious superficial resemblance to epiphyllous fungi, and might readily be mistaken for the fructifications of certain Ascomycetous species. As examples of scale-insects simulating fungi, reference may be made to such genera of the Coccineae as *Aspidiotus*, *Diaspis*, *Lecanium*, *Coccus*, and others. The female insects lying on the surface of a leaf, if preserved as a fossil impression, might easily be mistaken for perithecia[1].

Another pitfall in fossil mycology may be illustrated by a description of a supposed fungus, *Sclerotites Salisburiae*[2], Mass. on a Tertiary *Ginkgo* leaf. The figure given by Massalongo represents a *Ginkgo* leaf with well marked veins, the lamina between the veins being traversed by short discontinuous and longitudinally-running lines; the latter are referred to as the fungus. In a recent *Ginkgo* leaf one may easily detect with the naked eye a number of short lines between and parallel to the veins, which if examined in section are found to be secretory canals. There can be little doubt that *Sclerotites Salisburiae* owes its existence to the preservation of these canals.

The list of fossil fungi given by Meschinelli in Saccardo's *Sylloge Fungorum*[3] includes certain species which are of no botanical value, and should have no place in any list which claims to be authentic.

Among the numerous examples of fossil 'fungi' which have no claim to be classed with plants, there are some which are in all probability the galleries of wood-eating animals. The radiating grooves frequently found on the inner face of the bark of a pine tree made by species of the beetle *Bostrychus* might be mistaken for the impressions of the firm strands of mycelial tissue of some Basidiomycetous fungus.

In some notes on fossil fungi by J. F. James[4] contributed to the American Journal of Mycology in 1893, it is pointed out that a supposed fungus described by Lesquereux from the

---

[1] For figures of the Coccineae, see Comstock (88), Maskell (87), Judeich and Nitsche (95) &c.

[2] Massalongo (59) Pl. i. fig. 1, p. 87.

[3] Meschinelli (92).   [4] James, J. F. (93²).

Lower Coal-Measures as *Rhizomorpha Sigillariae*[1], bears a strong
likeness to some insect-burrows, such as those of *Bostrychus*.

"A new fungus from the Coal-Measures" described by
Herzer in 1893[2] may probably be referred to animal agency.
In any case there is no evidence as to the fungoid nature of
the object represented in the figure accompanying Herzer's
description.

More trustworthy evidence of fossil fungi is afforded
by the marks of disease in petrified tissue and by the
presence of true mycelia. In examining closely the calcareous
and siliceous plant-tissues from the Coal-Measures and
other geological horizons, one occasionally sees fine thread-
like hyphae ramifying through the cells or tracheal cavities;
in many cases the hyphae bear no reproductive organs and
cannot as a rule be referred to a particular type of fungus.
If the hyphal filaments are unseptate, they most likely
belong to some Phycomycetous species; or if they are obviously
septate the Mesomycetes or the Mycomycetes are the more
probable groups. Occasionally there may be found indications
of the characteristic *clamp-connections* in the septate filaments;
a small semicircular branch, which is given off from a mycelium
immediately above a transverse wall, bends round to fuse with
the filament just below the septum, thus serving as a small
loop-line connecting the cell-cavity above and below a cross wall.
Such clamp-connections are usually confined to the hyphae
of Basidiomycetes and thus serve as a useful aid in identification. A good example of a clamp-connection in a fossil
mycelium is figured by Conwentz[3] in his monograph on the
Baltic amber-trees of Oligocene age. The stout and thick type
of hypha found in some fossil woods agrees closely with that
of *Polyporus, Agaricus melleus* and other well-known recent
Basidiomycetes.

In a section of a piece of lignified coniferous wood recently
brought by Col. Feilden from Kolguev island[4], the brown and

---

[1] Lesquereux (87).   [2] Herzer (93).   [3] Conwentz (90) Pl. xii. fig. 5.
[4] Feilden, H. W. (96); Seward (96²) p. 62, appendix to Feilden's paper. I am
indebted to Dr Bonney for an opportunity of examining the plant remains from
the Feilden collection.

stout hyphae of a fungus are clearly seen as distinct dark lines traversing the tracheal tissue. The occurrence of septa and the large diameter of the mycelial branches at once suggest a comparison with such recent forms as *Agaricus melleus, Polyporus* and other Basidiomycetes. The age of the Kolguev wood is not known with any certainty.

The vesicular swellings such as those represented in fig. 41, A, B, D and E, may easily be misinterpreted. Such spherical expansions in a mycelium, either terminal or intercalary, may be sporangia, oogonia or large resting-spores, or non-fungal cell-contents, and it is usually impossible in the absence of the contents to determine their precise nature. Hartig[1] and others have drawn attention to the occurrence of such bladder-like swellings in the mycelia of recent fungi, which have nothing to do with reproductive purposes; under certain conditions the hyphae of a fungus growing in the cavity of a cell or trachea may form such vesicles, and these, as in fig. 42, D, *m* may completely fill up the cavity of a large tracheid.

Some good examples of bladder-like swellings, such as occur in the mycelium of *Agaricus melleus* and other recent fungi, have been figured by Conwentz[2] in fossil wood of Tertiary age from Karlsdorf. The swellings in this fossil fungus might easily be mistaken for oogonia or sporangia; especially as they are few in number and spherical in form.

A similar appearance is presented by a mass of tyloses in the cavity of an old vessel or tracheid; and vesicular cell-contents, as in the cells of fig. 41, A, 2–5, may closely simulate a number of thin-walled fungal spores or sporangia.

A good example of such a vesicular tissue, in addition to that already quoted, is afforded by a specimen of an Eocene fern, *Osmundites Dowkeri* Carr.[3] described by Carruthers in 1870. The ground-tissue cells contain traces of distinct fungal hyphae (fig. 41, B), and in many of the parenchymatous elements the cavity is completely filled with spherical vesicles; in other cases one finds hyphae in the centre of the cell while vesicles line the wall, as shewn in fig. 41, B. Carruthers refers to these

[1] Hartig (78).   [2] Conwentz (80) Pl. v. fig. 17.
[3] Carruthers (70) Pl. xxv. fig. 3.

bladders as starch grains, and this may be their true nature; their appearance and abundant occurrence in the parenchyma certainly suggest vesicular cell-contents rather than fungal cells. I could detect no proof of any connection between the hyphae and bladders, and the absence of the latter in the cavities of the tracheids, fig. 41, C, favoured the view of their being either starch-grains or other vacuolated contents similar to that in the cells of the Portland Cycad (fig. 41, A) referred to on p. 88.

The vacuolated cell-contents partially filling the cells in fig. 41, D, present a striking resemblance to the contents of the cells 2–5 in fig. 41, A. In fig. D the frothy and contracted substance might be easily mistaken for a parasitic or saprophytic fungus, but this resemblance is entirely misleading. It is by no means uncommon to find the cells of recent plants occupied by such vacuolated contents, especially in diseased tissues in which a pathological effect produces an appearance which has more than once misled the most practised observers.

In the important work recently published by Renault on the Permo-Carboniferous flora of Autun, there is a small spore-like body described as a teleutospore, and classed with the Puccineae[1]. We have as yet no satisfactory evidence of the existence of this section of Fungi in Palaeozoic times, and Renault's description of *Teleutospora Milloti* from Autun might be seriously misleading if accepted without reference to his figure. The fragment he describes cannot be accepted as sufficient evidence for the existence of a Palaeozoic *Puccinia*.

The same author refers another Palaeozoic fungus to the Mucorineae under the name of *Mucor Combrensis*[2]; this identification is based on a mycelium having a resemblance to the branched thallus of *Mucor*, but in the absence of reproductive organs such resemblance is hardly adequate as a means of recognition.

The occurrence of hyphal cells in calcareous shells and corals has already been alluded to.[3] In addition to the examples referred to above, there is one which has been

---

[1] Renault (96) p. 427, fig. 80, *d*.
[2] *ibid.* p. 427, fig. 80, *a—c*.   [3] p. 127.

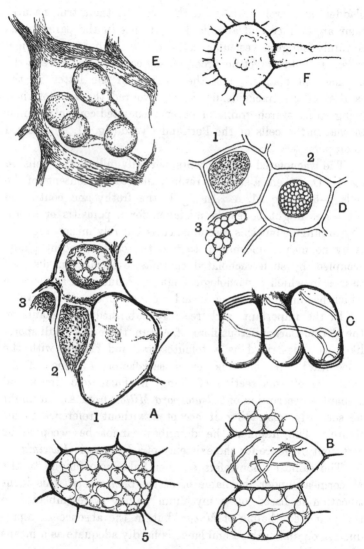

FIG. 41.   A. Cells of *Cycadeoidea gigantea* Sew.   × 355.   B and C. Parenchymatous cells and scalariform tracheids of *Osmundites Dowkeri* Carr.   × 230.
D. Epidermal cells of *Memecylon* (*Melastomaceae*) with vacuolated contents.
E. *Peronosporites antiquarius* Smith, (No. 1923 in the Williamson collection).
× 230.   F. *Zygosporites*.   × 230.   (A, B, C and E drawn from specimens
in the British Museum; D from a drawing by Prof. Marshall Ward; F from
a specimen in the Botanical Laboratory Collection, Cambridge.)

described by Etheridge[1] from a Permo-Carboniferous coral. This observer records the occurrence of tubular cavities in the calices of *Stenopora crinita* Lonsd., and attributes their origin to a fungus which he names *Palaeoperone endophytica*; he mentions one case in which a tube contains fine spherical spore-like bodies which he compares with the spores of a *Saprolegnia*. As pointed out above (p. 128), it is almost impossible to decide how far these tubes in shells and corals should be attributed to fungi, and how far to algae.

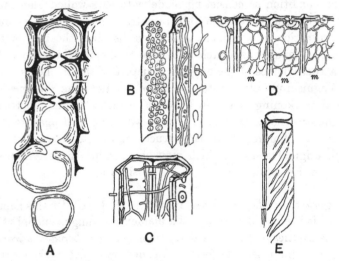

FIG. 42. A, B, C. Tracheids of coniferous wood attacked by *Trametes radiciperda* Hart. (*Polyporus annosus* Fr.) D and E. Tracheids attacked by *Agaricus melleus* Vahl. A, × 650, B—E, × 360. (After Hartig.)

Passing from the direct evidence obtained from the presence of fungal hyphae in petrified tissues, we must draw attention to the indirect evidence of fungal action afforded by many fossil plants. It is important to be familiar with at least the more striking effects of fungal ravages in recent wood in order that we may escape some of the mistakes to which pathological phenomena may lead us in the case of fossils[2].

The gradual dissociation of the elements in a piece of

[1] Etheridge (92) Pl. VII.
[2] Hartig (78) and (94), Göppert and Menge (83).

fossil wood owing to the destruction of the middle lamellae, the occurrence of various forms of slit-like apertures in the walls of tracheids (fig. 42, E) and the production of a system of fine parallel striation on the walls of a vessel are among the results produced by parasitic and saprophytic fungi. With the help of a ferment secreted by its hyphae, a fungus is able to eat away either the thickening cell layers or the middle lamellae or both, and if, as in fig. 42, A, only the middle lamellae are left one might easily regard such tissue in a fossil condition as consisting of delicate thin-walled elements. The oblique striae on the walls of a tracheid may often be due to the action of a ferment which has dissolved the membrane in such a manner as to etch out a system of spiral lines, probably as a consequence of the original structure of the tracheids. In distinguishing between the woods of Conifers the presence of spiral thickening layers in the wood element is an important diagnostic character, and it is necessary to guard against the confusion of purely secondary structures, due to fungal action, with original features which may be of value in determining the generic affinity of a piece of fossil wood.

Oochytrium Lepidodendri, Ren. Fig. 43, 1. Under this name Renault has recently described a filamentous fungus endophytic in the cavities of the scalariform tracheids of a *Lepidodendron*[1]. The mycelium has the form of slender branched hyphae with transverse septa. Numerous ovoid and more or less spherical sporangia occur as terminal swellings of the mycelial threads. The long axis of the ovoid forms measures 12—15 μ, and the shorter axis 9—10 μ; the contents may be seen as a slightly contracted mass in the sporangial cavity. In some of the sporangia one sees a short apical prolongation in the form of a small elongated papilla, as shown in fig. 43, 1. Renault refers this fungus to the Chytridineae, and compares it with *Cladochytrium, Woronina, Olpidium,* and other recent genera.

In the immediate neighbourhood of two of the sporangia shown in the uppermost tracheid of fig. 43, 1, there are seen a few minute dark dots which are described as spores petrified

[1] Renault (96) p. 425, fig. 78.

in the act of escaping from a lateral pore.　This interpretation strikes one as lacking in scientific caution.

The sporangia of *Hyphochytrium infestans*[1], as figured by Fischer in Rabenhorst's work bear a close resemblance to those of the fossil.　It would seem very probable that Renault's species may be reasonably referred to the Chytridineae, as he proposes.

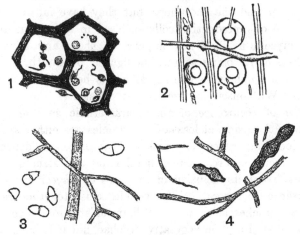

Fig. 43.　1.　*Oochytrium Lepidodendri*, Ren.　(After Renault.)　2.　*Polyporus vaporarius* Fr. var. *succinea*.　(After Conwentz.)　3.　*Cladosporites bipartitus* Fel.　(After Felix.)　4.　*Haplographites cateniger* Fel.　(After Felix.)

Peronosporites antiquarius W. Smith.　Fig. 41, E.

In an address to the Geologists' Association delivered by Mr Carruthers in 1876 a brief reference, accompanied by a small-scale drawing, is made to the discovery of a fungus in the scalariform tracheids of a *Lepidodendron* from the English Coal-Measures[2].　In the following year Worthington Smith published a fuller account of the fungus, and proposed for it the above name[3], which he chose on the ground of a close similarity between the mycelium and reproductive organs of the fossil form and recent members of the

[1] Fischer in Rabenhorst, vol. i. (92) p. 144.
[2] Carruthers (76) p. 22, fig. 1.　　　[3] Smith, W. G. (77) p. 499.

Peronosporeae. In Smith's description the mycelium is described as bearing spherical swellings containing zoospores. These spherical organs are fairly abundant and not infrequently met with in sections of petrified plant-tissues from the English Coal-Measures; they may be oogonia or sporangia, or in some cases mere vesicular expansions of a purely vegetative hypha. No confirmation has been given to the supposed spores referred to by Smith. Prof. Williamson and others have carefully examined the specimens, but they have failed to detect any trace of reproductive cells enclosed in the spherical sacs[1]. The mycelium does not appear to show any satisfactory evidence of its being septate as figured by Smith.

The example shown in fig. 41 E has been drawn from one of the Williamson specimens: it illustrates the form and manner of occurrence of the characteristic swellings. It is probable that some at least of the vesicles are either sporangia or oogonia, but we cannot speak with absolute confidence as to their precise nature. The general habit and structure of the fungus favour its inclusion in the class of *Phycomycetes*. The occurrence of several of the vesicles close together on short hyphal branches, as shown in Williamson's figures, suggests the spherical swellings on vegetative hyphae, but it is impossible to speak with absolute confidence. There is a close resemblance between this English form and one recently described by Renault as *Palaeomyces gracilis* Ren.[2]; the two fossils should probably be placed in the same genus.

The examples referred to below and originally recorded by Cash and Hick no doubt belong to the same type as Smith's *Peronosporites*.

The sketches reproduced in fig. 44 have been drawn from specimens originally described by Cash and Hick in 1878[3] The sections were cut from a calcareous nodule from the Halifax Coal-Measures containing fragments of various plants and among others a piece of cortical tissue, probably of a *Lepidodendron* or *Stigmaria*. In a transverse section of this

---

[1] Williamson (81) Pl. xlviii. p. 301.
[2] Renault (96) p. 439, figs. 88 and 89.
[3] Cash and Hick (78²).

tissue one sees under a moderately high power that the cells
have become partially separated from one another by the

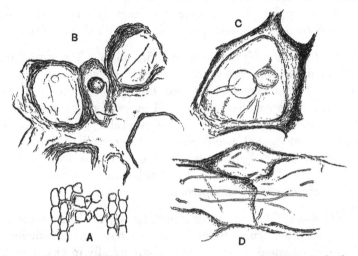

FIG. 44. Cells with fungal hyphae. A. A piece of disorganised tissue,
showing the separation of the cells. B. Part of A more highly magnified.
C. A single cell containing two swollen hyphae. D. Partially destroyed
cell-membranes pierced by fungal hyphae. (Drawn from sections in the
Edinburgh Botanical Museum, originally described by Cash and Hick.)

destruction of the middle lamella (fig. 44 A). The cell-
cavities and the spaces between the isolated cells contain
numerous fine fungal hyphae, which here and there terminate
in spherical swellings. One such swelling is shown under a low
power in fig. 44 A, in the middle uppermost cell, and more
highly magnified in fig. 44 B. In fig. C there are two such
swellings (the larger one having a diameter of ·003 mm.) in
contact, but the connection does not appear to be organic.
The cell-walls of the infected tissue present a ragged and
untidy appearance, and in places (*e.g.* fig. 44 D) the membrane
has been pierced by some of the mycelial branches.

This fungus bears a close resemblance to *Peronosporites
antiquarius*, but it is impossible to determine its precise
botanical position without further data. In Cash and Hick's
paper in which the above fungus is briefly dealt with, some

small spore-like bodies are figured which the authors speak of as possibly a Myxomycetous fungus[1] There is however no sound reason for such a supposition.

As examples of Ascomycetous fungi found in silicified wood of Tertiary age, two species may be quoted from Felix.

Cladosporites bipartitus Felix[2], fig. 43, 3. The mycelium and conidia of this form were discovered in some Eocene silicified wood from Perekeschkul near Baku, on the shores of the Caspian. The conidia are elliptical or pyriform in shape and divided by a transverse septum into two cells. No traces were found of any special conidiophores. The mycelium consists of septate branched hyphae, rendered conspicuous by a brown colouration. Felix compares the fossil with the recent genera *Cephalothecium* and *Cladosporium*.

Haptographites canteniger Felix[3], fig. 43, 4. The conidia of this form were found to be fairly abundant in the silicified tissue investigated by Felix; they occur usually in chains of 2 to 6 conidia having an ovoid or flask-shaped form, with a thick membrane (fig. 43, 4). The mycelium consists of branched hyphae divided into long cylindrical cells by transverse septa; occasional instances were found of an H-shaped fusion between lateral branches of parallel hyphae.

Felix compares this species with examples of the genera *Haptographium* and *Dematium* of the family Sphaeriaceae; it was found in the woody tissue of a dicotyledonous stem from Perekeschkul.

Zygosporites sp. The object represented in fig. 41 F consists of a stalked spherical sac bearing a number of radiating arms which are divided distally into delicate terminations. We find similar bodies figured by Williamson[4] in his IXth and Xth Memoirs on the Coal-Measure plants; he includes some of them under the generic term *Zygosporites*, and

---

[1] Cash and Hick, Pl. vi. fig. 3.
[2] Felix (94) p. 276, Pl. xix. fig. 1.
[3] *ibid.* p. 274, Pl. xix. figs. 5 and 6.
[4] Williamson (78) and (80).

compares them with the zygospores of the freshwater algae Desmideae. Hitherto these spore-like fossils have only been recorded as isolated spheres, but in the example shown in fig. 41 F there is a distinct tubular and thin-walled stalk attached to the *Zygosporites*. The specimen was found in the partially disorganised cortical tissue of a *Lyginodendron* stem from the English Coal-Measures. It is difficult to decide as to the precise nature of the fossil, but the presence of the hyphal stalk points to a fungus rather than an alga as the most probable type of plant with which to connect it. It may possibly be a sporangium of a fungus comparable with the common mould *Mucor*, or it may be a zygospore formed by the conjugation of two hyphae of which only one has been preserved.

For an example of a fossil representative of the Basidiomycetes we may turn to the excellent monograph by Conwentz on the Baltic amber trees, and quote one of the forms which he has described.

Polyporus vaporarius Fr. *f. succinea*[1], fig. 43, 2. In several preparations of the wood preserved by petrifaction in amber Conwentz found distinct indications of the ravages of a fungus, which suggested the presence of the recent species *Polyporus vaporarius* Fr. With the help of the indirect evidence afforded by the pathological effects as seen in the tissues of the host-plant, and the direct evidence of the fungal mycelium Conwentz was led to this identification.

The mycelium is brown in colour, in part thick-walled, and in part with thin walls, transversely septate and not much branched. In the portion of one of Conwentz' figures reproduced in fig. 43, 2, the rents and holes in the tracheid walls are clearly shown; they afford the indirect evidence of fungal attacks, and are of the same nature as those shown in fig. 42, B, C and E.

Enough has been said to call attention to the paucity of exact data on which to generalise as to the geological history

---

[1] Conwentz (90) p. 119, Pl. xi. pp. 2, 3, Pl. xv. fig. 8.

of fungi. The types selected for description or passing allusion have not been chosen in each case because of their special intrinsic value, but rather as convenient examples by which to illustrate authentic records or to serve as warnings against possible sources of error.

It would seem that we have fairly good and conclusive evidence of the existence in Permo-Carboniferous times of Phycomycetous fungi, but it is not until we pass to post-Palaeozoic or even Tertiary plants that we discover satisfactory representatives of the higher fungi or Mycomycetes. If special attention were paid to the investigation of fossil fungi, it is quite possible that our knowledge of the past history of the group might be considerably extended. It is essential that the greatest caution should be exercised in the identification of forms and in their reference to definite families; otherwise our lists of fossil species will serve to mislead, and to emphasize the untrustworthy character of palaeobotanical data. Unless we feel satisfied as to the position of a fossil fungus it is unwise to use a generic term suggestive of a definite family or recent genus. Such a name as Renault has used in one instance, *Palaeomyces*, might be employed as a useful and comprehensive designation.

VII. CHAROPHYTA.

CHARACEÆ. NITELLEÆ.

It has been the general custom to include the Characeæ or Stoneworts among the Chlorophyceæ (green algae), of which they form a distinctly isolated family. On the whole, it would seem better to follow the course lately adopted by Migula[1] and allow the Characeæ to rank as a family of a distinct group, Charophyta. While agreeing in many respects with plants higher in the scale than Thallophytes, the Stoneworts do not sufficiently resemble the Bryophyta to be included in that group.

---

[1] Migula (90) in Rabenhorst's *Kryptogamen Flora*, vol. v.

The Charophyta are plants containing chlorophyll, living in fresh and brackish water; the stem is jointed, and bears at the nodes whorls of leaves, on which are borne the reproductive organs. The antheridia are spherical in shape and of complex structure, containing numerous biciliate antherozoids. The oogonia are oval in form and contain a single large egg-cell. The Chara-plant is developed from a *protonema* formed from the germinating oospore. Vegetative reproduction is effected by means of bulbils, accessory shoots, etc.

The Nitelleæ have not been recognised in a fossil condition. The absence or feeble development of a calcareous incrustation renders the genera of this family less likely to be preserved than such a genus as *Chara*.

Chareæ.

Leaves and stems with or without a cortical investment. Fruit with a five-celled *corona*. The envelope of the 'fruit' and other parts of the plant are frequently encrusted with carbonate of lime.

In the genus *Chara*, the best known member of the family, the plant as a whole resembles in its general habit and external differentiation of parts the higher plants. The stem consists of long internodes separated by short nodes bearing whorls of leaves. Each internode consists of a long cylindrical cell, which becomes enclosed by a cortical sheath composed of rows of cells which have grown upwards and downwards from the peripheral nodal cells. The cortical cells are usually spirally twisted and impart to the stem a characteristic appearance; they are divided by transverse walls into numerous cells some of which occasionally grow out into short processes (fig. 45 c). The leaves repeat on a smaller scale the structural features of the stem, but possess a limited growth, whereas the stem has an unlimited power of growth by means of a large hemispherical apical cell. Branches arise in the axils of the leaves. The plants are either monoecious or dioecious. The oogonium is elliptical in shape, and is borne on a short stalk-cell, it contains a single oosphere. The wall of the oogonium is formed of five spirally twisted cells which have grown over it from the five peripheral cells of a

leaf-node. The tips of the investing cells project at the apex
in the form of a terminal crown or *corona* (fig. 45, *E, c*). The
antheridia have a complex structure, and produce a very large
number of motile antherozoids.

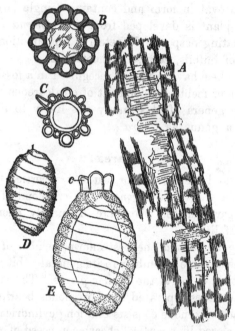

Fɪɢ. 45. *A* and *B*. *Chara Knowltoni* Sew. From a section in the British
Museum. *C*. Stem of *Chara foetida* A. Br. in transverse section (after
Migula. × 18). *D*. Interior of oogonium of *C. foetida*. *E*. Oogonium
of *C. foetida* (*D* and *E* after Migula. × 50).

After fertilisation, the egg-cell becomes surrounded by a
membrane, at first colourless, but afterwards yellow or brown.
The inner cell-walls of the cells surrounding the oospore become
thicker and darker in colour; the outer walls remain thin and
eventually fall away. The lateral walls may or may not
become thickened. In most of the Chareae a calcareous deposit
is formed between the hard shell and the outer walls of the
cells enveloping the oospore. This calcareous shell is developed
subsequently to the thickening and hardening of the inner

walls of the fruit-case. The cells of the corona and stalk do not
become calcareous. In the fossil Charas, it is this calcareous
shell that is preserved. In the members of the Chareae the
stems are usually encrusted with carbonate of lime, and thus
have a much better chance of preservation than the slightly
calcareous Nitelleae.

Chara.

The generic characters have already been described in the
brief account of the family Chareae.

The generic name was proposed by Vaillant in 1719[1], and
adopted by Linnaeus, who classed the Stoneworts with aquatic
phanerogams. As long ago as 1623[2] a figure of *Chara* was
published by Caspar Bauhin as a form of *Equisetum*. The
generic name *Chara* has usually been applied to recent and
fossil species alike. The existing species have a wide distribution; *Chara foetida*, A. Br., a common British form, occurs
in practically all parts of the world. Stems and calcareous
'fruit-cases' occur fairly commonly in a fossil state, and differ
but little from recent species, at least as regards essential
features.

It is difficult to say at what geological horizon the Stoneworts are first represented. The first certain traces of *Chara*
occur in Jurassic rocks, but certain spirally marked subspherical
bodies have been recorded from Devonian and Carboniferous
strata, which closely resemble Chara oogonia, and may be
Palaeozoic representatives of the genus.

In 1889 Mr Knowlton[3] of the American Geological Survey
described some 'problematic organisms' found in Devonian
rocks at the falls of the Ohio. Examples of these fossils are
shown in fig. 46 *b* and *c*; the spirally grooved body measures
from 1·50 to 1·80 mm. in diameter, and about 1·70 mm. in
length. The Chara-like character of the fossils had been
previously suggested by Meek[4] in 1873. Without going into
the arguments for or against placing these fossils in the Chareae,

[1] Vaillant (1719) p. 17. [2] Migula (90) p. 53.
[3] Knowlton (89²). [4] Meek (73) p. 219.

they may at least be mentioned as possible but not certain Palaeozoic forms of *Chara* or an allied genus.

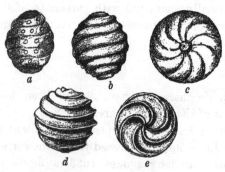

Fig. 46. *a. Chara Bleicheri* Sap. ×30. *b* and *c.* Devonian *Chara?* sp. *circa* ×12. *d* and *e. Chara Wrighti* Forbes. *circa* ×12.

1. *Chara Bleicheri*, Saporta. Fig. 46, *a.* In this form the fruits' are minute and subspherical, ·39—·44 mm. long, and 35—·40 mm. broad, showing in side view 5—6 slightly oblique spiral bands. Each spiral band bears a row of slightly projecting tubercles.

This species was first described by Saporta[1] from the Oxfordian (Jurassic) rocks of the Department of Lot in France; it is compared by the author of the species with *Chara Jaccardi* Heer, described by Heer from the Upper Jurassic rocks of Switzerland.

2. *C. Knowltoni*, Seward. Fig. 45, *a* and *b*, and Fig. 47. The Oogonia are broadly oval, about ·5 mm. in length, and at the broadest part of about the same breadth. The surface is marked by eleven or twelve bands in the form of a flattened spiral. The stems possess investing cortical cells.

This species was founded on specimens from the Wealden beds of Sussex[2], but numerous examples of Chara 'fruits' and stems have long been known from the uppermost Jurassic rocks of the Dorset coast and the Isle of Wight, which may probably be included in this species. These fossil Charas are

[1] Saporta (73) p. 214, Pl. IX. figs. 8–11.
[2] Seward (94²) p. 13, fig. 1.

abundant[1] in the Chert beds of Purbeck age seen in the cliffs near Swanage. Pieces of corticated stems from this locality are represented in fig. 45 *A* and *B*.

The cortical cells surrounding a large internodal cell are very clearly seen in the section shown in fig. 45 *B*, and in the longitudinal view in fig. 45 *A*. The resemblance of these specimens to the stems of recent Stoneworts is very striking.

FIG. 47.   *Chara Knowltoni*, Sew.   × 30.

The single oogonium of fig. 47 was found in the Wealden beds near Hastings.

3. *Chara Wrighti*, Forbes.   Fig. 46, *d* and *e*.   This species is characterised by globular or somewhat elliptical oogonia, with six or seven spiral bands.

It is very abundant in the Lower Headon beds of Hordwell Cliffs on the Hampshire coast[2]. Various species of *Chara* are commonly met with in the Oligocene beds of the Isle of Wight and Hampshire, as well as in the Paris basin beds, and elsewhere. Well preserved 'fruits' and stem fragments are met with in a siliceous rock of Upper Oligocene age imported from Montmorency in the Paris basin, and used as a stone for grinding phosphates at some chemical works near Upware, a few miles from Cambridge.

Many other species of fossil Charas are known from various horizons and localities, but the above examples suffice as illustrative types. In Post-Tertiary deposits masses of *Chara* and plant fragments occasionally occur forming blocks of Travertine. Examples of such Chara beds have been recorded by Sharpe from Northampton[3], by Lyell[4] from Forfarshire, and

---

[1] Woodward, H. B. (95) pp, 234, 261, *etc.*
[2] Forbes, E. (56) p. 160, Pl. VII.
[3] *Vide* p. 69, fig. 10.
[4] Lyell (29).

by other writers from several other districts. Beds of calcareous marl are occasionally seen as whitish streaks in the peat of the Fenland[1]; these often consist in great part of Charas. A season's growth of *Chara* in a shallow lake or mere in the Fens may appear as a white line in a section of peaty and other material which has been formed on the site of old pools or lakes.

The recognition of specific characters in the isolated Chara 'fruits' usually met with in a fossil state is exceedingly unsatisfactory; the features usually relied on in the living species are not preserved, and great care should be taken in the separation of the various forms.

---

[1] Skertchly (77) p. 60.

CHAPTER VIII.

BRYOPHYTA (Muscineae).

I. HEPATICAE (Liverworts). II. MUSCI (Mosses).

THE Bryophyta are small plants, varying in size from 1 mm. to about 30 cm., creeping or erect, having a thalloid, or more usually a foliose body, consisting of a cell-mass exhibiting in most cases a distinct internal differentiation. They possess no true roots and no true vascular tissue. The life-history of the members of the group is characterised by a well-marked and definite alternation of generations. The Moss or Liverwort plant is the sexual generation (*gametophyte*), and as a result of the fertilisation of an egg-cell the asexual or spore-bearing generation (*sporophyte*) is produced. The sporophyte never exhibits a differentiation into stem and leaves. Asexual and vegetative reproduction are effected by means of spores, bulbils, or detached portions of the plant-body. Sexual reproduction is by means of biciliate antherozoids produced in *antheridia* and egg-cells formed singly in *archegonia*.

In the Bryophytes the distinguishing characteristics are more constant and well-defined than in the Thallophytes. In the former the plant never consists of a single cell or coenocyte, but is always multicellular, and exhibits in most cases a definite physiological division of labour as expressed in the histological differentiation of distinct tissue-systems. In the Thallophytes there is no true alternation of generation in the same sense as in the Mosses and Liverworts and in the higher plants. In the Bryophytes the sexual reproduction has reached a higher stage of development and a much greater constancy as regards the nature of the reproductive organs.

On the germination of the spore there is usually formed a fairly distinct structure known as the *protonema*, from which the Moss or Liverwort developes as a bud[1].

I. HEPATICAE. $\begin{cases} \text{MARCHANTIALES.} \\ \text{ANTHOCEROTALES.} \\ \text{JUNGERMANNIALES.} \end{cases}$

The vegetative plant-body possesses a different organisation on the ventral and dorsal side; it has the form of a thalloid creeping plant (Thalloid Liverworts), or of a delicate stem with thin appendages or leaves without a midrib (Foliose Liverworts). In most cases the body of the plant is made up of parenchymatous tissue, showing but little internal differentiation; in one or two genera a few strengthening or mechanical fibres occur among the thinner walled ground-tissue. On the germination of the spore, a feebly developed protonema is produced, which gives rise to the Liverwort plant. Reproduction as in the group Bryophyta.

The Liverworts have a very wide geographical distribution, and are specially abundant in moist shady situations; they grow on stones or damp soil, and occur as epiphytes on other plants. *Marchantia, Pellia,* and *Jungermannia* are among the better known British representatives of the class.

Considering the soft nature of the body of recent Liverworts, it is not surprising that they are poorly represented in a fossil state. In the absence of the sexual reproductive organs, or of the sporophytes, which have scarcely ever been preserved, exact identification is almost hopeless. The difficulties already referred to in dealing with the algae, as regards the misleading similarity between the form of the thallus and the bodies of other plants, have to be faced in the case of the Liverworts. Many of the thalloid Liverworts, if preserved in the form of a cast or impression without internal structure or reproductive organs, could hardly be distinguished from various genera of algae in which the thallus has the form of a forked plate-like

---

[1] Schiffner and Müller in Engler and Prantl (95), Campbell (95), Dixon and Jameson (96) are among the best of modern writers on the Bryophyta.

body. Such genera as *Pellia, Marchantia, Lunularia, Reboulia,* and others bear a striking resemblance to *Fucus, Chondrus* and many other algae.

Imperfect specimens of certain Lichens, not to mention some of the Polyzoa, might easily be mistaken for Liverworts. Among the higher plants, there are some forms of the *Podostemaceae* which simulate in habit both thalloid and foliose Liverworts as well as Mosses[1]. The members of this Dicotyledonous family are described as water-plants with a Moss- or Liverwort-like form; they occur on rocks in quickly-flowing water in the tropics. In one instance a recent Podostemaceous genus has been described as a member of the Anthocerotales; the genus *Blandowia*[2], referred to by Willdenow as a Liverwort, has since been recognised as one of the *Podostemaceae*. The resemblance between some of the foliose *Hepaticae* and genera of Mosses is often very close. In certain Mosses, such as *Hookeria pennata*[3], the large two-ranked leaves suggest the branches of a *Selaginella*.

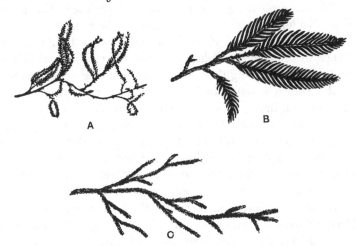

Fig. 48. A. *Tristichia hypnoides* Spreng. From a specimen in the British
Museum. B. *Podocarpus cupressina* Br. and Ben. (After Brown and
Bennett[4].) C. *Selaginella Oregana* Eat. From a plant in the Cambridge
Botanic Garden. A, B and C very slightly reduced.

[1] Hooker, J. D. (91) p. 513.   [2] Schiffner (95) p. 140.
[3] Hooker, W. J. (20) Pl. CLXIII.   [4] Bennett and Brown (38), Pl. v.

The plant reproduced in fig. 48 A (*Tristichia*), one of the
Podostemaceae, might easily be mistaken for a foliose Liverwort
if found as a fragmentary fossil.   Such species of *Selaginella*
as *S. Oregana* Eat. and *S. rupestris* Spring (fig. 48 C) have a
distinctly moss-like habit and do not present a very obvious
resemblance to the more typical and better known Selaginellas.
The twig of a *Podocarpus* (*P. cupressina*)[1] in fig. 48 B affords
an instance of a conifer which simulates to some extent certain
of the larger-leaved Liverworts; it bears a resemblance also to
some fossil fragments referred to *Selaginellites* or *Lycopodites*.
A small fossil specimen figured by Nathorst[2] from Japan as
possibly a *Lycopodites* may be compared with a coniferous twig,
and with some of the larger Liverworts, *e.g.* species of *Plagiochila*[3].
Podocarpus cupressina is, however, chiefly instructive as an
example of the striking differences which are met with among
species of the same genus; it differs considerably from the
ordinary species of *Podocarpus*, and might well be identified
as a member of some other group than that of the Coniferae.

We have no records of Palaeozoic Hepaticae.   The fossils
which Zeiller has figured in his *Flore de Brive* as *Schizopteris
dichotoma* Gümb.[4] and *S. trichomanoides* Göpp. bear a resemblance to some forms of hepatics, but there is no satisfactory
evidence for removing them from the position assigned to them
by the French writer.   In Mesozoic rocks a few specimens are
known which bear a close resemblance as regards the form of the
thalloid body to recent Liverworts, but the identification of such
fossils cannot be absolutely trusted.   Two French authors,
MM. Fliche and Bleicher[5], have described a plant from Lower
Oolite rocks near Nancy as a species of *Marchantia*, *M. oolithius*,
but they point out the close agreement of such forked laminar
structures to algae and lichens.   From Tertiary and Post-
Tertiary beds a certain number of fossil species have been
recorded, but they possess no special botanical interest.

[1] Bennett and Brown (38) p. 35.
[2] Nathorst (90) Pl. ii. fig. 3.
[3] Lindenberg (39) Pl. ix. fig. 1.
[4] Zeiller (92²) Pl. i. figs. 7 and 8.
[5] Fliche and Bleicher (81).

ORDER MARCHANTIALES.

The plant-body is always thalloid, bearing rhizoids on the lower surface, and having an epidermis with pores limiting the upper or dorsal surface.

Marchantites.

This convenient generic name was proposed by Brongniart in 1849[1]; it may be briefly defined as follows:

Vegetative body of laminar form, with apparently dichotomous branches, and agreeing in habit with the recent thalloid Hepaticae, as represented by such a genus as *Marchantia*.

The name *Marchantites* is preferable to *Marchantia*, as the latter implies identity with the recent genus, whereas the former is used in a wide sense and refers rather to a definite form of vegetative body than to a particular generic type.

1. *Marchantites erectus* (Leckenby). Fig. 49. This species may be described as follows: The thalloid body is divided into spreading dichotomously branched segments, obtusely pointed apically. The slightly wrinkled surface shows a distinct and comparatively broad darker and shorter median band, with lighter coloured and thinner margins.

In 1864 Leckenby described this plant from the Lower Oolite beds of the Yorkshire coast near Scarborough, as *Fucoides erectus*, regarding it as a fossil alga. I recently pointed out

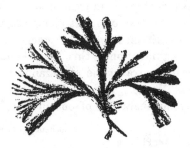

FIG. 49. *Marchantites erectus* (Leck.). From the type-specimen in the Woodwardian Museum. Nat. size.

[1] Brongniart (49) p. 12.

that the general appearance and mode of occurrence of the specimens suggest a liverwort rather than an alga, and proposed the substitution of the genus *Marchantites*[1]. It would, however, be unwise to speak with any great confidence as to the real affinities of the fossil.

The example shown in the figure is the type-specimen of Leckenby[2]: the breadth of the branches is about 3 mm. Under a low magnifying power the surface shows distinct and somewhat oblique wrinklings, the general appearance being very similar to that of some recent forms of the genus *Marchantia*.

A closely allied species has recently been described from the Wealden beds of Ecclesbourne, near Hastings, on the Sussex coast, as *Marchantites Zeilleri* Sew.[3].

In a recent monograph on Jurassic plants from Poland, apparently containing much that is of the greatest value, but which is unfortunately written in the Polish language, Raciborski[4] describes a new species of thalloid Liverwort under the name of *Paleohepatica Rostafinski*. The specimens are barren plants larger than any Jurassic species hitherto described; they agree closely in habit with Saporta's Tertiary species *Marchantites Sezannensis*.

2. *Marchantites Sezannensis* Saporta. Fig. 50. The body is broadly linear and dichotomously branched, with a somewhat undulating margin. Midrib on the dorsal surface depressed, but more prominent on the ventral surface. The upper surface is divided into hexagonal areas, in each of which occurs a central pore. There are two rows of scales along the median line on the lower surface. Stalked male receptacles.

Brongniart[5] first mentioned this fossil hepatic, which was found in the calcareous travertine of Sézanne of Oligocene age in the Province of Marne. The specimens figured by Saporta[6] show very clearly the characters of one of the *Marchantiaceae*,

[1] Seward (94[2]) p. 17.   [2] Leckenby (64) Pl. XI. fig. 3.
[3] Seward *loc. cit.* p. 18, Pl. I. fig. 3.
[4] Raciborski (94) p. 10, Pl. VII. figs. 1—3.
[5] Brongniart (49) p. 12.
[6] Saporta (68) p. 308, Pl. I. figs. 1—8. *Vide* also Watelet (66) p. 40, Pl. XI. fig. 6

and in this case we have the additional evidence of the characteristic male receptacles which are given off from a point
towards the apex of the lobes, and arise from a slight median

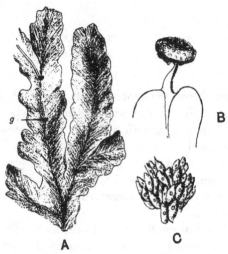

Fig. 50. *Marchantites Sezannensis* Sap. *A.* Surface view of the thallus;
g, ? cups with gemmae. *B* A male branch. *C.* A portion of *A*
magnified to show the surface features. (After Saporta.)

depression. In one of Saporta's figures (reproduced in fig. 50 *A*)
there are represented some median scars which may mark the
position of cups similar to those which occur on recent species
of *Marchantia,* and in which gemmæ or bulbils are produced.

The collection of Sézanne fossils in the Sorbonne includes
some very beautiful casts of *Marchantites* in which the
structural details are preserved much more perfectly than in
the examples described by Saporta. In a few specimens which
Prof. Munier-Chalmas recently showed me the reproductive
branches were exceedingly well shown. The fossils occur as
moulds in the travertine, and the museum specimens are in the
form of plaster-casts taken from the natural moulds.

Several species of Liverworts belonging to the Marchantiales
and Jungermanniales have been recorded from the amber of
North Germany, of Oligocene age. These appear to be represented by small fragments, such as are figured by Göppert and

Berendt[1] in their monograph on the amber plants, published in 1845. The determinations have since been revised by Gottsche[2], who recognises species of *Frullania*, *Jungermannia*, and other genera.

II. MUSCI. { SPHAGNALES.
             { ANDREAEALES.
             { BRYALES.

The plant-body (gametophyte) in the Musci consists of a stem bearing thin leaves, usually spirally disposed, rarely in two rows. The internal differentiation of the stem is generally well marked, and in some cases is comparable in complexity with the structure of the higher plants. A protonema arises from the spore, having the form of a branched filamentous, or more rarely a thalloid structure. Reproduction as in the group Bryophyta.

Mosses like Liverworts have an extremely wide distribution, and occur in various habitats. In many districts vast tracts of country are practically monopolised by peat-forming genera, such as *Sphagnum* and other Mosses. Some genera are found on rocks at high altitudes in dry regions, a few grow as saprophytes, and many occur either as epiphytes on the leaves and stems of other plants, or carpeting the ground under the shade of forest trees.

In the simpler Mosses, the stem consists of a parenchymatous ground-tissue with a few outer layers of thicker-walled and smaller cells. In others there is a distinct central cylinder which occupies the axis of the stem, and consists of long and narrow cells; in the more complex forms the structure of the axial tissues suggests the central cylinder or *stele* of higher plants. The genus *Polytrichum*, so abundant on English moors, illustrates this higher type of stem differentiation. In a transverse section of the stem the peripheral tissue is seen to be composed of thick-walled cells, passing internally into large parenchymatous tissue. The axial part is occupied by a

---

[1] Göppert and Berendt (45) Pl. VI. and (53).
[2] Gottsche (86).

definite central cylinder consisting in the centre of elongated elements with dark-coloured and thick walls having thin transverse septa; surrounding this central tissue there are thinner walled elements, of which some closely agree in form with the sieve-tubes of the higher plants. The central tissue may be regarded as a rudimentary type of xylem, and the surrounding tissue as a rudimentary phloem. Each leaf is traversed by a median conducting strand which passes into the stem and eventually becomes connected with the axial cylinder.

The fertilisation of the egg-cell gives rise to the development of a long slender stalk terminating distally in a large spore-capsule. In section the stalk or seta closely resembles the leafy axis of the moss plant. Considering the fairly close approach of some of the mosses to the higher plants as regards histological characters, it is conceivable that imperfectly petri-fied stems of fossil mosses might be mistaken for twigs of Vascular Cryptogams.

Like Liverworts, Mosses have left very few traces of their existence in plant-bearing rocks. Without the aid of the characteristic moss-' fruit' or sporogonium it is almost impos-sible to recognise fossil moss-plant fragments. In species of the tropical genera *Spiridens* and *Dawsonia*, e.g. *S. longi-folius*[1] Lind. or *D. superba*[2] Grev. and *D. polytrichoides*[3] R. Br., the plant reaches a considerable length, and resembles twigs of plants higher in the scale than the Bryophytes. The finer branches of species of the extinct genus *Lepidodendron* are extremely moss-like in appearance. Again, *Cyathophyllum bulbosum* Muell[4], with its two kinds of leaves arranged in rows, is not at all unlike species of *Selaginella* or the hepatic genus *Gottschea*. It is by no means improbable that some of the Palaeozoic specimens described as twigs of *Lycopodites*, *Selaginites*, or *Lepidodendron*, may be portions of mosses. The fertile branches of *Lycopodium phlegmaria* in a fossil condition might be easily mistaken for fragments of a moss. In some conifers with small and crowded scale-leaves there is a certain resemblance to the stouter forms of moss stems.

---

[1] Schimper (65) Pl. III.   [2] Greville (47) Pl. XII.
[3] Brown, R. (11) Pl. XXIII.   [4] Hooker, W. J. (20) Pl. CLXII.

Such possible sources of error should be prominently kept in view when we are considering the value of negative evidence as regards the geological history of the Musci.

A recent writer[1] on mosses has expressed the opinion that no doubt the Musci played an exceedingly important rôle in past time. Although we have no proof that this was so, yet it is far from improbable, and the absence of fossil mosses must no doubt be attributed in part to their failure to be preserved in a fossil state.

In the numerous samples of Coal-Measure vegetation preserved in extraordinary perfection in the calcareous nodules of England, no certain trace of a moss has so far been discovered. The most delicate tissue in the larger Palaeozoic plants has often been preserved, and in view of such possibilities of petrifaction it might appear strange that if moss-like plants existed no fragments had been preserved. Their absence is, however, no proof of the non-existence of Palaeozoic mosses, but it is a fact which certainly tends towards the assumption that mosses were probably not very abundant in the Coal Period forests. Epiphytic mosses frequently occur on the stems and leaves of ferns and other plants in tropical forests. Such small and comparatively delicate plants would, however, be easily rubbed off or destroyed in the process of fossilisation, and it is extremely rare to find among petrified Palaeozoic plants the external features well preserved. It is probable that the forests extended over low lying and swampy regions, and that, in part, the trees were rooted in a submerged surface. Under such conditions of growth there would not be the same abundance of Bryophytes as in most of our modern forests.

To whatever cause the absence of mosses may be best attributed, it is a fact that should not be too strongly emphasised in discussions on plant-evolution.

Muscites.

This comprehensive genus may be defined as follows :—
Stem filiform, simple or branched, bearing small sessile

---

[1] Limpricht (90) p. 67.

leaves, with a delicate lamina, without veins or with a single median vein, arranged in a spiral manner on the stem.

Muscites[1] is one of those convenient generic designations which limited knowledge and incomplete data render necessary in palaeontology. Fossil plants which in their general habit bear a sufficiently striking resemblance to recent mosses, may be included under this generic name.

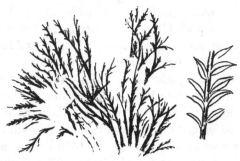

FIG. 51. *Muscites polytrichaceus* Ren. and Zeill. (after Renault and Zeiller).

1. *Muscites polytrichaceus* Renault and Zeiller. In this species the stems are about 3—4 cm. long and 1·3 m. broad, usually simple, but sometimes giving off a few branches, and marked externally by very delicate longitudinal grooves. The leaves are alternate, closely arranged, lanceolate, with an acute apex, gradually narrowed towards the base, 1—2 mm. long, traversed by a single median vein.

One of the French specimens, on which the species was founded[2], is shown in fig. 51, and the form of the leaves is more clearly seen in the small enlarged piece of stem. The authors of the species point out that the tufted habit of the specimens, their small size, and the membranous character of the leaves, all point to the Musci as the Class to which the plant should be referred in spite of the absence of repro-ductive organs.

Among recent mosses, the genus *Rhizogonium,*—one of the *Mniaceae,*—and *Polytrichum* are spoken of as offering a close

---

[1] Brongniart (28[2]) p. 93.
[2] Renault and Zeiller (88) p. 34, Pl. xli. figs. 2—4.

resemblance to the fossil form. The type-specimen was found in
the Coal-Measures of Commentry, and is now in the Museum of
the École des Mines in Paris; the figure given by MM. Renault
and Zeiller faithfully represents the appearance of the plant.

It has been suggested[1] that some small twigs figured by
Lesquereux[2] from the Coal-Measures of North America as
Lycopodites Meeki Lesq., may possibly be mosses. The specimens do not appear to be at all convincing, and cannot well
be included as probable representatives of Palaeozoic Musci.
Lycopodites Meeki Lesq. bears a close resemblance to the
recent *Selaginella Oregana* shown in fig. 48, C.

From Mesozoic rocks we have no absolutely trustworthy
fossil mosses. The late Prof. Heer[3] has quoted the occurrence
of certain fossil Caterpillars in Liassic beds as indicative of
the existence of mosses, but evidence of this kind cannot be
accepted as scientifically sound. In 1850 Buckman[4] described
and figured a few fragments of plants from a freshwater limestone at the base of the Lias series near Bristol. Among
others he described certain specimens as examples of a fossil
Monocotyledon, under the generic name *Najadita*. Mr Starkie
Gardner[5] subsequently examined the specimens, and suggested
that the Lias fragments referred to *Najadita* should be compared with the recent freshwater moss *Fontinalis*. In this
opinion he was supported by Mr Carruthers and Mr Murray of
the British Museum. In a footnote to the memoir in which
this suggestion is made, Gardner refers to a moss-capsule from
the same beds, which he had received from Mr Brodie. Through
the kindness of the latter gentleman, I have had an opportunity
of examining the supposed capsule, and have no hesitation in
describing it as absolutely indeterminable. It is in the form of
an irregularly oval brown stain on the surface of the rock, with
the suggestion of a stalk at one end, but there are no grounds
for describing the specimen as a moss-capsule, or indeed anything
else. The type-specimens figured by Brodie and subsequently
referred to a moss are now in the British Museum; they are

---

[1] Solms-Laubach (91) p. 186.   [2] Lesquereux (79) Pl. LXII. fig. 1.
[3] Heer (65) p. 89.   [4] Buckman (50) 1.   [5] Gardner (86) p. 203.

small and imperfect fragments of slender stems bearing rather long oval leaves which might well have belonged to a moss. The material is however too fragmentary to allow of accurate diagnosis or determination.

2. *Muscites ferrugineus* (Ludg.). This species possesses a slender stem bearing crowded ovate-acuminate leaves. The capsules are cup-shaped, borne on a short stalk, with a circular opening without marginal teeth. This fossil was first figured and described by Ludwig[1] from a brown ironstone of Miocene age at Dernbach in Nassau. The author of the species placed it in the recent genus *Gymnostomum*, and Schimper[2] afterwards changed the generic name to *Sphagnum*, at the same time altering the specific name to *Ludwigi*. The evidence is hardly strong enough to justify a generic designation which implies identity with a particular recent genus, and it is a much safer plan to adopt the non-committal term *Muscites*, at the same time retaining Ludwig's original specific name. Without having examined the type-specimen it is impossible to express a definite opinion as to the accuracy of the description given by Ludwig; if the capsule is correctly identified it is the oldest example hitherto recorded of a fossil moss-sporogonium.

[1] Ludwig (59) p. 165, Pl. LXIII. fig. 9.
[2] Schimper and Schenk (90) p. 75.

CHAPTER IX.

PTERIDOPHYTA (Vascular Cryptogams).

I. EQUISETALES. II. SPHENOPHYLLALES.
III. LYCOPODIALES. IV. FILICALES.

THE Pteridophytes include plants which vary in size from a few millimetres[1] to several metres in height. The spore on germination gives rise to a small thalloid structure, the *prothallium*, on which the sexual organs are developed; this is the *gametophyte* or sexual generation. The sexual organs have the form of typical archegonia and antheridia. From the fertilised egg-cell there is developed the Pteridophyte plant or *sporophyte*, which bears the spores. This asexual generation shows a well-marked external differentiation into stem and leaves, and bears true roots. Internally the tissues exhibit a high degree of differentiation into distinct tissue-systems. True vascular bundles occur, which may or may not be capable of secondary thickening by means of a *cambium*, *i.e.* a definitely localised zone of meristematic tissue. The sporangia are borne either on the ordinary foliage leaves or on special spore-bearing leaves called *sporophylls*, which differ in a greater or less degree from the sterile leaves.

The majority of the best known and most important Palaeozoic genera are either true Vascular Cryptogams, or possess certain of the pteridophytic characteristics combined with those of higher plants. It is not merely the commoner and more familiar recent genera with which the student of extinct types must be acquainted, but it is extremely important

---

[1] *e.g.* the Fern *Trichomanes Goebelianum* Gies. Giesenhagen (92) p. 157.

that he should make himself familiar with the rarer, less known and more isolated recent forms, which often throw most light on the affinities of the older representatives of the group. It is often the case, the more isolated living plants are, the more likely are they to afford valuable assistance in the interpretation of genera representing a class, which reached its maximum development in the earlier periods of the earth's history. The importance of paying special attention to such recent plants as may be looked upon as survivals of a class now tending towards extinction, will be more thoroughly realised after the extinct vascular cryptogams have been dealt with.

A comparison of the Pteridophyta and Bryophyta brings out certain points of divergence. In the first place, the sporophyte assumes in the former class a much more prominent rôle, and the gametophyte has suffered very considerable reduction. The gametophyte, *i.e.* the structure which is formed on the germination of the asexually-produced spore, is usually short-lived, small, and more or less dependent on the sporophyte for its nutrition. In a few cases only is it capable of providing itself with the essential elements of food. On the other hand, the sporophyte, at a very early stage of its development becomes free from the gametophyte and is entirely self-supporting. Reproduction is effected as in the Bryophyta by sexual reproductive organs and by asexual methods. Not only have we in the Pteridophytes a much more complete external division of the plant-body into definite members, which subserve distinct functions, and behave as well-defined physiological organs adapted for taking a certain share in the life-functions of the individual, but the internal differentiation has reached a much higher stage. True vascular tissue, consisting of xylem and phloem, occurs for the first time in this class. The whole plant is traversed by one or more vascular strands composed of xylem and phloem elements, which are respectively concerned with the distribution of inorganic and organic food substances.

The Pteridophyta include the most important fossil plants. It is from a study of the internal structure of various extinct representatives of this class, that palaeobotanists have been able to contribute facts of the greatest interest and importance towards the advancement of botanical science.

The botanist's chief aim in the anatomical investigation of Palaeozoic genera is to discover data which point the way to a solution of the problems of plant-evolution. In the abundant material afforded by the petrified remnants of ancient floras we have the means of tracing the past history of existing groups or individual forms, and it is from the Palaeozoic Pteridophytes that our most valuable results have been so far obtained.

In this and the following chapters of Volume I. two divisions of the Pteridophyta are dealt with in such detail as the nature of the book allows. In the earlier chapters of Volume II. the remaining representatives of this class will be described. As in the preceding chapters such recent plants will be described as are most essential for the correct interpretation of the fossil forms.

It is impossible to do more than confine our attention to a few only of the genera of living plants which directly concern us; some acquaintance with the general facts of plant morphology must be assumed. Among the most useful text-books or books of reference on the Pteridophyta the student may consult those mentioned in the footnote[1].

I. EQUISETALES.

Leaves usually small in proportion to the size of the whole plant, arranged in whorls at the nodes. Sporangia borne on specially modified sporophylls or sporangiophores, which are aggregated to form a definite strobilus or spore-bearing cone.

EQUISETACEAE. (Recent Species.)

The leaves are in whorls, coherent in the form of a sheath, and traversed by longitudinal veins which do not fork or anasto-

---

[1] Scott (96) a text-book for elementary students ; a full account is given of *Equisetum* and other genera of primary importance. Vines (95) Part iii. Campbell (95), Luerssen (89) in Rabenhorst's *Kryptogamen-Flora*, vol. iii., Van Tieghem (91), de Bary (84), Baker (87).

mose. The stem is divided into comparatively long internodes separated by the leaf-bearing nodes, and the branches arise in the leaf-axils at the nodes. The fertile leaves or sporophylls differ from the sterile leaves, and usually occur in definite aggregations or strobili containing spores of one kind (*isosporous*). In the single living genus *Equisetum*, the outer coat of the mature spore forms two hygroscopically sensitive filamentous structures or *elaters*. On the germination of the spore the gametophyte is developed in the form of a small lobed prothallium 1—2 cm. in length. In most cases there are distinct male and female prothallia.

The genus *Equisetum* L., the common Horse-tail, is the sole living representative of this Family. It occurs as a common native plant in Britain, and has a wide geographical distribution. Species of *Equisetum* are abundant in the temperate zones of both hemispheres, and occur in arctic as well as tropical latitudes. Wallace[1] speaks of Horse-tails, "very like our own species," growing at a height of 5000 feet on the Pangerango mountain in Java. In favourable situations the large British Horse-tail, *Equisetum maximum* Lam. (= *E. Telmateia* Erhb.), occasionally reaches a height of about six feet, and growing in thick clusters forms miniature forests of trees with slender erect stems and regular circles of long and thin branches. A tropical species, *Equisetum giganteum* Linn.[2] living in the marshes of Mexico and Cuba[2], and extending southward to Buenos Ayres and Chili, reaches a height of twenty to forty feet, but the stem always remains slender, and does not exceed an inch in diameter. Groves of such tall slender plants on the eastern slopes of the Andes[3] suggest to the palaeobotanist an enfeebled forest-growth recalling the arborescent Calamites of a Palaeozoic vegetation. The twenty-five existing species of *Equisetum* are remnants of various generic types of former epochs, and possess a special interest from the point of view of the geological history of plants. A brief description of the

---

[1] Wallace (86) p. 117.

[2] Baker (87) p. 4. Hooker, W. J. (61) Pl. LXXIV. *Vide* also Milde (67) for figures of *Equisetum*.

[3] Seeman (65).

main characters of the recent genus will enable the student to
appreciate the points of difference and agreement between the
extinct and present representatives of the Equisetales.

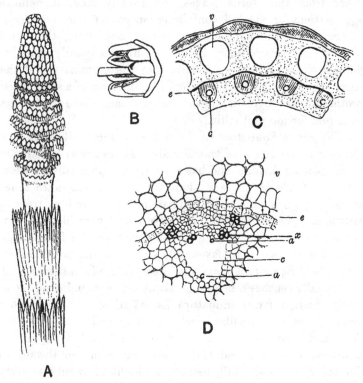

Fɪɢ. 52. *Equisetum maximum* Lam. A. Fertile shoot with strobilus and
sterile leaf-sheaths [after Luerssen (89); slightly less than nat. size].
B. Sporophyll bearing open sporangia (after Luerssen; slightly enlarged).
C. Part of a transverse section (diagrammatic); *v*, vallecular canals, *e*, endodermis, *c*, carinal canals (after Luerssen; × 20). D. *Equisetum arvense*
L. Part of a transverse section of an internode of a sterile shoot.
v, cortex, *e*, endodermis, *x*, xylem tracheids, *a* remains of annular tracheids
of the protoxylem, *c*, carinal canal (after Strasburger; × 90).

Equisetum.

The plant consists of a perennial underground creeping
rhizome, branching into secondary rhizomes, divided into well-
marked nodes and internodes. From the nodes are given off

two sets of buds, which may develope into ascending aerial
shoots or descending roots. At each node is a leaf-sheath more
or less deeply divided along the upper margin into teeth
representing the tips of coherent leaves (fig. 52, A).

In some species one or more internodes of underground
branches become considerably swollen and assume the form of
ovate or elliptical starch-storing tubers, which are capable of
giving rise to new plants by vegetative reproduction. Tubers,
either singly or in chains, occur in *E. arvense* Linn., *E. silvaticum*
Linn., *E. maximum* Lam., among British species.

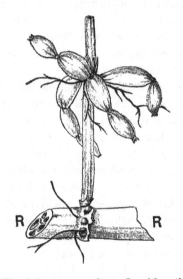

Fig. 53.　Rhizome (R) of *Equisetum palustre* L. with a thin shoot giving off
roots and tuberous branches from a node [after Duval-Jouve (64)].

In the example shown in fig. 53 (*Equisetum palustre* L.[1])
the stout rhizome R gives off from its node, marked by a small
and irregular leaf-sheath, two thin roots and a single shoot.
The latter has a leaf-sheath at its base, and from the second
node, with a larger leaf-sheath, there have been developed
branches with tuberous internodes; the constrictions between

[1] Duval-Jouve (64) Pl. i. fig. 5.

the tubers and the tips of the terminal tubers bear small leaf-sheaths. Branched roots are also given off from the upper node of the erect shoot.

Near the surface of the ground the buds on the rhizome nodes develope into green erect shoots. The shoot axis is marked out into long internodes separated by nodes bearing the leaf-sheaths. The surface of each internode is traversed by regular and more or less prominent longitudinal ridges and grooves; each ridge marking the position of an internal longitudinal vascular strand. In the axil of each leaf, that is in the axil of each portion of a leaf-sheath corresponding to a marginal uni-nerved tooth, there is produced a lateral bud which may either remain dormant or break through the leaf-sheath and emerge as a lateral branch. At the base of each branch an adventitious root may be formed from a cell immediately below the first leaf-sheath, but in aerial shoots the roots usually remain undeveloped. The lateral branches repeat on a smaller scale the general features of the main axis. In some species, the shoots are unbranched, and in others the slender branches arise in crowded whorls from each node. Leaves, roots and branches are given off in whorls, and the whorls from each node alternate with those from the node next above and next below.

In some species of *Equisetum* the aerial stem terminates in a conical group of sporophylls, while in others the strobilus is formed at the apex of a pale-coloured fertile shoot, which never attains any considerable length and dies down early in the season of growth (fig. 52, A). Below the terminal cone or strobilus there occur one or two modified leaf-sheaths. Such a ring of incompletely developed leaves intervening between the cone of sporangiophores and the normal leaves, is known as the *annulus*. The annulus is seen in fig. 52, A, immediately below the lowest whorl of sporophylls; it has the form of a low sheath with a ragged margin. In the region of the cone the internodes remain shorter, and the whorls of appendages, known as sporophylls or sporangiophores, have the form of stalked structures terminating distally in a hexagonal peltate disc, which bears on its inner face a ring of five to ten oval sporangia (fig. 52, B). Each sporangium contains numerous spores which eventually escape by the longitudinal dehiscence of the sporangial wall. The opening of the sporangia is probably assisted by the movements of the characteristic elaters formed from the outer wall of each spore.

The spores, which are capable of living only a short time, grow into aerial green prothallia, 1—2 cm. in length; these have the form of irregularly and more or less deeply lobed structures. On the larger and more deeply lobed prothallia the archegonia or female reproductive organs are borne, and the smaller or male prothallia bear the antheridia. On the fertilisation of an egg-cell, the *Equisetum* plant is gradually developed. For a short time parasitic on the female prothallium or gametophyte, the young plant soon takes root in the ground and becomes completely independent.

As seen in transverse section through a young stem near the apex, the axis consists of a mass of parenchyma, in which may be distinguished a central larger-celled tissue, surrounded by a ring of smaller-celled groups marking the position of a circle of embryonic vascular strands. In each young vascular strand, a few of the cells next the pith may be seen to have thicker walls and to be provided with a ring-like internal thickening; these have passed over into the condition of annular tracheids and represent the *protoxylem* elements. At a later stage, a transverse section through the stem shows a central hollow pith, formed by the tearing apart and subsequent disappearance of the medullary parenchymatous cells, which were unable to keep pace with the growth in thickness of the stem. The pith cavity is bridged across at each node by a multi-layered plate of parenchyma, which forms the so-called nodal *diaphragm*. The inner edge of each vascular strand is now found to be occupied by a small irregularly circular canal (fig. 52, C, c, and D, c) in which may be seen some of the rings of protoxylem tracheids (D, a) which have been torn apart and almost completely destroyed. These canals, known as *carinal canals*, have arisen by the tearing and disruption of the thin-walled cells in the immediate neighbourhood of the protoxylem. Each carinal canal is bounded by a layer of elongated parenchymatous cells which form part of the xylem of the vascular bundle, and is succeeded internally by the general ground-tissue of the stem. The xylem parenchyma next a carinal canal is succeeded externally by phloem tissue, consisting of short protoplasmic cells and longer elements, without nuclei and poor in contents; the latter may be regarded as sieve-tubes. On either side of the phloem, the xylem occurs in two separate bands or groups of annular and reticulately thickened tracheids. In some species, e.g. *Equisetum xylochaetum* Metten.[1] and *E. giganteum*[2] L. a native of South America, the xylem has the form of two bands composed of fairly numerous tracheids, but in most species three xylem tracheids occur in small groups, as shown in the figure of *E. maximum* (fig. 52, D). In the shape of the vascular bundle, and in the formation of the carinal canal, there is a distinct resemblance between the vascular bundles of *Equisetum* and those of a monocotyledonous stem. These collateral stem-bundles of xylem and phloem traverse each internode as distinct strands, and at the nodes each strand forks into two branches (fig. 54, A), which anastomose with the alternating bundles passing into the stem from the leaf-sheath.

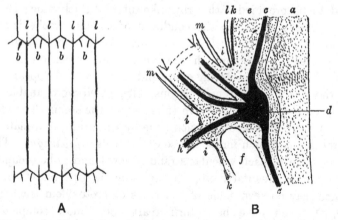

A   B

Fig. 54. A. Plan of the vascular bundles in the stem of an *Equisetum*; b, branches passing out to buds (after Strasburger); l, vascular strands passing to the leaf-segments. B. Longitudinal section through a node of *E. arvense* L. (after Duval-Jouve; × 20). Explanation in the text.

<hr/>

[1] Milde (67) Pl. xix. fig. 8.   [2] *ibid.* Pl. xxxi. fig. 3.

Thus the vascular strands of each internode alternate in position with those of the next internode.

There are certain points connected with the vascular bundles in the nodal region of a shoot, which have an important bearing on the structure of fossil equisetaceous stems. Fig. 54 B represents a diagrammatic longitudinal section through the node of a rhizome of *Equisetum arvense* from which a root *h* is passing off in a downward direction, and a branch in an upward direction. The black band *c* in the parent stem shows the position of the vascular strands; in the region of the node the vascular tissue attains a considerable thickness, as seen at *d* in the figure. The bands passing out to the left from *d* go to supply the branch and root respectively. The increased breadth of the xylem strands at the node is due to the intercalation of a number of short tracheids. Fig. 55, 4 shows a transverse section through a mature node of *Equisetum maximum*; *px* marks the position of the protoxylem and *e* that of the endodermis. On comparing this section with that of the internodal vascular bundle in fig. 52, D, the much greater development of wood in the former is obvious; the carinal canal of the internodal bundle is absent in the section through a node. The disposition of the xylem tracheids in fig. 55, 4 shows a certain regularity which, though not very well marked, suggests the development of wood elements as the result of cambial activity. Longitudinal sections through the nodal region demonstrate the existence of "cells similar to those of an ordinary cambium, and a cell-formation resulting from their division which is similar to that in an ordinary secondary thickening."[1] The short tracheids which make up this nodal mass of xylem differ from those in the internodal bundle in their smaller size, and in being reticulately thickened. There is, therefore, evidence that in the nodes of some *Equisetum* stems additional xylem elements are produced by a method of growth comparable with the cambial activity which brings about the growth in thickness of a forest-tree[2]. The

---

[1] Cormack (93) p. 71.

[2] Williamson and Scott (94) p. 877. These authors, in referring to Cormack's description of the secondary nodal wood of *E. maximum*, express doubts as to the existence of such secondary growth in *all* species of the genus.

significance of these statements will be realised when the
structure of the extinct genus *Calamites* is described and
compared with that of *Equisetum*.

The small drawing in fig. 55, 3 shows part of the ring of
thick nodal wood; the section cuts through two bundles about
their point of bifurcation, the strand *x* is passing out in a
radial direction to a lateral branch, the strand to the right of
x and the separate fragment of a strand to the left of *x* are
portions of leaf-trace bundles on their way to the leaf-sheath.
Reverting to fig. 54, B, the other structures seen in the section
are the leaf-sheaths (*l* and *m*), the vallecular canal (*f*), the
epidermis, cortex and pith (*k, e* and *a*) of the stem. The
epidermis which has been ruptured by the root and branch is
indicated at *i, i*; the dotted lines traversing the upper part of
the pith of the lateral branch mark the position of a nodal
diaphragm.

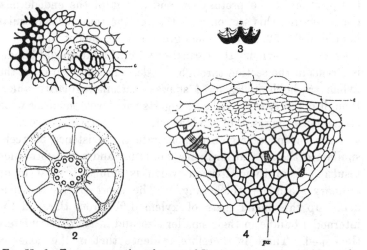

FIG. 55. 1. Transverse section of a root of *Equisetum variegatum* Schl., *e* endodermis, or outer layer of the phloeoterma (after Pfitzer; × 160). 2. Transverse section of rhizome of *E. maximum*, slightly enlarged. 3. Transverse
section through a node of *E. maximum*, *x*, branch of vascular strand (slightly
enlarged). 4. Transverse section through a node of *E. maximum* showing
the mass of xylem, *px* protoxylem (× 175). (Figs. 3 and 4 after Cormack.)

Immediately external to each vascular strand, as seen in
transverse section, there is a layer of cells containing starch,

and this is followed by a distinct endodermis, of which the cells show the characteristic black dot in the cuticularised radial walls (fig. 52, D). Beyond the endodermis there is the large-celled parenchyma of the rest of the cortex. Tannin cells occur here and there scattered among the ground tissue. On the same radius on which each vascular strand occurs, the cortical parenchyma passes into a mass of sub-epidermal thick-walled mechanical tissue or stereome. Alternating with the ridges of stereome, the grooves are occupied by thin-walled chlorophyll-containing tissue which carries on most of the assimilating functions, and communicates with the external atmosphere by means of stomata arranged in vertical rows down each internode. The continuity of the cortical tissue is interrupted by the occurrence of large longitudinal *vallecular canals* alternating in position with the stem ridges and vascular strands (fig. 52, C, *v*). The epidermis consists of a single layer of cells, containing stomata, and with the outer cell-walls impregnated with silica.

In certain species of *Equisetum, e.g. E. palustre* L., the whole circle of vascular strands is enclosed by an endodermis, and has the structure typical of a monostelic stem. In others *e.g. E. litorale* Kühl. each vascular strand is surrounded by a separate endodermis, and in some forms *e.g. E. silvaticum* L. there is an inner as well as an outer endodermal layer[1]. Without discussing the explanation given to this variation in the occurrence of the endodermis, it may be stated that in all species of *Equisetum* the stem may be regarded as monostelic[2].

In the rhizome the structure agrees in the main with that of the green shoots, but the vallecular canals attain a larger size, and the pith is solid. A slightly enlarged transverse section of a rhizome of *Equisetum maximum* is shown in fig. 55, 2, the small circles surrounding the pith mark the position of the vascular bundles and carinal canals; the much larger spaces between the central cylinder and the surface of the stem are the vallecular canals.

The central cylinder or stele of the root is of the diarch, triarch or tetrach type; *i.e.* there may be 2, 3 or 4 groups of

---

[1] Pfitzer (67).   [2] Strasburger (91) p. 443.

protoxylem in the xylem of the root stele. The axial portion is occupied by large tracheids, and the smaller tracheids of the xylem occur as radially disposed groups, alternating with groups of phloem. External to the xylem and phloem strands there occur two layers of cells, usually spoken of as a double endodermis, but it has been suggested that it is preferable to describe the double layer as the *phloeoterma*[1], of which the inner layer has the functions of a pericycle, and the outer that of an endodermis. A transverse section of a root is seen in fig. 55, 1, the dark cells on the left are part of a thick band of sclerenchyma in the cortex of the root, the layer *e* is the outer layer of the phloeoterma.

Without describing in detail the development[2] of the sporangia, it should be noted that the sporangial wall is at first 3 to 4 cells thick, but it eventually consists of a single layer. The cells have spiral thickening bands on the ventral surface, and rings on the cells where the longitudinal splitting takes place. Each sporangium is supplied by a vascular bundle which is given off from that of the sporangiophore axis. The strobili are isosporous.

FOSSIL EQUISETALES.
{
I.   EQUISETITES.
II. PHYLLOTHECA.
III. SCHIZONEURA.
IV. CALAMITES.
V.   ARCHAEOCALAMITES.
}

In dealing with the fossil Equisetales, we will first consider the genera *Equisetites*, *Phyllotheca* and *Schizoneura*, and afterwards describe the older and better known genera *Calamites* and *Archaeocalamites*. A thoroughly satisfactory classification of the members of the Equisetales is practically impossible without more data than we at present possess. It has been the custom to include *Equisetites*, *Phyllotheca* and *Schizoneura* in the family Equisetaceae, and to refer *Calamites* and *Archaeo-*

---

[1] Strasburger (91) p. 435.
[2] Bower (94) p. 495.

calamites to the Calamarieae; such a division rests in part on assumption, and cannot be considered final. When we attempt to define the Equisetales and the two families Equisetaceae and Calamarieae, we find ourselves seriously hampered by lack of knowledge of certain important characters, which should be taken into account in framing diagnoses. There is little harm in retaining provisionally the two families already referred to, if we do not allow a purely arbitrary classification to prejudice our opinions as to the affinities of the several members of the Equisetales.

The Equisetaceae might be defined as a family including plants which were usually herbaceous but in some cases arborescent, bearing verticils of leaves in the form of sheaths more or less deeply divided into segments or teeth. The strobili were isosporous and consisted of a central axis bearing verticils of distally expanded sporophylls with sporangia, as in *Equisetum*. The genus *Equisetites* might be included in this family, but it must be admitted that we know next to nothing as to its anatomy, and we cannot be sure that the strobili were always isosporous.

The genus *Schizoneura* is too imperfectly known to be defined with any approach to completeness, or to be assigned to a family defined within certain prescribed limits. *Phyllotheca* is another genus about which we possess but little satisfactory knowledge; we are still without evidence as to its structure, and the descriptions of the few strobili that are known are not consistent. Recent work points to a probability of *Phyllotheca* being closely allied to *Annularia*, a genus included in the Calamarieae, and standing for a certain type of Calamitean foliage-shoots.

In comparing the Calamarieae with the Equisetaceae, the alternation of sterile and fertile whorls in the strobilus, and the free linear leaves at the nodes instead of leaf-sheaths are two characters made use of as distinguishing features of the genus *Calamites* as the type of the Calamarieae. On the other hand, the strobili of *Phyllotheca* appear to agree with those of *Calamites* rather than with those of *Equisetum*, and strobili of *Archaeocalamites* have been found exhibiting the typical

Equisetum characters.   The sheath-like form of the leaves is not necessarily peculiar to the Equisetaceae, and we have evidence that leaf-sheaths occurred on the nodes of Calamitean plants. In *Archaeocalamites* the leaves possess characteristic features, and can hardly be said to agree more closely with those of Calamites than with the leaves of *Phyllotheca* or *Sphenophyllum*, a genus belonging to another class of Pteridophytes.

On the whole, then, without discussing further the possibilities of a subdivision of the Equisetales, we may regard the genera *Calamites, Archaeocalamites, Equisetites, Equisetum, Phyllotheca* and *Schizoneura* as so many members of the Equisetales, without insisting on a classification which cannot be supported by satisfactory evidence.

Our knowledge of *Calamites* is fairly complete.   Abundant and well-preserved material from the Coal-Measures of England, and from Permo-Carboniferous rocks of France, Germany and elsewhere, has enabled palaeobotanists to investigate the anatomical characters of both the vegetative and reproductive structures of this genus.   We are in a position to give a detailed diagnosis of Calamitean stems, roots and strobili, and to determine the place of this type of plant in a system of classification.   *Calamites* not only illustrates the possibilities of palaeobotanical research, but it demonstrates the importance of fossil forms as foundations on which to construct the most rational classification of existing plants.   The close alliance between *Calamites* and the recent Equisetaceae has been clearly established, and certain characteristics of the former genus render necessary an extension and modification of the definition of the class to which both *Calamites* and *Equisetites* belong.   The Calamites broaden our conception of the Equisetaceous alliance, and by their resemblance to other extinct Palaeozoic types they furnish us with important links towards a phylogenetic series, which the other members of the Equisetales do not supply.

From the Upper Devonian to the Permian epoch *Calamites* and other closely related types played a prominent part in the vegetation of the world.   We have no good evidence for the existence of *Calamites* in Triassic times: in its place there were gigantic Equisetums which resembled modern Horse-tails in a remarkable degree. In the succeeding Jurassic period tree-like Equisetums were still in existence, and species of *Equisetites* are met with in rocks of this age in nearly all parts of the world. A few widely distributed species are known from Wealden rocks, but as we ascend the geologic series from the Jurassic strata, the Equisetums become less numerous and the individual plants gradually assume proportions practically identical with those of existing forms.

I. *Equisetites.*

The generic name *Equisetites* was proposed by Sternberg in 1838[1] as a convenient designation for fossil stems bearing a close resemblance to recent species of *Equisetum*. Some authors have preferred to apply the name *Equisetum* to fossil and recent species alike, but in spite of the apparent identity in the external characters of the fossil stems with those of existing Horse-tails, and a close similarity as regards the cones, there are certain reasons for retaining Sternberg's generic name. It is important to avoid such nomenclature as might appear to express more than the facts admit. If the custom of adding the termination -*ites* to the root of a recent generic term is generally followed, it at once serves to show that the plants so named are fossil and not recent species. Moreover, in the case of fossil Equisetums we know nothing of their internal structure, and our comparisons are limited to external characters. Stems, cones, tubers, and leaves are often very well preserved as sandstone casts with distinct surface-markings, but we are still in want of petrified specimens. There is indeed evidence that some of the Triassic and Jurassic species of *Equisetites*, like the older Calamites, possessed the power of secondary growth in thickness, but our deductions are based solely on external characters.

In the following pages a few of the better known species of *Equisetites* are briefly described, the examples being chosen

---

[1] Sternberg (38) p. 43.

partly with a view to illustrate the geological history of the genus, and partly to contribute something towards a fuller knowledge of particular species. One of the most striking facts to be gleaned from a general survey of the past history of the Equisetaceae is the persistence since the latter part of the Palaeozoic period of that type of plant which is represented by existing Equisetums. There is perhaps no genus in existence which illustrates more vividly than *Equisetum* the survival of an extremely ancient group, which is represented to-day by numerous and widely spread species. The Equisetaceous characteristics mark an isolated division of existing Vascular Cryptogams, and without reference to extinct types it is practically impossible to do more than vaguely guess at the genealogical connections of the family. When we go back to Palaeozoic plants there are indications of guiding lines which point the way to connecting branches between the older Equisetales and other classes of Pteridophytes. The recently discovered genus *Cheirostrobus*[1] is especially important from this point of view.

The accurate description of species, and the determination of the value of such differences as are exhibited in the surface characters of structureless casts, are practically impossible in many of the fossil forms. In certain living Horse-tails we find striking differences between fertile and sterile shoots, and between branches of different orders. The isolated occurrence of fragments of fossil stems often leads to an artificial separation of 'species' largely founded on differences in diameter, or on slight variations in the form of the leaf-sheaths. It is wiser to admit that in many cases we are without the means of accurate diagnosis, and that the specific names applied to fossil Equisetums do not always possess much value as criteria of taxonomic differences.

The specimens of fossil Equisetums are usually readily recognised by the coherent leaf-segments in the form of nodal sheaths resembling those of recent species. The tissues of the cortex and central cylinder are occasionally represented by a thin layer

---

[1] Scott (97). This genus will be described in Volume II.

of coal pressed on to the surface of a sandstone cast, or covering
a flattened stem-impression on a piece of shale. It is sometimes
possible under the microscope to recognise on the carbonised
epidermal tissues the remains of a surface-ornamentation
similar to that in recent species, which is due to the occurrence
of siliceous patches on the superficial cells. Longitudinal rows
of stomata may also be detected under favourable conditions of
preservation. The nodal diaphragms of stems have occasionally
been preserved apart, but such circular and radially-striated
bodies may be misleading if found as isolated objects. Casts
of the wide hollow pith of *Equisetites*, with longitudinal ridges
and grooves, and fairly deep nodal constrictions, have often
been mistaken for the medullary casts of *Calamites*.

Several species of *Equisetites* have been recorded from the
Upper Coal-Measures and overlying Permian rocks, but these
present special difficulties. In one instance described below,
(*Equisetites Hemingwayi* Kidst.), the species was founded on a
cast of what appeared to be a strobilus made up of sporophylls
similar to those in an *Equisetum* cone. In other Permo-
Carboniferous species the choice of the generic name *Equisetites*
has been determined by the occurrence of leaf-sheaths either
isolated or attached to the node of a stem. The question to
consider is, how far may the Equisetum-like leaf-sheath be
regarded as a characteristic feature of *Equisetites* as distinct
from *Calamites*? In the genus *Calamites* the leaves are
generally described as simple linear leaves arranged in a whorl
at the nodes, but not coherent in the form of a sheath (fig. 85).
The fusion of the segments into a continuous sheath or collar
is regarded as a distinguishing characteristic of *Equisetites* and
Equisetum. The typical leaf-sheath of a recent Horse-tail has
already been described. In some species we have fairly large
and persistent free teeth on the upper margin of the leaf-sheath,
but in other Equisetums the rim of the sheath is practically
straight and has a truncated appearance, the distal ends of the
segments being separated from one another by very slight
depressions, as in a portion of the sheath of *Equisetum ramosissimum* Desf. of fig. 58, *C*. In other leaf-sheaths of this
species there are delicate and pointed teeth adherent to the

17—2

margin of the coherent segments; the teeth are deciduous, and after they have fallen the sheath presents a truncated appearance. This difference between the sheaths to which the teeth are still attached and those from which they have fallen is illustrated by fig. 58, *B* and *C*; it is one which should be borne in mind in the description of fossil species, and has probably been responsible for erroneous specific diagnoses. In some recent Horse-tails the sheath is occasionally divided in one or two places by a slit reaching to the base of the coherent segments[1]; this shows a tendency of the segments towards the free manner of occurrence which is usually considered a Calamitean character. In certain fossils referred to the genus *Annularia*, the nodes bear whorls of long and narrow leaves which are fused basally into a collar (fig. 58, *D*). There are good grounds for believing that at least some Annularias were the foliage shoots of true Calamites. Again, in some species of *Calamitina*, a sub-genus of *Calamites*, the leaves appear to have been united basally into a narrow sheath. We see, then, that it is a mistake to attach great importance to the separate or coherent character of leaf-segments in attempting to draw a line between the true *Calamites* and *Equisetites*. Potonié[2]

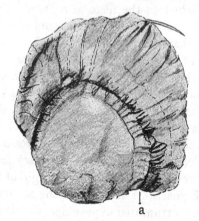

a

Fig. 56.　Calamitean leaf-sheath.　From a specimen in the Woodwardian Museum.　*a*, base of leaf-sheath; (very slightly reduced).

[1] Potonié (93) Pl. xxv. fig. 1*a*.
[2] *ibid.* p. 179. *Vide* also Potonié (92).

while pointing out that this distinction does not possess much value as a generic character, retains the genus *Equisetites* for certain Palaeozoic Equisetum-like leaf-sheaths.

Fig. 56 represents a rather faint impression of a leaf-sheath and nodal diaphragm. The specimen is from the Coal-Measures of Ardwick, Manchester. The letter *a* probably points to the attachment of the sheath to the node of the stem. The flattened sheath is indistinctly divided into segments, and at the middle of the free margin there appears to be a single free tooth. The lower part of the specimen, as seen in the figure, shows the position of the nodal diaphragm. Between the diaphragm and the sheath there are several slight ridges converging towards the nodal line ; these agree with the characteristic ridges and grooves of Calamite casts which are described in detail in Chapter X. There is another specimen in the British Museum which illustrates, rather more clearly than that shown in fig. 56, the association of a fused leaf-sheath with a type of cast usually regarded as belonging to a Calamitean stem. Some leaf-sheaths of Permian age described by Zeiller[1] as *Equisetites Vaujolyi* bear a close resemblance to the sheath in fig. 58 E. The nature of the true Calamite leaves is considered more fully on a later page.

The examples of supposed *Equisetites* sheaths referred to below may serve to illustrate the kind of evidence on which this genus has been recorded from Upper Palaeozoic rocks. I have retained the name *Equisetites* in the description of the species, but it would probably be better to speak of such specimens as 'Calamitean leaf-sheaths' rather than to describe them as definite species of *Equisetites*. We have not as yet any thoroughly satisfactory evidence that the *Equisetites* of Triassic and post-Triassic times existed in the vegetation of earlier periods.

In Grand'Eury's *Flore du Gard*[2] a fossil strobilus is figured under the name *Calamostachys tenuissima* Grand'Eury, which consists of a slender axis bearing series of sporophylls and

[1] Zeiller (92[2]) p. 56, Pl. xii. Other similar leaf-sheaths have been figured by Germar (44) Pl. x., Schimper (74) Pl. xvii. and others.
[2] Grand'Eury (90) p. 223, Pl. xv. fig. 16.

sporangia apparently resembling those of an *Equisetum*. There are no sterile appendages or bracts alternating with the sporophylls; and the absence of the former suggests a comparison with *Equisetites* rather than *Calamites*. Grand'Eury refers to the fossil as "parfois à peine perceptible," and a recent examination of the specimen leads me to thoroughly endorse this description. It was impossible to recognise the features represented in Grand'Eury's drawing. Setting aside this fossil, there are other strobili recorded by Renault[1] and referred by him to the

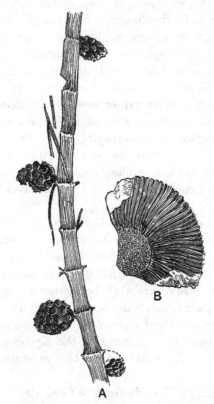

Fig. 57. A. *Equisetites Hemingwayi* Kidst. From a specimen in the British Museum. ⅔ nat. size. B. Diaphragm and sheath of an Equisetaceous plant, from the Coal-Measures. ⅔ nat. size. From a specimen in the British Museum.

[1] Renault (93) Pl. XLII. figs. 6 and 7.

genus *Bornia* (*Archaeocalamites*), which also exhibit the Equisetum-like character; the axis bears sporophylls only and no sterile bracts. It would appear then that in the Palaeozoic period the Equisetaceous strobilus, as we know it in *Equisetum*, was represented in some of the members of the Equisetales.

1. *Equisetites Hemingwayi* Kidst. Fig. 57, *A*.

Mr Kidston[1] founded this species on a few specimens of cones found in the Middle Coal-Measures of Barnsley in Yorkshire. The best example of the cone described by Kidston has a length of 2·5 cm., and a breadth of 1·5 cm.; the surface is divided up into several hexagonal areas 4 mm. high and 5 mm. wide. Each of these plates shows a fairly prominent projecting point in its centre; this is regarded as the point of attachment of the sporangiophore axis which expanded distally into a hexagonal plate bearing sporangia. An examination of Mr Kidston's specimens enabled me to recognise the close resemblance which he insists on between the fossils and such a recent Equisetaceous strobilus as that of *Equisetum limosum* Sm. Nothing is known of the structure of the fossils beyond the character of the superficial pattern of the impressions, and it is impossible to speak with absolute confidence as to their nature. The author of the species makes use of the generic name *Equisetum*; but in view of our ignorance of structural features it is better to adopt the more usual term *Equisetites*.

Since Kidston's description was published I noticed a specimen in the British Museum collection which throws some further light on this doubtful fossil. Part of this specimen is shown in fig. 57, *A*. The stem is 21 cm. in length and about 5 mm. broad; it is divided into distinct nodes and internodes; the former being a little exaggerated in the drawing. The surface is marked by fine and irregular striations, and in one or two places there occur broken pieces of narrow linear leaves in the neighbourhood of a node. Portions of four cones occurring in contact with the stem, appear to be sessile on the nodes, but the preservation is not sufficiently good to enable one to speak with certainty as to the manner of attachment. Each cone consists of regular hexagonal depressions, which agree exactly with the surface characters of Kidston's type-specimen. The manner of occurrence of the cones points to a lateral and not a terminal attachment. The stem does not show any traces of Equisetaceous leaf-sheaths at the nodes, and such fragments of leaves as occur appear to have the form of separate linear segments; they are not such as are met with on *Equisetites.* It agrees with some of the slender foliage-shoots of Calamitean plants often described under the generic name *Asterophyllites.* As regards the cones; they differ from the known Calamitean strobili in the absence of sterile bracts, and appear to consist entirely of distally expanded sporophylls as in *Equisetum.* The general impression afforded by the fossil is that we have not sufficient evidence for definitely associating this stem and cones with a true *Equisetites.* We may, however, adhere to this generic title until more satisfactory data are available.

2. *Equisetites spatulatus* Zeill. Fig. 58, *A.*

This species is chosen as an example of a French *Equisetites* of Permian age. It was recently founded by Zeiller[1] on some specimens of imperfect leaf-sheaths, and defined as follows :—

Sheaths spreading, erect, formed of numerous uninerved coherent leaves, convex on the dorsal surface, spatulate in form, 5—6 cm. in length and 2—3 mm. broad at the base, and 5—10 mm. broad at the apex, rounded at the distal end.

The specimen shown in fig. 58, *A*, represents part of a flattened sheath, the narrower crenulated end being the base of the sheath. The limits of the coherent segments and the position of the veins are clearly marked. Zeiller's description accurately represents the character of the sheaths. They agree closely with an Equisetaceous leaf-sheath, but as I have already pointed out, we cannot feel certain that sheaths of this kind were not originally attached to a Calamite stem.

[1] Zeiller (95).

The portion of a leaf-sheath and a diaphragm represented in fig. 57, *B*, agrees closely with Zeiller's examples. This specimen is from the English Coal-Measures, but it is not advisable to

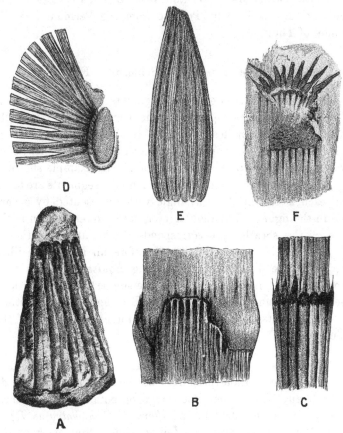

Fig. 58. *A. Equisetites spatulatus*, Zeill. Leaf-sheath. ¼ nat. size. (After Zeiller.)

　　　B. E. columnaris, Brongn. From a specimen in the British Museum. ¾ nat. size.

　　　C. Equisetum ramosissimum, Desf. × 2.

　　　D. Annularia stellata (Schloth.). Leaf-sheath. Slightly enlarged. (After Potonié.)

　　　E. Equisetites zeaeformis (Schloth.). Leaf-sheath. ¼ nat. size. (After Potonié.)

　　　F. E. lateralis, Phill. From a specimen in the Scarborough Museum. Nat. size.

attempt any specific diagnosis on such fragmentary material. It is questionable, indeed, if these detached fossil leaf-sheaths should be designated by specific names. Another similar form of sheath, hardly distinguishable from Zeiller's species, has recently been described by Potonié from the Permian (Rothliegende) of Thuringia.

3. *Equisetites zeaeformis* (Schloth.)[1]　Fig. 58, *E*.

The sheaths consist of linear segments fused laterally as in *Equisetum*. In some specimens the component parts of the sheath are more or less separate from one another, and in this form they are apparently identical with the leaves of *Calamites* (*Calamitina*) *varians*, Sternb. The example shown in fig. 58, *E* is probably a young leaf-sheath; the segments are fused, and each is traversed by a single vein represented by a dark line in the figure. The regular crenulated lower margin is the base of the sheath, and corresponds to the upper portion of fig. 58, *A*. This species affords, therefore, an interesting illustration of the difficulty of separating *Equisetites* leaves from those of true *Calamites*. Potonié has suggested that the leaf-sheath of a young Calamite might well be split up into distinct linear segments as the result of the increase in girth of the stem.

Other Palaeozoic species of *Equisetites* have been recorded, but with one exception these need not be dealt with, as they do not add anything to our knowledge of botanical importance. The specimen described in the *Flore de Commentry* as *Equisetites Monyi*, by Renault and Zeiller[2], differs from most of the other Palaeozoic species of *Equisetites*, in the fact that we have a stem with short internodes bearing a leaf-sheath at each node divided into comparatively long and distinct teeth. This species presents a close agreement with specimens of *Calamitina*, but Renault and Zeiller consider that it is generically distinct. They suggest that the English species, originally

---

[1] Potonié (93) p. 179, Pl. xxv. figs. 2—4.
[2] Renault and Zeiller (88) p. 396, Pl. LVII. fig. 7.

described and figured by Lindley and Hutton[1] as *Hippurites gigantea*, and now usually spoken of as *Calamitina*, should be named *Equisetites*. It would probably be better to adopt the name *Calamitina* for the French species. The type-specimen of this species is in the Natural History Museum, Paris.

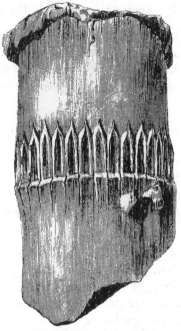

F𝗂𝗀. 59.   *Equisetites platyodon* Brongn.   (After Schoenlein, slightly reduced.)

When we pass from the Permian to the Triassic period, we find large casts of very modern-looking Equisetaceous stems which must clearly be referred to the genus *Equisetites*. The portion of a stem represented in fig. 59 known as *Equisetites platyodon* Brongn.[2] affords an example of a Triassic Equisetaceous stem with a clearly preserved leaf-sheath. The stem measures about 6 cm. in diameter. One of the oldest known Triassic species is *Equisetites Mougeoti*[3] (Brongn.) from the Bunter series of the Vosges.

[1] Lindley and Hutton (31) Pl. cxiv.
[2] Schoenlein and Schenk (65) Pl. v. fig. 1.
[3] Schimper and Mougeot (44) p. 58, Pl. xxix.

The Keuper species *E. arenaceus* is, however, more completely known. The specimens referred to this species are very striking fossils; they agree in all external characters with recent Horse-tails but greatly exceed them in dimensions.

4. *Equisetites arenaceus* Bronn.

This plant has been found in the Triassic rocks of various parts of Germany and France; it occurs in the Lettenkohl group (Lower Keuper), as well as in the Middle Keuper of Stuttgart and elsewhere. The species may be defined as follows:—

Rhizome from 8—14 cm. in diameter, with short internodes, bearing lateral ovate tubers. Aerial shoots from 4—12 cm. in diameter, bearing whorls of branches, and leaf-sheaths made up of 110—120 coherent uni-nerved linear segments terminating in an apical lanceolate tooth. Strobili oval, consisting of crowded sporangiophores with pentagonal and hexagonal peltate terminations.

The casts of branches, rhizomes, tubers, buds and cones enable us to form a fairly exact estimate of the size and general appearance of this largest fossil Horse-tail. The Strassburg Museum contains many good examples of this species, and a few specimens may be seen in the British Museum. In the École des Mines, Paris, there are some exceptionally clear impressions of cones of this species from a lignite mine in the Vosges.

It is estimated that the plant reached a height of 8 to 10 meters, about equal to that of the tallest recent species of *Equisetum*, but in the diameter of the stems the Triassic plant far exceeded any existing species.

It is interesting to determine as far as possible, in the absence of petrified specimens, if this Keuper species increased in girth by means of a cambium. There are occasionally found sandstone casts of the pith-cavity which present an appearance very similar to that of Calamitean medullary casts[1]. The

---

[1] Jäger (27).

nodes are marked by comparatively deep constrictions, which probably represent the projecting nodal wood. The surface of the casts is traversed by regular ridges and grooves as in an ordinary Calamite, and it is probable that in *Equisetites arenaceus*, as in *Calamites*, these surface-features are the impression of the inner face of a cylinder of secondary wood (*cf.* p. 310). Excellent figures of this species of *Equisetites* are given by Schimper in his Atlas of fossil plants[1], also by Schimper and Koechlin-Schlumberger[2], and by Schoenlein and Schenk[3].

5. *Equisetites columnaris* Brongn. Figs. 11 and 58, *B*.

This species, which is by far the best known British *Equisetites*, was founded by Brongniart[4] on some specimens from the Lower Oolite beds of the Yorkshire coast. Casts of stems are familiar to those who have collected fossils on the coast between Whitby and Scarborough; they are often found in an erect position in the sandstone, and are usually described as occurring in the actual place of growth. As previously pointed out (p. 72), such stems have generally been deposited by water, and have assumed a vertical position (fig. 11). Young and Bird[5] figured a specimen of this species in 1822, and in view of its striking resemblance to the sugar-cane, they regarded the fossil as being of the same family as *Saccharum officinarum*, if not specifically identical.

A specimen was described by König[6] in 1829, from the Lower Oolite rocks of Brora in the north of Scotland under the name of *Oncylogonatum carbonarium*, but Brongniart[7] pointed out its identity with the English species *Equisetites columnaris*.

Our acquaintance with this species is practically limited to the casts of stems. A typical stem of *E. columnaris* measures 3 to 6 cm. in diameter and has fairly long internodes. The

[1] Schimper (74) Pls. IX—XI.
[2] Schimper and Koechlin-Schlumberger (62).
[3] Schoenlein and Schenk (65) Pls. I—IV.
[4] Brongniart (28) p. 115, Pl. XIII.
[5] Young and Bird (22) p. 185, Pl. III. fig. 3.
[6] König, in Murchison (29) p. 293, Pl. XXXII.
[7] Murchison (29) p. 368.

largest stem in the British Museum collection has internodes about 14 cm. long and a diameter of about 5 cm. In
some cases the stem casts show irregular lateral projections in
the neighbourhood of a node, but there is no evidence that the
aerial shoots of this species gave off verticils of branches. In
habit *E. columnaris* probably closely resembled such recent
species as *Equisetum hiemale* L., *E. trachyodon* A. Br. and
others.

The stems often show a distinct swelling at the nodes; this
may be due, at least in part, to the existence of transverse
nodal diaphragms which enabled the dead shoots to resist
contraction in the region of the nodes. The leaf-sheaths
consist of numerous long and narrow segments often truncated
distally, as in fig. 58, *B*, and as in the sheath of such a recent
Horse-tail as *E. ramosissimum* shown in fig. 58, *C*. In some
specimens one occasionally finds indications of delicate acuminate teeth extending above the limits of a truncated sheath.
Brongniart speaks of the existence of caducous acuminate teeth
in his diagnosis of the species, and the example represented
in fig. 58, *B*, demonstrates the existence of such deciduous
appendages. There is a very close resemblance between the
fossil sheath of fig. 58, *B*, with and without the teeth, and
the leaf-sheath of the recent *Equisetum* in fig. 58, *C*. In
some specimens of *E. columnaris* in which the cast is covered
with a carbonaceous film, each segment in a leaf-sheath is
seen to be slightly depressed in the median portion, which
is often distinctly marked by numerous small dots, the edges
of the segment being flat and smooth. The median region
is that in which the stomata are found and on which deposits
of silica occur.

6. *Equisetites Beani* (Bunb.). Figs. 60—62.

Bunbury[1] proposed the name *Calamites Beani* for some
fossil stems from the Lower Oolite beds of the Yorkshire coast,
which Bean had previously referred to in unpublished notes as
C. giganteus. The latter name was not adopted by Bunbury

[1] Bunbury (51) p. 189.

on account of the possible confusion between this species
and the Palaeozoic species *Calamites gigas* Brong. The generic
name *Calamites* must be replaced by *Equisetites* now that we
are familiar with more perfect specimens which demonstrate
the Equisetean characters of the plant.

Schimper[1] speaks of this species as possibly the pith-cast
of *Equisetites columnaris*, but his opinion cannot be maintained; the species first described by Bunbury has considerably
larger stems than those of *E columnaris*. It is not impossible,
however, that *E. columnaris* and *E. Beani* may be portions
of the same species. The chief difference between these forms
is that of size; but we have not sufficient data to justify the
inclusion of both forms under one name. Zigno[2], in his work
on the Oolitic Flora, figures an imperfect stem cast of *E. Beani*
under the name of *Calamites Beani*, but the species has

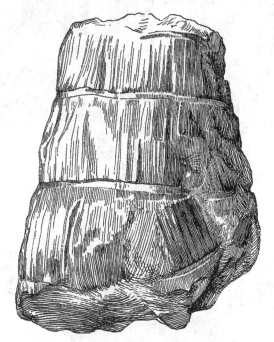

Fig. 60. *Equisetites Beani* (Bunb.). ⅔ nat. size. [After Starkie Gardner (86)
Pl. IX. fig. 2.]

[1] Schimper (69) p. 267.   [2] Zigno (56) Pl. III. fig. 1, p. 45.

Fig. 61.   *Equisetites Beani* (Bunb.).   From a specimen in the British
Museum, $\frac{2}{3}$ nat. size.   (No. V. 2725.)

received little attention at the hands of recent writers. In 1886 Starkie Gardner[1] figured a specimen which was identified by Williamson as an example of Bunbury's species; but the latter pointed out the greater resemblance, as regards the external appearance of the Jurassic stem, to some of the recent arborescent Gramineae[2] than to the Equisetaceae. Williamson, with his usual caution, adds that such appearances have very little taxonomic value. Fig. 60 is reproduced from the block used by Gardner in his memoir on Mesozoic Angiosperms; he quotes the specimen as possibly a Monocotyledonous stem. The fossil is an imperfect cast of a stem showing two clearly marked nodal regions, but no trace of leaf-sheaths. A recent examination of specimens in the museums of Whitby, Scarborough, York and London has convinced me that the plant named by Bunbury *Calamites Beani* is a large *Equisetites*. As a rule the specimens do not show any indications of the leaf-sheaths, but in a few cases the sheaths have left fairly distinct impressions.

In the portion of stem shown in fig. 61 the impressions of the leaf segments are clearly marked. This specimen affords much better evidence of the Equisetaceous character of the plant than those which are simply internal casts. The narrow projecting lines extending upwards from the nodes in the figured specimen probably represent the divisions between the several segments of each leaf-sheath.

In the museums of Whitby and Scarborough there are some long specimens, in one case 44 cm. in length, and 33 cm. in circumference, which are probably casts of the broad pith-cavity. These casts are often transversely broken across at the nodes, so that they consist of three or four separate pieces which fit together by clean-cut faces. This manner of occurrence is most probably due to the existence of large and resistant nodal diaphragms which separated the sand-casts of adjacent internodes. In the York museum there are some large diaphragms, 10 cm. in diameter, preserved separately in a piece of rock containing a cast of *Equisetites Beani*. The nodal diaphragms of some of the Carboniferous Calamites were the seat of cork development[3],

---

[1] Gardner (86) Pl. ix. fig. 3.   [2] Williamson (83) p. 4.
[3] Williamson and Scott (94) p. 889, Pl. lxxix. fig. 19.

and it may be that the frequent preservation of Equisetaceous diaphragms in Triassic and Jurassic rocks is due to the protection afforded by a corky investment.

The stem shown in fig. 62 appears to be a portion of a shoot of *E. Beani* not far from its apical region. From the lower nodes there extend clearly marked and regular lines or slight grooves tapering gradually towards the next higher node; these are no doubt the impressions of segments of leaf-sheaths. The sheaths themselves have been detached and only their impressions remain. The flattened bands at the node of the stem in fig. 60, and shown also in fig. 61, mark the place of attachment of the leaf-sheaths. On some of these nodal bands one is able to recognise small scars which are most likely the casts of outgoing leaf-trace bundles.

Some of the internal casts of this species are marked by numerous closely arranged longitudinal lines, which are probably the impressions of the inner face of a central woody cylinder. In the smaller specimen shown in fig. 62 we have the apical

Fig. 62. *Equisetites Beani* (Bunb.). From a specimen in the Scarborough Museum. Very slightly reduced.

portion of a shoot in which the uppermost internodes are in an unexpanded condition.

It is impossible to give a satisfactory diagnosis of this species without better material. The plant is characterised chiefly by the great breadth of the stem, and by the possession of leaf-sheaths consisting of numerous long and narrow segments. *Equisetites Beani* must have almost equalled in size the Triassic species, *E. arenaceus*, described above.

7.   *Equisetites lateralis* Phill.   Figs. 58, *F*, 63, and 64.

This species is described at some length as affording a useful illustration of the misleading character of certain features which

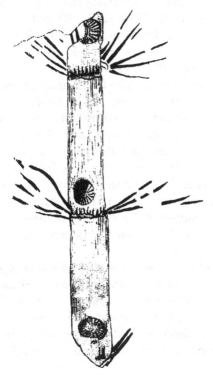

Fɪɢ. 63.   *Equisetites lateralis* Phill.   From a specimen in the British Museum. Slightly reduced.

are entirely due to methods of preservation. The specific name was proposed by Phillips in his first edition of the *Geology of*

the Yorkshire Coast for some very imperfect stems from the
Lower Oolite rocks near Whitby[1]. The choice of the term
lateralis illustrates a misconception; it was given to the plant
in the belief that certain characteristic wheel-like marks on
the stems were the scars of branches. Lindley and Hutton[2]
figured a specimen of this species in their *Fossil Flora*, and
quoted a remark by " Mr Williamson junior" (afterwards Prof.
Williamson) that the so-called scars often occur as isolated
discs in the neighbourhood of the stems. Bunbury[3] described
an example of the same species with narrow spreading leaves
like those of a Palaeozoic *Asterophyllites*, and proposed this
generic name as more appropriate than *Equisetites*. In all probability the example shown in fig. 63 is that which Bunbury
described. It is certainly the same as one figured by Zigno[4] as
Calamites lateralis in his *Flora fossilis formationis Oolithicae*.

This specimen illustrates a further misconception in the
diagnosis of the species. The long linear appendages spreading
from the nodes are, I believe, slender branches and not leaves ;
they have not the form of delicate filmy markings on the rock
face, but are comparatively thick and almost woody in appearance. The true leaves are distinctly indicated at the nodes,
and exhibit the ordinary features of toothed sheaths.

Heer[5] proposed to transfer Phillips' species to the genus
Phyllotheca, and Schimper[6] preferred the generic term *Schizoneura*. The suggestion for the use of these two names would
probably not have been made had the presence of the *Equisetum*
sheaths been recognised.

The circular depressions a short distance above each node
are the 'branch scars' of various writers. Schimper suggested
that these radially marked circles might be displaced nodal
diaphragms. Andrae[7] figured the same objects in 1853
but regarded them as branch scars, although in the specimen he describes, there are several of them lying apart from

[1] Phillips J. (29) Pl. x. fig. 13.   [2] Lindley and Hutton (31) Pl. CLXXXVI.
[3] Bunbury (51) p. 189.   [4] Zigno (56) Pl. III. fig. 3, p. 46.
[5] Heer (77) p. 43, Pl. IV.
[6] Schimper (69) p. 284. *Vide* also Nathorst (80) p. 54.
[7] Andrae (53) Pl. VI. figs. 1—5.

the stems, and to one of them is attached a portion of a leaf-sheath. Solms-Laubach[1] points out that the internodal position of these supposed scars is an obvious difficulty; we should not expect to find branches arising from an internode. After referring to some specimens in the Oxford museum, he adds—"In presence of these facts the usual explanation of these structures appears to me, as to Heer, very doubtful....We are driven to the very arbitrary assumption that they represent the lowest nodes of the lateral branches which were inserted above the line of the nodes of the stem." Circular discs similar to those of *E. lateralis* have been found in the Jurassic rocks of Siberia[2] and elsewhere. There are one or two examples of such discs from Siberia in the British Museum. If the nodal diaphragms were fairly hard and stout, it is easy to conceive that they might have been pressed out of their original position when the stems were flattened in the process of fossilisation. It is not quite clear what the radial spoke-like lines of the discs are due to; possibly they mark the position of bands of more resistant tissue or of outgoing strands of vascular bundles. A detached diaphragm is seen in fig. 64 C; in the centre it consists of a flat plate of tissue, and the peripheral region is traversed by the radiating lines. In the stem of fig. 64, A the deeply divided leaf-sheaths are clearly seen, and an imperfect impression of a diaphragm is preserved on the face of the middle internode. In fig. 64 B a flattened leaf-sheath is shown with the free acuminate teeth fused basally into a continuous collar[3]. The short piece of stem of *Equisetites lateralis* shown in fig. 58, *F*, shows how the free teeth may be outspread in a manner which bears some resemblance to the leaves of *Phyllotheca*, but a comparison with the specimens already described, and a careful examination of this specimen itself, demonstrate the generic identity of the species with *Equisetites*. The carbonaceous film on the surface of such stems as those of fig. 58, F, and 64, A, shows a characteristic shagreen texture which may possibly be due to the presence of silica in the epidermis as in recent Horse-tails.

---

[1] Solms-Laubach (91) p. 180.   [2] *cf.* p. 283.
[3] There is a similar specimen in the Oxford Museum.

There is another species of *Equisetites, E. Münsteri*, Schk.,
from a lower geological horizon which has been compared with

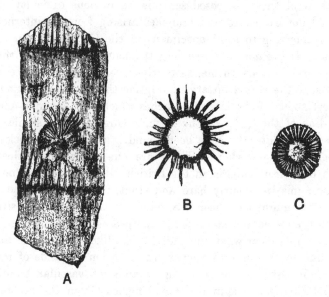

FIG. 64. *Equisetites lateralis* Phill. A. Part of a stem showing leaf-sheaths
and an imperfect diaphragm. B. A single flattened leaf-sheath. C. A
detached nodal diaphragm. From a specimen in the York Museum.
Slightly reduced.

E. lateralis, and lends support to the view that the so-called
branch-scars are nodal diaphragms[1]. This species also affords
additional evidence in favour of retaining the generic name
Equisetites for Phillips' species. *Equisetites Münsteri* is a typical
Rhaetic plant; it has been found at Beyreuth and Kuhnbach,
as well as in Switzerland, Hungary and elsewhere. A specimen
of *Equisetites* originally described by Buckman as *E. Brodii*[2],
from the Lower Lias of Worcestershire, may possibly be identical with *E. Münsteri*. The leaf-sheaths of this Rhaetic species
consist of broad segments prolonged into acuminate teeth; some

---

[1] Since this was written I have found a specimen of *Equisetites lateralis* in
the Woodwardian Museum, in which a diaphragm like that in fig. 64, C, occurs
in the centre of a flattened leaf-sheath similar to that of fig. 64, B.

[2] Buckman (50) p. 414.

of the examples figured by Schenk[1] show clearly marked impressions of displaced nodal diaphragms exactly as in *E. lateralis*.
Another form, *Equisetum rotiferum* described by Tenison-Woods[2]
from Australia, is closely allied to, or possibly identical with
E. lateralis.

8. *Equisetites Burchardti* Dunker[3]. Fig. 65.

This species of *Equisetites* is fairly common in the Wealden
beds of the Sussex coast near Hastings, and also in Westphalia.

Fɪɢ. 65. *Equisetites Burchardti* Dunk. Showing a node with two tubers and
a root. From a specimen in the British Museum. Nat. size.

It is characterised by having long and slender internodes,
bearing at the nodes leaf-sheaths with five or six pointed segments, and by the frequent formation of branch-tubers. These
tuberous branches closely resemble those which are formed on
the underground shoots of *Equisetum arvense* L., *E. sylva-
ticum* L. and others; they occur either singly or in chains[4]
In the specimen shown in the figure the left-hand tuber is
remarkably well preserved, its surface is somewhat sunk and
shrivelled, and the apex is surrounded by a nodal leaf-sheath.
A thin branched root is given off just below the point of
insertion of the oval tuber.

No other species of *Equisetites* affords such numerous

---

[1] Schenk (67).

[2] Tenison-Woods (83), Pl. vɪ. figs. 5 and 6. Specimen no. V. 3358 in the
British Museum.

[3] Dunker (46) p. 2, Pl. v. fig. 7.   [4] Seward (94²) p. 30.

examples of tubers as this Wealden plant. By some of the
earlier writers the detached tubers of *E. Burchardti* were
described as fossil seeds under the name *Carpolithus*.

FIG. 66. *Equisetites Yokoyamae* Sew. From specimens in the British
Museum. Nat. size.

The specimens shown in fig. 66 have been referred to
another species, *E. Yokoyamae* Sew.[1]; they were obtained from
the Wealden beds of Sussex, but according to Mr Rufford,
who discovered them, the smaller tubers of this species are not
found in association with those of *E. Burchardti*. The stems
are very narrow and the tubers have a characteristic elliptical
form; the species is of little value botanically, but it affords
another instance of the common occurrence of these tuberous
branches in the Wealden Equisetums.

Similar fossil tubers, on a much larger scale, have been
found in association with the Triassic *Equisetites arenaceus*;
with *E. Parlatori* Heer[2], a Tertiary species from Switzerland,
and with other Mesozoic and Tertiary stems. *E. Burejensis*[3],
described by Heer from the Jurassic rocks of Siberia, bears a
close resemblance to the Wealden species.

The description of the above species by no means exhausts
the material which is available towards a history of fossil

[1] Seward (94²) p. 33.   [2] Heer (55) vol. III. p. 158, Pl. CXLV.
[3] Heer (77) p. 99, Pl. XXII.

Equisetums. The examples which have been selected may serve to illustrate the kind of specimens that are usually met with, as well as some of the possible sources of error which have to be borne in mind in the description of species.

Such Tertiary species as have been recorded need not be considered; they furnish us with no facts of particular interest from a morphological point of view. The wide distribution of *Equisetites* especially during the Jurassic period, is one of the most interesting lessons to be learnt from a review of the fossil forms. No doubt a detailed comparison of the several species from different parts of the world would lead us to reduce the number of specific names; and at the same time it would emphasize the apparent identity of fossils which have been described from widely separated latitudes under different names.

Specimens of *Equisetites* are occasionally found in plant-bearing beds apart from the other members of a Flora; this isolated manner of occurrence suggests that the plant grew in a different station from that occupied by Cycads and other elements of the vegetation[1].

A selection of Triassic and Jurassic species arranged in a tabular form demonstrates the world-wide distribution of this persistent type of plant[2].

II. *Phyllotheca.*

The generic name *Phyllotheca* was proposed by Brongniart[3] in 1828 for some small fossil stems from the Hawkesbury river, near Port Jackson, Australia. The stems of this genus are divided into nodes and internodes and possess leaf-sheaths as in *Equisetum*, but *Phyllotheca* differs from other Equisetaceous plants in the form of the leaves and in the character of its sporophylls. We may define the genus as follows:—

Plants resembling in habit the recent Equisetums. Stems simple or branched, divided into distinct nodes and internodes,

---

[1] *Vide* Saporta (73) p. 227.

[2] The distribution will be dealt with in Volume II.

[3] Brongniart (28) p. 151.

the latter marked by longitudinal ridges and grooves; from the nodes are given off leaf-sheaths consisting of linear-lanceolate uninerved segments coherent basally, but having the form of free narrow teeth for the greater part of their length. The long free teeth are usually spread out in the form of a cup and not adpressed to the stem, the tips of the teeth are often in-curved.

The sporangia are borne on peltate sporangiophores attached to the stem between whorls of sterile leaves.

Our knowledge of *Phyllotheca* is unfortunately far from complete. The chief characteristic of the vegetative shoots consists in the cup-like leaf-sheaths; these are divided up into several linear segments, which differ from the teeth of an *Equisetum* leaf-sheath in their greater length and in their more open and spreading habit of growth. The large loose sheaths of the fertile shoots of some recent Horse-tails bear a certain resemblance to the sheaths of *Phyllotheca*. The diagnosis of the fertile shoots is founded principally on some Permian specimens of the genus described by Schmalhausen from Russia[1] and redescribed more recently by Solms-Laubach[2]. Prof. Zeiller[3] has, however, lately received some examples of *Phyllotheca* from the Coal-Measures of Asia Minor which bear strobili like those of the genus *Annularia*, a type which is dealt with in the succeeding chapter. A description of a few species will serve to illustrate the features usually associated with this generic type, as well as to emphasize the unsatisfactory state of our knowledge as to the real significance of such supposed generic characteristics.

There are a few fossil stems from Permian rocks of Siberia, from Jurassic strata in Italy, and from Lower Mesozoic and Permo-Carboniferous beds in South America, South Africa, India and Australia which do not conform in all points to the usually accepted definition of *Equisetites*, and so justify their inclusion in an allied genus. On the other hand there are numerous instances of stems or branches which have been

---

[1] Schmalhausen (79) p. 12, Pl. i. figs. 1—3.
[2] Solms-Laubach (91) p. 181.   [3] Zeiller (96).

referred to *Phyllotheca* on insufficient grounds. Our knowledge of this Equisetaceous plant has recently been extended
by Zeiller[1], who has recorded its occurrence in the Coal-
Measures of Asia Minor associated with typical Upper
Carboniferous plants. The same author[2] has also brought
forward good evidence for the Permian age of the beds
in Siberia and Altai, where *Phyllotheca* has long been
known. It is true that Zigno's species of the genus occurs in
Italian Jurassic rocks, but on the whole it would seem that
this genus is rather a Permian than a Jurassic type. The
species which Zeiller describes under the name *Phyllotheca
Rallii* from the Coal-Measures of Herakleion (Asia Minor)
shows some points of contact with *Annularia*. It is much
to be desired, however, that we might learn more as to the
reproductive organs of this member of the Equisetales; until
we possess a closer acquaintance with the fructification we
cannot hope to arrive at any satisfactory conclusion as to the
exact position of the genus among the Calamarian and Equisetaceous forms. M. Zeiller[3] informs me that his specimens of
P. Rallii, which are to be fully described in a forthcoming work,
include fossil strobili resembling those of *Annularia radiata*.
The verticils of linear leaves fused basally into a sheath agree
in appearance with the star-like leaves of *Annularia*, but in
Phyllotheca Rallii the segments appear to spread in all directions
and are not extended in one plane as in the typical *Annularia*[4].

1. *Phyllotheca deliquescens* (Göpp.).

In an account of some fossil plants collected by Tchikatcheff
in Altai, Göppert[5] describes and figures two imperfect stems of
an Equisetum-like plant. Owing to the apparent absence of
nodal lines on the surface of the stem the generic name *Anarthrocanna* is proposed for the fossils; and the manner in which the
main axis appears to break up into slender branches suggested
the specific name *deliquescens*. Schmalhausen[6] afterwards

---

[1] Zeiller (95²).   [2] *ibid.* (96).   [3] Letter, July 30, 1897.
[4] On this character of Annularian leaves, *vide* p. 337.
[5] Göppert (45) p. 379, Pl.xxv. figs. 1, 2.   [6] Schmalhausen (79) p. 12.

recognised the generic identity of Göppert's fragments with the Indian and Australian stems referred to the genus *Phyllotheca* by McCoy[1] and Bunbury[2].

We may define the species as follows:—

Stem reaching a diameter of 2—3 cm. with internodes as much as 4 cm. long, the surface of which is traversed by longitudinal ridges and grooves which are continuous and not alternate at the nodes. Branches arise in verticils from the nodes. The leaves have the form of funnel-shaped sheaths split up into narrow and spreading linear segments, each of which is traversed by a median vein. The fertile shoot terminates in a loose strobilus bearing alternating whorls of sterile bracts and sporangiophores.

The specimens on which this diagnosis is founded are for the most part fragments of sterile branches. Some of these present the appearance of Calamitean stems in which the ridges and grooves continue in straight lines from one internode to the next. Similar stem-casts have been referred by some writers to the allied genus *Schizoneura*, and it would appear to be a hopeless task to decide with certainty under which generic designation such specimens should be described. The portion of stem shown in fig. 67 affords an example of an Equisetaceous plant, probably in the form of a cast of a hollow pith, which might be referred to either *Phyllotheca* or *Schizoneura*. The specimen was found in certain South African rocks which are probably of Permo-Carboniferous age[3]. It agrees closely with some stems from India described by Feistmantel[4] as *Schizoneura gondwanensis*, and it also resembles equally closely the Australian specimens referred by Feistmantel[5] to *Phyllotheca australis* and some stems of *Phyllotheca indica* figured by Bunbury[6].

The longitudinal ridges and grooves shown in fig. 67 probably represent the broad medullary rays and the projecting wedges of secondary wood surrounding a large hollow

[1] McCoy (47) Pl. XI. fig. 7.   [2] Bunbury (61) Pl. XI. fig. 1.

[3] Seward (97²) p. 324, Pl. XXIV. fig. 1.

[4] Feistmantel (81) Pl. IX. A. fig. 7, &c.

[5] *ibid.* (90) Pl. XIV. fig. 5.   [6] Bunbury (61) Pl. XI. fig. 1.

pith, as in *Calamites*. In the Calamitean casts the ridges and grooves of each internode usually alternate in position with those of the next, as in *Equisetum* (fig. 54, A), but in *Phyllotheca*, *Schizoneura* and *Archaeocalamites* there is no such regular alternation at the nodes of the internodal vascular strands.

FIG. 67. *Phyllotheca ?* ¾ nat. size. From a South African specimen of Permo-Carboniferous age in the British Museum.

In *Phyllotheca* and *Schizoneura* there are no casts of 'infra-nodal canals' below each nodal line, but these are by no means always found in true Calamites. It is therefore practically impossible to determine the generic position of such fossils as that shown in fig. 67 without further evidence than is afforded by leafless casts.

A few examples of *Phyllotheca deliquescens* have been

described by Schmalhausen in which a branch bears clusters
of sporangiophores, alternating with verticils of sterile bracts.
The sporangiophores appear to have the form of stalked peltate
appendages bearing sporangia, very similar to the sporangiophores of *Equisetum*. Solms-Laubach[1] has examined the best of
Schmalhausen's specimens, and a carefully drawn figure of one
of the fertile branches is given in his *Fossil Botany*.

The significance of this manner of occurrence of sporangiophores and whorls of sterile bracts on the fertile branch will
be better understood after a description of the strobilus of
Calamites. In *Phyllotheca* the sporangiophores appear to have
been given off in whorls, which were separated from one another
by whorls of sterile bracts, whereas in *Equisetum* there are no
sterile appendages associated with the sporangiophores of the
strobilus, with the exception of the annulus at the base of the
cone. Heer[2] first drew attention to the fact that in *Phyllotheca*
we have a form of strobilus or fertile shoot to a certain extent
intermediate in character between *Equisetum* and *Calamites*.

In abnormal fertile shoots of *Equisetum*, sporophylls occasionally occur above and below a sterile leaf-sheath. Potonié[3]
has figured such an example in which an apical strobilus is
succeeded at a lower level by a sterile leaf-sheath, and this again
by a second cluster of sporophylls. As Potonié points out, this
alternation of fertile and sterile members affords an interesting
resemblance between *Phyllotheca* and *Equisetum*. It suggests
a partial reversion towards the Calamitean type of strobilus.

2. *Phyllotheca Brongniarti* Zigno. Fig. 68, A.

This species of *Phyllotheca* from the Lower Oolite rocks of
Italy is known only in the form of sterile branches. The leaves
are fused basally into an open cup-like sheath which is dissected
into several spreading and incurved linear segments. The
internodes are striated longitudinally; they are about 2 mm.
in diameter and 10 mm. in length.

---

[1] Solms-Laubach (91) p. 181, fig. 17.
[2] Heer (82) p. 9.
[3] Potonié (96²) p. 115, fig. 3.

The specimen represented in fig. 68, A, was originally described by the Italian palaeobotanist Zigno[1]; it serves to illustrate the points of difference between this genus and the ordinary *Equisetum*. The open and spreading sheaths clasping the nodes and the erect solitary branches give the plant a distinctive appearance.

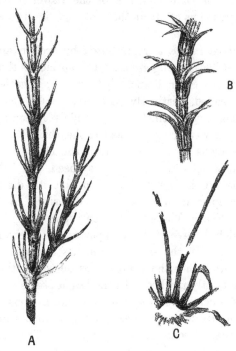

Fig. 68. A. *Phyllotheca Brongniarti*, Zigno. Nat. size. (After Zigno.)
B. *Calamocladus frondosus*, Grand'Eury. (After Grand'Eury.) Slightly enlarged.
C. *Phyllotheca indica*, Bunb. Part of a leaf-sheath. From a specimen in the Museum of the Geological Society. Slightly enlarged.

3. *Phyllotheca indica* Bunb. and *P. australis* Brongn. Fig. 68, C.

Sir Charles Bunbury[2] described several imperfect specimens from the Nagpur district of India under this name, but he

[1] Zigno (56) Pl. vii. p. 59. [2] Bunbury (61).

expressed the opinion that it was not clear to him if the plant was specifically distinct from the *Phyllotheca australis* Brongn. previously recorded from New South Wales. Feistmantel[1] subsequently described a few other Indian specimens, but did not materially add to our knowledge of the genus. Bunbury's specimens were obtained from Bharatwádá in Nagpur, in beds belonging to the Damuda series of the Lower Gondwana rocks, usually regarded as of about the same age as the Permian rocks of Europe.

Phyllotheca indica is represented by broken and imperfect fragments of leaf-bearing stems. The species is thus diagnosed by Bunbury:—"Stem branched, furrowed; sheaths lax, somewhat bell-shaped, distinctly striated; leaves narrow linear, with a strong and distinct midrib, widely spreading and often recurved, nearly twice as long as the sheaths." An examination of the specimens in the Museum of the Geological Society of London, on which this account was based, has led me to the opinion that it is practically impossible to distinguish the Indian examples from *P. australis* described by Brongniart[2] from New South Wales. The few specimens of the latter species which I have had an opportunity of examining bear out this view. In the smaller branches the axis of *P. indica* is divided into rather short internodes on which the ridges and grooves are faintly marked. In the larger stems the ridges and grooves are much more prominent, and continuous in direction from one internode to the next; a few branches are given off from the nodes of some of the specimens. The leaves are not very well preserved; they consist of a narrow collar-like basal sheath divided up into numerous long and narrow segments, which are several times as long as the breadth of the sheath, and not merely twice as long as Bunbury described them. Each leaf-sheath has the form of a very shallow cup-like rim clasping the stem at a node, with long free spreading segments which are often bent back in their distal region. The general habit of the leafy branches appears to be identical with that of *P. australis* as figured by McCoy.

---

[1] Feistmantel (81), Pl. xii. A.   [2] Brongniart (28) p. 152.

Prof. Zeiller informs me that in the type-specimen on which Brongniart founded the species, *P. australis*, the sheath appears to be closely applied to the stem with a verticil of narrow spreading segments radiating from its margin. It may be, therefore, that in the Australian form there was not such an open and cup-like sheath as in *P. indica*; but it would be difficult, without better material before us, to feel confidence in any well marked specific distinctions between the Indian and Australian Phyllothecas.

On the broader stems, such as that of fig. 67, we have clearly marked narrow grooves and broader and slightly convex ridges, which present an appearance identical with that of some Calamitean stems. In the specimen figured by Bunbury[1] in his Pl. X, fig. 6, there is a circular depression on the line of the node which represents the impression of the basal end of a branch; on the edges of the node there are indications of two other lateral branches. The nature of this stem-cast points unmistakeably to a woody stem like that of *Calamites*. The precise meaning of the ridges and grooves on the cast is described in the Chapter dealing with Calamitean plants.

Grand'Eury[2] in his monograph on the coal-basin of Gard, has recently described under the name of *Calamocladus frondosus* what he believes to be the leaf-bearing axes of a Calamitean plant. The thicker branches are almost exactly identical in appearance with the broader specimens of *Phyllotheca*. The finer branches of *Calamocladus* bear cup-like leaf-sheaths which are divided into long and narrow recurved segments (fig. 67, B), precisely as in *Phyllotheca*. These comparisons lead one to the opinion that the *Phyllotheca* of Australia and India may be a close ally of the Permo-Carboniferous Calamitean plants. The form of the leaf-whorls of *Annularia* (Calamarian leaf-bearing branches) and of *Calamocladus* is of the same type as in *Phyllotheca*; the character of the medullary casts is also the same. The nature of the fertile shoot of *Phyllotheca* described by Schmalhausen from Siberia, with its alternating whorls of sterile and fertile leaves, is another point of agreement between this genus and

[1] Bunbury (61).   [2] Grand'Eury (90) p. 221.

Calamitean plants. An Equisetaceous species has been described
from the Newcastle Coal-Measures of Australia by Etheridge[1]
in which there are two forms of leaves, some of which closely
resemble those of *Phyllotheca indica*, while others are compared
with the sterile bracts of *Cingularia*, a Calamitean genus
instituted by Weiss[2].

When we turn to other recorded forms of *Phyllotheca* many
of them appear on examination to have been placed in this
genus on unsatisfactory grounds. Heer figures several stem
fragments from the Jurassic rocks of Siberia as *P. Sibirica*
Heer[3], and it was the resemblance between this form and the
English *Equisetites lateralis* which led to the substitution of
Phyllotheca for *Equisetites* in the latter species. Without
examining Heer's material it is impossible to criticise his
conclusions with any completeness, but several of his specimens
appear to possess leaf-sheaths more like those of *Equisetum* than
of *Phyllotheca*.

The frequent occurrence of isolated diaphragms and the
comparatively long acuminate teeth of the leaf-sheath afford
obvious points of resemblance to *Equisetites lateralis*. Some
of the examples figured by Heer appear to be stem fragments,
with numerous long and narrow filiform leaves different in
appearance from those of other specimens which he figures.
It may be that some of the less distinct pieces of stems are
badly torn specimens in which the internodes have been
divided into filiform threads. Heer also figures a fertile axis
associated with the sterile stems, and this does not, as Heer
admits, show the alternating sterile bracts such as Schmalhausen
has described. So far as it is possible to judge from an examination of Heer's figures and a few specimens from Siberia in
the British Museum—and this is by no means a safe basis on
which to found definite opinions—there appears to be little
evidence in favour of separating the fossils described as *Phyllotheca Sibirica* from *Equisetites*. This Siberian form may
indeed be specifically identical with *Equisetites lateralis* Phill.

Various species of *Phyllotheca* have been described from

[1] Etheridge (95).   [2] Weiss (76) p. 88.
[3] Heer (77) p. 43, Pl. IV. (78) p. 4, Pl. I.

Jurassic and Upper Palaeozoic rocks in Australia. Some of these possess cup-like leaf-sheaths, and in the case of the thicker specimens they show continuous ridges and grooves on the internodes, as well as a habit of branching similar to that in some of the Italian Phyllothecas. In some of the stems it is however difficult to recognise any characters which justify the use of the term *Phyllotheca*. A fragment figured by Tenison-Woods[1] as a new species of *Phyllotheca, P. carnosa*, from Ipswich, Queensland, affords an example of the worthless material on which species have not infrequently been founded. The author of the species describes his single specimen as a "faint impression"; the figure accompanying his description suggests a fragment of some coniferous branch, as Feistmantel has pointed out in his monograph on Australian plants.

It is important that a thorough comparative examination should be made of the various fossil Phyllothecas with a view to determine their scientific value, and to discover how far the separation of *Phyllotheca* and *Equisetites* is legitimate in each case. There is too often a tendency to allow geographical distribution to decide the adoption of a particular generic name, and this seems to have been especially the case as regards several Mesozoic and Palaeozoic Southern Hemisphere plants.

The geological and geographical range of *Phyllotheca* is a question of considerable interest, but as already pointed out it is desirable to carefully examine the various records of the genus before attempting to generalise as to the range of the species. *Phyllotheca* is often spoken of as a characteristic member of the *Glossopteris* Flora of the Southern Hemisphere, and its geological age is usually considered to be Mesozoic rather than Palaeozoic.

III. *Schizoneura.*

The plants included under this genus were originally designated by Brongniart[2] *Convallarites* and classed as Monocotyledons. Some years later Schimper and Mougeot[3] had

---

[1] Tenison-Woods (83) Pl. ix. fig. 2.　　　　[2] Brongniart (28) p. 128.
[3] Schimper and Mougeot (44) p. 48, Pls. xxiv—xxvi.

the opportunity of examining more perfect material from the Bunter beds of the Vosges, and proposed the new name *Schizoneura* in place of Brongniart's term, on the grounds that the specimens were in all probability portions of Equisetaceous stems, and not Monocotyledons. Our knowledge of this genus is very limited, but the characteristics are on the whole better defined than in the case of *Phyllotheca*. The following diagnosis illustrates the chief features of *Schizoneura*.

Hollow stems with nodes and internodes as in *Equisetum*; the surface of the internodes is traversed by regular ridges and grooves, which are continuous and not alternate in their course from one internode to the next. The leaf-sheaths are large and consist of several coherent segments; the sheaths are usually split into two or more elongate ovate lobes, and each lobe contains more than one vein. Fertile shoots are unknown.

Two of the best known and most satisfactory species are *Schizoneura gondwanensis* Feist. and *S. paradoxa* Schimp. and Moug.

Schizoneura gondwanensis Feist. Fig. 69, A and B.

This species is represented by numerous specimens from the Lower Gondwana rocks of India[1]; it is characterised by narrow articulated stems which bear large leaf-sheaths at the nodes. The sheaths may have the form of two large and spreading elongate-oval lobes, each of which is traversed by several veins (fig. 69, B), or the lobes may be further dissected into long linear single-veined segments, as in fig. 69, A. It is supposed that in the young condition each node bears a leaf-sheath consisting of laterally coherent segments which, as development proceeds, split into two or more lobes. Feistmantel records this species from the Talchir, Damuda and Panchet divisions of the Lower Gondwana series of India; these divisions are regarded as equivalent to the Permo-Carboniferous and Triassic rocks of Europe. The two specimens shown in fig. 69 are from the Lower Gondwana rocks of the Raniganj Coal-field, India.

---

[1] Feistmantel (81) p. 59, Pls. I. A—x. A.

As already pointed out[1], some of the specimens of flat and broader stems referred by Feistmantel to *Schizoneura* are

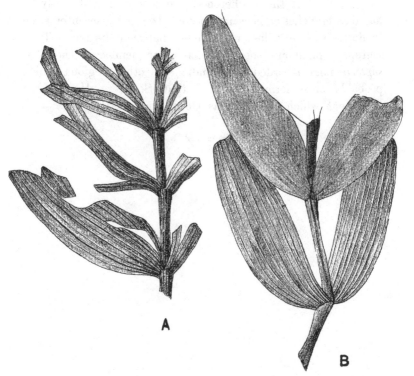

FIG. 69. *Schizoneura gondwanensis* Feist. (After Feistmantel; slightly reduced.)

identical in appearance with stems which have been described from India and elsewhere as species of *Phyllotheca*.

There are a few specimens of *S. gondwanensis* in the British Museum, but the genus is poorly represented in European collections.

A similar plant was described in 1844 by Schimper and Mougeot[2] from the Bunter rocks of the Vosges as *Schizoneura paradoxa*. This species bears a very close resemblance to the Indian forms, and indeed it is difficult to point to any distinction of taxonomic importance. Feistmantel considers

[1] *ante*, p. 284.   [2] Schimper and Mougeot (44) p. 50, Pls. XXIV.—XXVI.

that the European plant has rather fewer segments in the leaf-sheaths, and that the Indian plant had somewhat stronger stems. Both of these differences are such as might easily be found on branches of the same species. It is, however, interesting to notice the very close resemblance between the Lower Trias European plant and the somewhat older member of the *Glossopteris* flora recorded from India and other regions, which probably once formed part of that Southern Hemisphere Continent which is known as Gondwana Land[1].

[1] Seward (97[2]).

CHAPTER X.

I. EQUISETALES (*continued*).

(CALAMARIEAE.)

In order to minimise repetition and digression the following account of the Calamarieae is divided into sections, under each of which a certain part of the subject is more particularly dealt with.   After a brief sketch of the history of our knowledge of *Calamites*, and a short description of the characteristics of the genus, the morphological features are more fully considered. A description of the most striking features of the better known Calamitean types is followed by a short discussion on the question of nomenclature and classification, and reference is made to the manner of occurrence of *Calamites* and to some of the possible sources of error in identification.

IV. *Calamites.*

I. **Historical Sketch.**

In the following account of the Calamarieae the generic name *Calamites* is used in a somewhat comprehensive sense. As previous writers have pointed out, it is probable that under this generic name there may be included more than one type of plant worthy of generic designation.   Owing to the various opinions which have been held by different authors, as to the relationship and botanical position of plants now generally

included in the Calamarieae, there has been no little confusion
in nomenclature. Facts as to the nature of the genus *Calamites* have occasionally to be selected from writings containing
many speculative and erroneous views, but the data at our
disposal enable us to give a fairly complete account of the
morphology of this Palaeozoic plant.

In the earliest works on fossil plants we find several figures
of *Calamites*, which are in most cases described as those of fossil
reeds or grasses. The *Herbarium diluvianum* of Scheuchzer[1]
contains a figure of a Calamitean cast which is described as
probably a reed. Another specimen is figured by Volkmann[2]
in his *Silesia subterranea* and compared with a piece of sugar-
cane. A similar flattened cast in the old Woodwardian collection at Cambridge is described by Woodward[3] as "part of a
broad long flat leaf, appearing to be of some *Iris*, or rather an
Aloe, but 'tis striated without." Schulze[4], one of the earlier
German writers, figured a Calamitean branch bearing verticils
of leaves, and described the fossil as probably the impression of
an Equisetaceous plant. It has been pointed out by another
German writer that the Equisetaceous character of *Calamites*
was recognised by laymen many years before specialists shared
this view.

One of the most interesting and important of all the older
records of *Calamites* is that published by Suckow[5] in 1784.
Suckow is usually quoted as the author of the generic name
Calamites; he does not attempt any diagnosis of the plant, but
merely speaks of the specimens he is describing as "Calamiten."
The examples figured in this classic paper are characteristic
casts from the Coal-Measures of Western Germany. Suckow
describes them as ribbed stems, which were found in an oblique
position in the strata and termed by the workmen Jupiter's
nails ("Nägel"). Previous writers had regarded the fossils
as casts of reeds, but Suckow correctly points out that the
ribbed character is hardly consistent with the view that the

[1] Scheuchzer (1723), p. 19, Pl. IV. fig. 1.
[2] Volkmann (1720), p. 110, Pl. XIII. fig. 7.
[3] Woodward, J. (1728), Pt. II. p. 10.   [4] Schulze, C. F. (1755), Pl. II. fig. 1.
[5] Suckow (1784), p. 363.

casts are those of reeds or grasses. He goes on to say that the material filling up the hollow pith of a reed would not have impressed upon it a number of ribs and grooves such as occur on the Calamites. He considers it more probable that the casts are those of some well-developed tree, probably a foreign plant. *Equisetum giganteum* L. is mentioned as a species with which *Calamites* may be compared, although the stem of the Palaeozoic genus was much larger than that of the recent Horse-tail. The tree of which the Calamites are the casts must, he adds, have possessed a ribbed stem, and the bark must also have been marked by vertical ribs and grooves on its *inner face*. It is clear, therefore, that Suckow inclined to the view that *Calamites* should be regarded as an *internal cast* of a woody plant. Such an interpretation of the fossils was generally accepted by palaeobotanists only a comparatively few years ago, and the first suggestion of this view is usually attributed to Germar, Dawes, and other authors who wrote more than fifty years later than Suckow.

One of the earliest notices of *Calamites* in the present century is by Steinhauer[1], who published a memoir in the Transactions of the American Philosophical Society in 1818 on *Fossil reliquia of unknown vegetables in the Carboniferous rocks*. He gives some good figures of Calamitean casts under the generic name of *Phytolithus*, one of those general terms often used by the older writers on fossils. Among English authors, Martin[2] may be mentioned as figuring casts of *Calamites*, which he describes as probably grass stems. By far the best of the earlier figures are those by Artis[3] in his *Antediluvian Phytology*. This writer does not discuss the botanical nature of the specimens beyond a brief reference to the views of earlier authors. Adolphe Brongniart[4], writing in 1822, expresses the opinion that the Calamites are related to the genus *Equisetum*, and refers to M. de Candolle as having first suggested this view. In a later work Brongniart[5] includes species of *Calamites* as figured by Suckow, Schlotheim, Sternberg and Artis in

[1] Steinhauer (18), Pls. v. and vi.   [2] Martin (09), Pls. viii. xxv. and xxvi.
[3] Artis (25).   [4] Brongniart (22), p. 218.
[5] Brongniart (28), p. 34.

the family *Equisetaceae*. Lindley and Hutton[1] give several
figures of Calamites in their *Fossil flora*, but do not commit
themselves to an Equisetaceous affinity.

An important advance was made in 1835 by Cotta[2], a
German writer, who gave a short account of the internal
structure of some Calamite stems, which he referred to a new
genus *Calamitea*. The British Museum collection includes some
silicified fragments of the stems figured and described by Cotta
in his *Dendrolithen*. Some of the specimens described by this
author as examples of *Calamitea* have since been recognised as
members of another family.

In 1840 Unger[3] published a note on the structure and
affinities of *Calamites*, and expressed his belief in the close
relationship of the Palaeozoic plant and recent Horse-tails.

An important contribution to our knowledge of *Calamites*
was supplied by Petzholdt[4] in 1841. His main contention was
the Equisetaceous character of this Palaeozoic genus. The
external resemblance between Calamite casts and *Equisetum*
stems had long been recognised, but after Cotta's account of
the internal structure it was believed that the apparent
relation between *Equisetum* and *Calamites* was not confirmed
by the facts of anatomy. Petzholdt based his conclusions on
certain partially preserved Permian stems from Plauenscher
Grund, near Dresden. Although his account of the fossils is
not accurate his general conclusions are correct. The specimens described by Petzholdt differ from the common Calamite
casts in having some carbonised remnants of cortical and woody
tissue. A transverse section of one of the Plauenscher Grund
fossils is shown in fig. 70. The irregular black patches were
described by Petzholdt as portions of cortical tissue, while he
regarded the spaces as marking the position of canals like the
vallecular canals in an *Equisetum*. Our more complete knowledge of the structure of a Calamite stem enables us to

---

[1] Lindley and Hutton (31).

[2] Cotta (50). I am indebted to Prof. Stenzel of Breslau for calling my
attention to the fact that Cotta's work appeared in 1832, but in 1850 the same
work was sold with a new title-page bearing this date.

[3] Unger (40).   [4] Petzholdt (41).

correlate the patches in which no tissue has been preserved
with the broad medullary rays, which separated the wedge-
shaped groups of xylem elements; the latter being more
resistant were converted into a black coaly substance, while the
cells of the medullary rays left little or no trace in the
sandstone matrix. The thin black line, which forms the limit

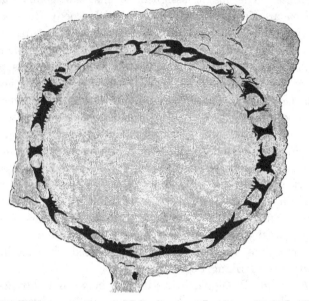

Fig. 70. Transverse section of a Calamite stem, showing carbonised remnants
of secondary wood. From a specimen (no. 40934), presented to the British
Museum by Dr Petzholdt from Plauenscher Grund, Dresden. ½ nat. size.

of the drawing in fig. 70, external to the carbonised wood, no
doubt marks the limit of the cortex, and the appendage indicated
in the lower part of the figure may possibly be an adventitious
root. It is interesting to note that Unger[1] in 1844 expressed
the opinion, which we now know to be correct, that the coaly
mass in the specimens described by Petzholdt represented the
wood, and that there was no proof of the existence of canals in
the cortex as Petzholdt believed.

Turning to Brongniart's later work[2] we find an important

[1] Unger (44).   [2] Brongniart (49), p. 49.

proposal which led to no little controversy. While retaining
the genus *Calamites* for such specimens as possess a thin bark
and a ribbed external surface, showing occasional branch-scars
at the nodes, and having such characters as warrant their
inclusion in the Equisetaceae, he proposes a second generic
name for other specimens which had hitherto been included in
Calamites. The fossils assigned to his new genus *Calamodendron* are described as having a thick woody stem, and as
differing from *Equisetum* in their arborescent nature. Brongniart's genus *Calamodendron* is made to include the plants for
which Cotta instituted the name *Calamitea*, and it is placed
among the Gymnosperms. This distinction between the Vascular Cryptogam *Calamites* and the supposed Gymnosperm
Calamodendron is based on the presence of secondary wood in
the latter type of stem. The prominence formerly assigned to
the power of secondary thickening possessed by a plant as a
taxonomic feature, is now known to have been the result of
imperfect knowledge. The occurrence of a cambium layer and
the ability of a plant to increase in girth by the activity of a
definite meristem, is a feature which some recent Vascular
Cryptogams[1] share with the higher plants; and in former
ages many of the Pteridophytes possessed this method of
growth in a striking degree.

Although Brongniart's distinction between *Calamites* and
Calamodendron has not been borne out by subsequent researches, the latter term is still used as a convenient designation for a special type of Calamitean structure. One of the
earliest accounts of the anatomy of *Calamodendron* stems is
by Mougeot[2], who published figures and descriptions of two
species, *Calamodendron striatum* and *C. bistriatum*.

Some years later Göppert[3], who was one of the greatest of
the older palaeobotanists, instituted another genus, *Arthropitys*[4], for certain specimens of silicified stems from the
Permian rocks of Chemnitz in Saxony, which Cotta had
previously placed in his genus *Calamitea* under the name of

---

[1] E.g. *Isoetes, Botrychium*, &c.   [2] Mougeot (52).
[3] Göppert (64), p. 183.   [4] ἄρθρον, joint; πίτυς, Pine-tree.

Calamitea bistriata[1]. Göppert rightly decided that the plants so named by Cotta differed in important histological characters from other species of *Calamitea*. The generic name *Arthropitys* has been widely adopted for a type of Calamitean stem characterised by definite structural features. The great majority of the petrified Calamite stems found in the English Coal-Measures belong to Göppert's *Arthropitys*.

The next proposal to be noticed is one by Williamson[2] in 1868; he instituted the generic name *Calamopitys* for a few examples of English stems, which differed in the structure of the wood and primary medullary rays from previously recorded types. We have thus four names which all stand for generic types of Calamitean stems. Of these *Calamodendron* and *Arthropitys* are still used as convenient designations for stems with well-defined anatomical characters. The genus *Calamitea* is no longer in use, and Williamson's name *Calamopitys* had previously been made use of by Unger[3] for plants which do not belong to the Calamarieae. As it is convenient to have some term to apply to such stems as those which Williamson made the type of *Calamopitys*, the name *Arthrodendron* is suggested by my friend Dr Scott[4] as a substitute for Williamson's genus.

The twofold division of the Calamites instituted by Brongniart has already been alluded to, and for many years it was generally agreed that both Pteridophytes and Gymnosperms were represented among the Palaeozoic fossils known as Calamites. The work of Prof. Williamson was largely instrumental in proving the unsound basis for this artificial separation; he insisted on the inclusion of all Calamites in the Vascular Cryptogams, irrespective of the presence or absence of secondary wood. By degrees the adherents of Brongniart's views acknowledged the force of the English botanist's contention. It is one of the many signs of the value of Williamson's work that there is now almost complete accord among palaeobotanical writers as to the affinities of Calamitean plants.

---

[1] The original specimens described by Göppert are in the rich palaeobotanical Collection of the Breslau Museum.

[2] Williamson (71[3]), p. 174.

[3] *vide* Solms-Laubach (96).   [4] Letter, November 1897.

In the following account of the Calamites, the generic name *Calamites* is used in a wide sense as including stems possessing different types of internal structure; when it is possible to recognise any of these structural types the terms *Calamodendron, Arthropitys* or *Arthrodendron* are used as subgenera. The reasons for this nomenclature are discussed in a later part of the Chapter.

Genus *Calamites*, Suckow, 1714 { This term was originally applied to the common pith-casts of Calamitean stems, *without reference to internal structure.*

Subgenera	*Calamodendron,*	Brongniart,	1849	These names have primarily reference to *internal structure.*
	Arthropitys	Göppert,	1864	
	Arthrodendron	Scott	1897	
	(= *Calamopitys*	Williamson,	1871)	

II. Description of the anatomy of Calamites.

a. Stems.   b. Leaves.   c. Roots.   d. Cones.

No fossils are better known to collectors of Coal-Measure plants than the casts and impressions of the numerous species of *Calamites*. In sandstone quarries of Upper Carboniferous rocks there are frequently found cylindrical or somewhat flattened fossils, varying from one to several inches in diameter, marked on the surface by longitudinal ridges and grooves, and at more or less regular intervals by regular transverse constrictions. Similar specimens are still more abundant as flattened casts in the blocks of shale found on the rubbish heaps of collieries. The sandstone casts are often separated from the surrounding rock by a loose brown or black crumbling material, and the specimens in the shale are frequently covered by a thin layer of coal.

Most of the earlier writers regarded such specimens as the impressions of the ribbed stems of plants similar to or identical with reeds or grasses. Suckow, and afterwards Dawes and others, expressed the opinion that the ordinary Calamite cast represented a hardened mass of sand or marl, which had filled up the

pith of a stem either originally fistular or rendered hollow by
decay. The investigation of the internal structure confirmed
this view, and proved that the surface-features of a Calamite
stem do not represent the external markings of the original
plant, but the form of the inner face of the cylinder of wood.
The ribs represent the medullary rays of the original stem or
branch, and the intervening grooves mark the position of the
strands of xylem which are arranged in a ring round a large
hollow pith[1].

With this brief preliminary account we may pass to a
detailed description of the anatomical characters of *Calamites*.

The genus *Calamites* may be briefly defined as follows:—

Arborescent plants reaching a height of several meters,
and having a diameter of proportional size. In habit of
growth the Calamites bore a close resemblance to *Equisetum*;
an underground rhizome giving off lateral branches and erect
aerial shoots bearing branches, either in whorls from regularly
recurring branch-bearing nodes, or two or three from each
node; and in some cases the stems bore occasional branches
from widely separated nodes. The leaves were disposed in
whorls either as star-shaped verticils on slender foliage shoots,
or in the form of a circle of long narrow leaves on the node of a
thicker branch. Adventitious roots were developed from the
nodal regions of underground and aerial stems. The cones had
the form of long and narrow strobili consisting of a central axis
bearing whorls of sterile and fertile appendages; the latter in
the form of sporangiophores bearing groups of sporangia. The
strobili were heterosporous in some cases, isosporous in others.
The stems had a large hollow pith bridged across by a
transverse diaphragm at the nodes in the centre of the single
stele; the latter consisted of a ring of collateral bundles
separated from one another by primary medullary rays. Each
group of xylem was composed of spiral, annular, scalariform and
occasionally reticulate tracheids, the position of the protoxylem
being marked by a longitudinal carinal canal. The shoots and
roots grew in thickness by means of a regular cambium layer.
The cortex consisted of parenchymatous and sclerenchymatous

[1] *Vide* p. 310.

cells, with scattered secretory sacs. The increase in girth of the central cylinder was often accompanied by a considerable development of cortical periderm. The roots differed from the shoots in having no carinal canals, and in the possession of a solid pith and centripetally developed primary xylem groups alternating with strands of phloem.

The above incomplete diagnosis includes only some of the more important structural features of the genus. Thanks to the researches begun by the late Mr Binney of Manchester and considerably extended by Carruthers, Williamson and later investigators, we are now in a position to give a fairly complete account of *Calamites*. The type of stem most frequently met with in a petrified condition in the English rocks is that to which Göppert applied the name *Arthropitys*, and it is this subgenus that forms the subject of the following description. Our knowledge of Calamitean anatomy is based on the examination of numerous fragments of petrified twigs and other portions of different specific types of the genus. It is seldom possible to differentiate specifically between the isolated fragments of stems and branches which are met with in calcareous or siliceous nodules. As so frequently happens in fossil-plant material, large specimens showing good surface features and broken fragments with well-preserved internal structure have to be dealt with separately.

a. Stems.

A transverse section of a young twig, such as is represented in fig. 71, illustrates the chief characteristics of the *primary structure* of a young branch of *Calamites*. The figure has been drawn from a section originally described by Hick[1] in 1894. A very young Calamite twig bears an exceedingly close resemblance to the stem of a recent *Equisetum*. The axial region of the stem may be occupied by parenchymatous cells, or the absence of cells in the centre may indicate the beginning of the gradual formation of the hollow pith, which is one of the characteristics of *Calamites*. The student of petrified Palaeozoic

[1] Hick (94), Pl. ix. fig. 1.

plants must constantly be on his guard against the possible
misinterpretation of Stigmarian 'rootlets,' which are frequently
found in intimate association with fossil tissues. The intrusion
of these rootlets is admirably illustrated by a section of a Calamite stem in the Williamson Collection (No. 1558) in which the
hollow pith, 2 cm. broad, contains more than a dozen Stigmarian
appendages.

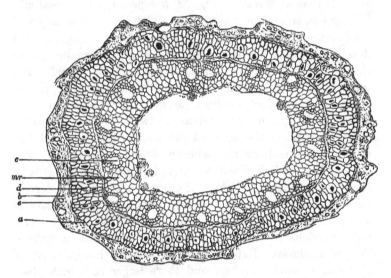

Fig. 71. Transverse section of a young Calamite stem. c, carinal canals;
mr, primary medullary rays; a, b, and d, cortex; e, epidermis. From a
section in the Manchester Museum, Owens College. × 60.

In the figured specimen of a Calamite twig (fig. 71)
there is a clearly marked differentiation into a cortical region
and a large stele or central cylinder. The pith-cells are already
partially disorganised, but there still remain a few fairly large
parenchymatous cells internal to the ring of vascular bundles.
The few irregular projections into the cavity of the large pith
consist of small fragments of cells, which may be the result
of fungal action. Mycelia of fungi are occasionally met with in
the tissues of older Calamite stems.

The position of the primary xylem groups is shown by the
conspicuous and regularly placed canals, c; these have been

formed in precisely the same manner as the corresponding
spaces in an *Equisetum* stem, and they are spoken of in both
genera as the carinal canals. Each canal owes its origin to the
disorganization and tearing apart of the protoxylem elements
and the surrounding cells. This may be occasionally seen in
examples of very young Calamites; the canals of a young twig
often contain apparently isolated rings which are coils of
elongated spiral threads. Fig. 72, *B* represents the canal of a
twig, cut in an oblique direction, in which the remains of spiral
tracheids are distinctly seen. In the stem of fig. 71 the
development has not advanced far enough to enable us to
clearly define the exact limits of each xylem strand. The
smaller elements bordering the canals constitute the primary
xylem, they are fairly distinct on the outer margin of some
of the canals seen in the section. Between the small patches
of primary xylem the outward extensions of the parenchyma
of the pith constitute the primary medullary rays; *mr*. The
distinct line encircling the canals and primary xylem has been
described by Hick as marking the position of the endodermis,
but it may possibly owe its existence to the tearing of the tissues
along the line where cambial activity is just beginning. This
layer of delicate dividing cells would constitute a natural
line of weakness. External to this line we have a zone of
tissue *a*, *d*, containing here and there larger cells with black
contents, which are no doubt secretory sacs. It is impossible to
distinguish with certainty any definite phloem groups, but in
other specimens these have been recognised immediately external to each primary xylem group; the bundles were typically
collateral in structure. Towards the periphery of the twig the
preservation is much less perfect; the outer portion of the
inner cortex, *d*, consists of rather smaller and thicker-walled
cells, but this is succeeded by an ill-defined zone containing a
few scattered cells, *b*, which have been more perfectly preserved.
The twig is too young to show any secondary tissue in the cortex;
but the tangential walls in some of the cortical cells afford evidence
of meristematic activity, which probably represents the beginning
of cork-formation. The limiting line, *e*, possibly represents the
cuticularised outer walls of an epidermal layer. The irregularly

wavy character of the surface of the specimen is probably the result of shrinking, and does not indicate original surface features.

In examining sections of calcareous nodules from the coal seams one meets with numerous fragments of small Calamitean twigs with little or no secondary wood; in some of these there is a small number of carinal canals, in others the canals are much more abundant. The former probably represent the smaller ramifications of a plant, and the latter may be regarded as the young stages of branches capable of developing into stout woody shoots[1]. Longitudinal sections of small branches teach us that the xylem elements next the carinal canals are either spiral or reticulate in character, the older tracheids being for the most part of the scalariform type, with bordered pits on the radial walls. This and other histological characters are admirably shown in the illustrations accompanying Williamson and Scott's memoir on *Calamites*. The student should treat the account of the anatomy of *Calamites* given in these pages as introductory to the much more complete description by these authors. They thus describe the course of the vascular bundles in a Calamitean branch :—

"The bundle-system of *Calamites* bears a general resemblance to that of *Equisetum*. A single leaf-trace enters the stem from each leaf, and passes vertically downwards to the next node. In the simplest cases the bundle here forks, its two branches attaching themselves to the alternating bundles which enter the stem at this node. In other cases both the forks attach themselves to the same bundle, so that, in this case, there is no regular alternation. In other cases, again, the bundle runs past one node without forking, and ultimately forms a junction with the traces of the second node below its starting-point. These variations may all occur in the same specimen. The xylem at the node usually forms a continuous ring, for where the regular dichotomous forks of the bundles are absent their place is usually taken by anastomoses[2]."

As in *Equisetum*, the xylem at the nodes possesses certain characteristic features which distinguish it from the internodal

---

[1] On this point *vide* Williamson and Scott (94), p. 869.
[2] Williamson and Scott, *loc. cit.* p. 876.

strands. It has already been pointed out that the xylem of
Equisetum increases in breadth at the nodes (p. 251, fig. 55, 4);

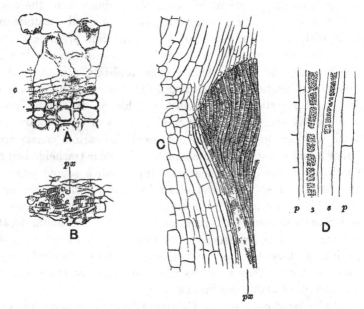

FIG. 72. *A.* External xylem elements and cambium, *c*, with imperfect
phloem. × 100.
B. Carinal canal containing protoxylem, *px.* × 65.
C. Radial longitudinal section through nodal xylem, *px.* × 35.
D. Phloem elements; *s*, sieve-tubes; *p, p*, parenchymatous cells.
(*A—C.* After Williamson and Scott. *D.* After Renault.)

the same is true of *Calamites.* In fig. 72, *C*, we have part of
a radial section of a Calamite twig in which the broad mass
of short nodal tracheids is clearly shown; this nodal wood
forms a prominent projection towards the pith. In the lower
part of the section the remains of some spiral protoxylem
tracheids are seen in a carinal canal.

The tracheids of the nodal wood are often reticularly pitted,
and so differ in appearance from the ordinary scalariform
elements.

It is rare to find the phloem clearly preserved, but in
specimens where it has been possible to examine this portion of
the vascular bundles, it is found to consist of elongated

cambiform cells and sieve-tubes. An unusually perfect specimen has been described by Renault[1] in which the phloem
elements are preserved in silica. Fig. 72, *D*, is copied
from one of Renault's drawings, the sieve-tubes, *s, s*, show
several distinct sieve-plates on the lateral walls of the tubes,
reminding one to some extent of the sieve-tubes in a Bracken
Fern. The cells, *p, p*, associated with the sieve-tubes are
square-ended elongated parenchymatous elements. Another
characteristic feature illustrated by longitudinal sections is
the nodal diaphragm; except in the smallest branches the
interior of each internode is hollow, and the ring of vascular
bundles is separated from the pith-cavity by a band of parenchymatous tissue. At each node this parenchyma extends
across the central cavity in the form of a nodal diaphragm, as
in the stem of *Equisetum.*

By far the greater number of the petrified fragments of
Calamites afford proof of cambial activity, and possess obvious
secondary tissues. In exceptionally perfect specimens the
xylem tracheids are found to be succeeded externally by a few
flattened thin-walled cells which are in a meristematic condition (fig. 72, *A, c*); these constitute the cambium zone, and
it is the *secondary structure* that results from the activity of
the meristematic cells that we have now to consider.

In petrified examples of branches in which the secondary
thickening has reached a fairly advanced stage, the wood is
usually the outermost tissue preserved, the more external
tissues having been detached along the line of cambium cells.
It is only in a few cases that we are able to examine all the
tissues of older examples.

The specimen represented in fig. 73 illustrates very clearly
the extension of the hollow pith up to the inner surface of
the vascular ring; the disorganisation of the pith-cells which
had already begun in the twig of fig. 71 has here advanced
much further. The bluntly rounded projections represent the
prominent primary xylem strands, each of which is traversed
by the characteristic carinal canal. Alternating with the
wedge-shaped groups of secondary xylem, *x*, we have the broad

[1] Renault (93), Pl. xlvii. fig. 4.

principal medullary rays, *mr*, which become slightly narrower towards the outside.   The inner face of each of these wide rays

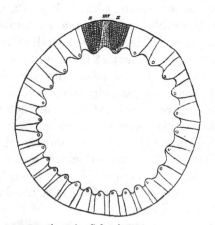

FIG. 73.   Transverse section of a Calamite stem.
mr, medullary ray.   After Williamson.
x, *x*, xylem.   (No. 1933 A.A. in the Williamson Collection.)

has a concave form, due to the less resistent nature of the medullary-ray cells as compared with the stronger xylem. The regularly sinuous form of the inner face of the vascular cylinder enables one to realise how the Calamite-casts (figs. 82, 99, and 101) have come to have the regular ridges and grooves on their surface.   The broad ridges on the cast mark the position of the wide medullary rays, while the grooves correspond to the more prominent ends of the vascular strands.   The tissues external to the wood have not been preserved in the example shown in fig. 73.   Some silicified specimens described by Stur[1] from Bohemia and now in the Museum of the Austrian Geological Survey, Vienna, admirably illustrate the connection between the surface features of a Calamite cast and the anatomy of the stem.

In the large section of a calcareous nodule diagrammatically shown in fig. 17 II. (p. 85) the secondary wood of a slightly flattened Calamite is the most prominent plant fragment. The pith-cavity has been almost obliterated by the lateral compression of the woody cylinder, but the presence of the carinal

[1] Stur (87).

canals along the inner edge of the wood may still be readily recognised. The appearance presented by a transverse section of the secondary wood of a Calamite is that of regular radial series of rather small rectangular tracheids, with occasional secondary medullary rays consisting of narrow and radially elongated parenchymatous cells. The principal rays[1] in the *Arthropitys* type of a Calamite stem are often found to gradually decrease in breadth as they pass into the secondary wood, until in the outer portion of the wood the primary medullary rays are practically obliterated by the formation of interfascicular xylem.

In fig. 74, *A*, we have a portion of a single xylem group of a thick woody stem. The stem from which the figure has been drawn was originally described by Binney[2] as *Calamodendron commune*; we now recognise it as a typical example of the subgenus *Arthropitys*. The specific term *communis* was used by Ettingshausen[3] in 1855 in a comprehensive sense to include more than twenty species of the genus *Calamites*, but since Binney's use of the term it has come to be associated with a definite type of *Arthropitys* stem, in which the primary medullary rays decrease rapidly in breadth towards the periphery of the wood. The wood of Binney's stem[4] measures 2·5 cm. across, but the pith-cavity has been crushed to the limits of a narrow band represented in the figure by the shaded portion. The strand of cells, *s*, in the pith is a portion of a Stigmarian appendage ("rootlet"), which penetrated into the hollow stem of the Calamite and became petrified by the same agency to which the preservation of the stem is due. These intruded Stigmarian appendages are of constant occurrence in the calcareous nodules; their intimate association with the tissues of other plants is often a serious source of error in the identification of petrified tissues. The inner portion of one of the

---

[1] The term *primary ray* may be conveniently restricted to the truly primary interfascicular tissue, and the term *principal ray* may be used for the outward extension of the primary rays by the cambium [Williamson and Scott (94), p. 878].

[2] Binney (68).   [3] Ettingshausen (55).

[4] The sections of fossil plants described by Binney were presented to the Woodwardian Museum, Cambridge, by his son (Mr J. Binney).

xylem groups is shown in fig. 74, A. External to the carinal
canal, the xylem tracheids are disposed in regular series and

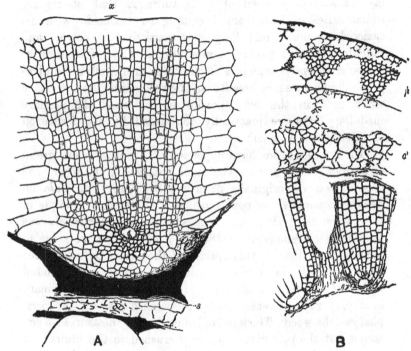

FIG. 74. A. Transverse section of part of a Calamite stem. [*Calamites*
                      (*Arthropitys*) *communis* (Binney).]
              s, Stigmarian appendage. x, xylem. From a specimen in the
                      Binney Collection, Cambridge. × 50.
          B. Transverse section of a stem.
              h, hypodermal tissue; c, inner cortex. From a specimen in the
                      Williamson Collection (no. 62). × 35.

associated with numerous narrow secondary medullary rays.
The width of the xylem wedge increases gradually as we pass
outwards, this is due to the formation of interfascicular xylem,
which in the more peripheral portion of the stem extends
across the primary medullary rays. The few primary medullary-
ray cells shown in the drawing illustrate the characteristic
tangentially elongated form and large size of the parenchymatous elements. Williamson and Scott have pointed out that
the tangentially elongated form of the medullary-ray cells is the

result of active growth, and not merely the expression of the tangential stretching of the stem consequent on secondary thickening.

A glance at the complete transverse section of the stem,—of which a small portion is shown in fig. 74 *A*,—suggests the existence of annual rings in the wood, but this appearance of rings is merely the result of compression. The secondary wood of a Calamite does not exhibit any regular zones of growth comparable with the annual rings of our forest trees.

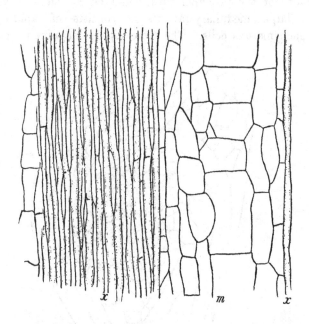

Fɪɢ. 75.   Longitudinal tangential section near the inner edge of the wood of
the Calamite of fig. 74.
x, *x*, secondary xylem and medullary rays; *m*, principal medullary
ray.   From a section in the Binney Collection.   × 50.

Before passing to other examples of Calamitean stems, reference may be made to the sections shown in figs. 75 and 76, which illustrate some further points in the structure of Binney's stems. In fig. 75 the xylem tracheids are shown at *x*, and between them the secondary medullary rays present the appearance of long and narrow parenchymatous cells; as the section is tangential the characteristic scalariform character of the tracheids is not shown, the ladder-like bordered pits being confined to the radial walls of the tracheal elements. The much greater length than breadth of the cells which form the rays associated with the xylem tracheids, is a characteristic feature in Calamitean stems. The breadth of the principal ray, *m*, shows that the section has passed through the wood a short distance from the pith; in a tangential section cut further into the wood the breadth of the principal rays would be considerably reduced. The large medullary-ray tissue consists of square-walled parenchymatous cells. The more highly magnified section, in

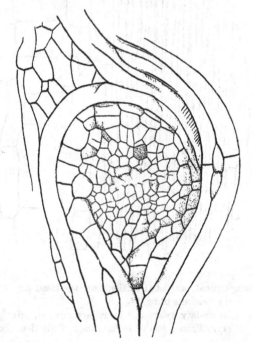

Fig. 76. Longitudinal tangential section of the same Calamite as that of figs. 74 and 75, showing a leaf-trace and curved tracheids at a node. From a section in the Binney Collection. × 100.

fig. 76, shows a central group of parenchyma containing a few transversely cut tracheids, but the two kinds of elements are not clearly differentiated in the figure; this group of cells is an outgoing leaf-trace which is enclosed by the strongly curved tracheids of the stem. The section is taken from the node of a stem where several leaf-trace bundles are passing out to a whorl of leaves; the few cells intercalated between the tracheids belong to the parenchyma of the secondary medullary rays.

In the small portion of a stem represented in fig. 74 *B*, the cortical tissues have been partially preserved; at the inner edge, next the hollow pith, there are two xylem groups, each with a carinal canal, and between them is part of a broad "principal" medullary ray[1]. The cambium has not been preserved, but beyond this region we have some of the large cells, *c*, of the inner cortex; these are followed by a few remnants of a smaller-celled tissue, and external to this part of the cortex there is a series of triangular groups, *h*, consisting of small thick-walled cells alternating with spaces which were originally occupied by more delicate parenchyma. The darker groups constitute hypodermal strands of mechanical tissue or stereome which lent support to the stem. The surface of a stem possessing such supporting strands would probably assume a longitudinally wrinkled or grooved appearance on drying; the intervening parenchyma, contracting and yielding more readily, would tend to produce shallow grooves alternating with the ridges above the stereome strands.

The complete section of the stem of which a small portion is shown in fig. 74 *B*, is figured by Williamson[2] in his 12th memoir on Coal-Measure plants. The section was obtained from Ashton-under-Lyne in Lancashire; it illustrates very clearly a method of preservation which is occasionally met with among petrified plants. The walls of the various tissue elements are black in colour and somewhat ragged, and the general appearance of the section is similar to that of a section of a charred piece of stem. It is possible that the Calamite twig was reduced to charcoal before petrifaction by a lightning flash or some other cause.

It is often said that the surface of a Calamite stem was probably marked by regular ridges and grooves similar to those

---

[1] *Vide* footnote, p. 311.
[2] Williamson (83[2]), Pl. xxxiii. fig. 19.

of the pith-cast, and that such external features are connected
with the arrangement of the tissues in the vascular cylinder..
The indication of grooves and ridges on the bark of fossil Calamites is no doubt the result of the existence in the hypoderm

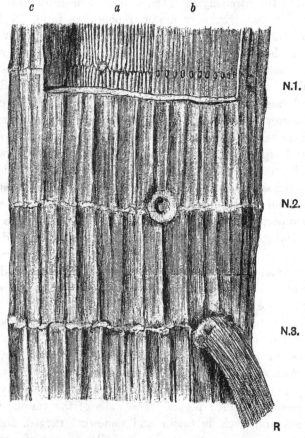

Fig. 77. Portion of a Calamite stem, showing the surface of the bark, *c* ; the
wood, *b*; the surface of the pith-cast, *a*. N.1—N.3. Nodes. R. Root.
(After Grand'Eury. Partially restored from a specimen in the École des
Mines, Paris.) ¾ nat. size.

of firm strands alternating with strands of less resistant cells.
It is very common to find Calamite pith-casts covered with
a layer of coal presenting a ribbed surface, but this is simply
due to the moulding of the coaly film on an internal pith-cast.

The broad grooves on such a specimen as that of fig. 77 are, on the other hand, probably an indication of the existence of hypoderm bands similar to those in fig. 74 *B, h.* The specimen from which fig. 77 is drawn shows many interesting features. The figure given by Grand'Eury, of which fig. 77 is a copy, is somewhat idealised, but the various surfaces can be made out in the fossil. The surface of the coaly envelope surrounding the pith-cast, *a*, is distinctly grooved, but the depressions have nothing to do with the surface features of the wood or the pith-cast; they are no doubt due to the occurrence of alternating bands of thick- and thin-walled tissue in the hypodermal region of the cortex; the peripheral strands of bast cells would stand out as prominent ribs as the stem tissue contracted during fos-silisation. At *b* (fig. 77) we have a view of the wood in which the position of the principal rays is indicated by fine longitudinal lines at regular intervals; the oval projections just below the nodal line are probably the casts of infranodal canals (*cf.* p. 324). At *a* the characteristic pith-cast is seen with a small branch-scar on the node. The scar on the middle node, *N* 2, is probably that of a root, and a root *R* is still attached to the node, *N* 3.

An interesting feature observed in some specimens of older Calamite branches is the development of periderm or cork. This is illustrated on a large scale by a unique specimen originally described by Williamson in 1878[1]. Figs. 78 and 79 represent transverse and longitudinal sections of this stem. This un-usually large petrified stem was found in the Coal-Measures of Oldham, in Lancashire. In the slightly reduced drawing, fig. 78, the large and somewhat flattened pith, *p*, 4·2 cm. in diameter, is shown towards the bottom of the figure. Sur-rounding this we have 58 or 59 wedge-shaped projecting xylem groups and broad medullary rays; the latter soon become indistinguishable as they are traced radially through the thick mass of secondary wood, 5 cm. wide, composed of scalariform tracheids and secondary medullary rays (fig. 78, 3). The secondary wood presents the features characteristic of *Cala-mites* (*Arthropitys*) *communis* (Binney). External to the wood there is a broken-up mass, about 5·5 cm. wide composed of

[1] Williamson (78), p. 323, Pl. xx. figs. 14 and 15.

regularly arranged (fig. 78, 2) and rather thick-walled cells; this consists of periderm, a secondary tissue, which has been

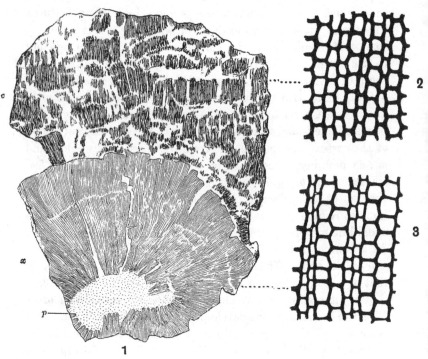

Fig. 78. 1. Transverse section of a thick Calamite stem.
                    p, pith; x, secondary wood; c, bark. (⅔ nat. size.)
             2. Periderm cells of bark.
             3. Xylem and medullary rays. (2 and 3, × 80.)
             From a specimen in the Williamson Collection (no. 79).

developed by a cork-cambium during the increase in girth of the plant. The more delicate cortical tissues have not been preserved, and the more resistant portion of the bark has been broken up into small pieces of corky tissue, among which are seen numerous Stigmarian appendages, pieces of sporangia and other plant fragments. These associated structures cannot of course be shown in the small-scale drawing of the figure.

In the radial longitudinal section (fig. 79) we see the pith with the projecting wood and the remains of a diaphragm at the

node. The mottled or watered appearance of the wood is due to
numerous medullary rays which sweep across the tracheids. The

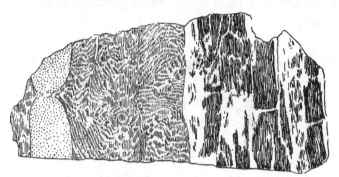

Fig. 79. Longitudinal section of the specimen stem in fig. 78.
    From a specimen in the Williamson Collection, British Museum (no. 80).
    ⅔ nat. size.

periderm elements, as seen in longitudinal section, are fibrous in
form.

The development of cork in a younger Calamite stem is
clearly shown in a specimen described by Williamson and Scott
in their Memoir of 1894. In a transverse section of the stem
several large cells of the inner cortex are seen to be in process
of division by tangential walls, and giving rise to radially
arranged periderm tissue [1].

The section diagrammatically sketched in fig. 80 is that of a
Calamite twig in which the wood appears to have been injured,
and the wound has been almost covered over by the formation
of callus wood. The young trees in a Palaeozoic forest might
easily be injured by some of the large amphibians, which were
the highest representatives of animal life during the Carboniferous period, just as our forest trees are often barked by deer,
rabbits, and other animals. Fissures might also be formed by
the expansion of the bark under the heating influence of the
sun's rays [2]. Such a specimen as that of fig. 80 gives an air of
living reality to the petrified fragments of the Coal period trees.

[1] Williamson and Scott (94), p. 888.
[2] Hartig (94), pp. 149, 297, *etc.*

It is well known how a wound on the branch of a forest tree becomes gradually overgrown by the activity of the cambium giving rise to a thick callus, which gradually closes over the

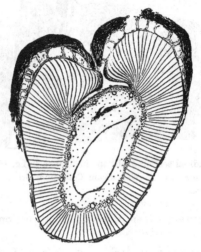

FIG. 80. Diagrammatic sketch of a transverse section of a Calamite twig, showing callus wood. From a specimen in the Cambridge Botanical Laboratory Collection. × *ca.* 10.

wounded surface in the form of two lips of wood which finally meet over the middle of the scar. The two lips of callus are clearly shown in the fossil branch arching over the tear in the wood just beyond the ring of carinal canals. The tissue external to the wood represents the imperfectly preserved cortex. A section which was cut parallel to that of fig. 80 shows a continuous band of wood beyond the wound, and the latter has the form of a small triangular gap; this section appears to have passed across the wound where it was narrower and has already been closed over by the callus. The formation of a rather different kind of callus wood has been described by Renault[1] and by Williamson and Scott[2], in stems where aborted or deciduous branches have been overgrown and sealed up by cambial activity.

[1] Renault (96), p. 91.

[2] Williamson and Scott, *loc. cit.* p. 893. *Vide* specimens 133*—135* in the Williamson Collection.

Some of the features to be noticed in longitudinal sections
of Calamite stems have already been de-
scribed, at least as regards younger
branches. The specimen shown in fig. 81
illustrates the general appearance of a stem
as seen in tangential and radial section. In
the lower portion, T, the course of the vascular bundles is shown by the black lines
which represent the xylem tracheids, bifur-
cating and usually alternating at each node.
Between the xylem strands are the broad
principal medullary rays. At b a branch
has been cut through on its passage out
from the parent stem, just above the nodal
line. In tangential sections of Calamite
stems one frequently sees both branches
and leaf-trace bundles (fig. 83, A), passing
horizontally through the wood and enclosed
by strongly curved and twisted tracheids.
In the upper part of the figure (81, R), the
section has passed through the centre of
the stem, and the wood is seen in radial
view; each node is bridged across by a
diaphragm of parenchymatous cells capable
of giving rise to a surface layer of periderm[1].

An outgoing branch, as seen in a tangential section of a stem, consists of a
parenchymatous pith surrounded by a ring
of vascular bundles, in which the characteristic carinal canals have not yet been
formed, but if the section has cut the branch
further from its base, there may be seen a circle of irregular
gaps marking the position of the carinal canals. Such gaps
are often occupied by thin parenchyma, and contain protoxylem
elements. The outgoing branches, as seen in a tangential section
of a Calamite stem, are seen to be connected with the wood of
the parent stem by curved and sinuous tracheids, which give

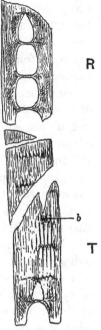

Fig. 81. *Calamites*.
Longitudinal sec-
tion (R, radial; T,
tangential) of a
small branch. b,
position of a lateral
branch. From a
specimen (no. 1937)
in the Williamson
Collection. Slightly
enlarged.

[1] *E.g.* specimen 132*** in the Williamson Collection.

to the stem-wood a curiously characteristic appearance[1], as if
the xylem elements had been pushed aside and contorted by
the pressure of the outgoing member. A tangential section
through a Pine stem[2] in the region of a lateral branch presents
precisely the same features as in *Calamites*. The branches are
given off from the stem immediately above a node and usually
between two outgoing leaf-trace bundles.

Specimens of pith-casts occasionally present the appearance
of a curved and rapidly tapered ram's horn, and the narrow
end of such a cast is sometimes found in contact with the node
of another cast. This juxtaposition of casts is shown unusually
well in fig. 82. In some of the published restorations of *Calamites*
the plant is represented as having thick branches attached to
the main stem by little more than a point. Williamson[3] clearly
explained this apparently unusual and indeed physically impossible method of branching, by means of sections of petrified stems.
The branches seen in fig. 82 are of course pith-casts, and in the
living plant the pith of each branch was surrounded by a mass
of secondary wood developed from as many primary groups
of xylem as there are grooves on the surface of the cast, each
of the grooves on an internode corresponding to the projecting
edge of a xylem group. At the junction of one branch with
another the pith was much narrower and the enclosing wood
thicker, so that the tapered ends of the cast merely show the
continuity by a narrow union between the pith-cavities of
different branches. Most probably the casts of fig. 82 are those
of a branched rhizome which grew underground, giving off
aerial shoots and adventitious roots. There is a fairly close
resemblance between the Calamite casts of fig. 82 and a stout
branching rhizome of a Bamboo, *e.g. Bambusa arundinacea*
Willd.; it is not surprising that the earlier writers looked upon
the Calamite as a reed-like plant.

Before leaving the consideration of stem structures there

---

[1] *Vide* Williamson (71), Pl. xxviii. fig. 38; (71²), Pl. iv. fig. 15; (78), Pl. xxi.
figs. 26—28. Williamson and Scott (94), Pl. lxxii. figs. 5 and 6. Renault (93),
Pl. xlv. figs. 4—6, etc. Felix (96), Pl. iv. figs. 2 and 3.

[2] Strasburger (91), Pl. ii. fig. 40.

[3] Williamson (78).

is another feature to which attention must be drawn. On the
casts shown in fig. 82 there is a circle of small oval scars
situated just below the nodes, these are clearly shown at
c, c, c. Each of the scars is in reality a slight projection
from the upper end of an internodal ridge. As the ridges
correspond to the broad inner faces of medullary rays, the

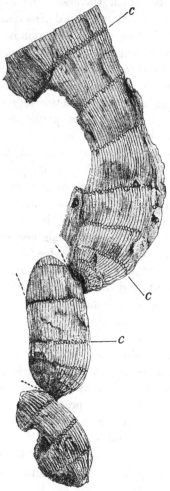

Fig. 82.   Branched rhizome of *Calamites*.   ½ nat. size.
C, C, nodes showing casts of infranodal canals.
From a specimen in the Manchester Museum, Owens College.

small projection at the upper end of each ridge is a cast of a
depression or canal which existed in the medullary tissue of
the living plant. There have been various suggestions as to
the meaning of these oval projections; several writers have
referred to them as the points of attachments of roots or other
appendages, but Williamson proved them to be the casts
of canal-like gaps which traversed the upper ends of principal medullary rays in a horizontal direction. In a tangential
section of a Calamite stem the summit of each primary
medullary ray often contains a group of smaller elements which
are in process of disorganisation, and in some cases these
cells give place to an oval and somewhat irregular canal.
In the diagrammatic tangential section represented in fig. 83, *A*
the upper end of each ray is perforated by a large oval
space, which has been formed as the result of the breaking
down of a horizontal band of cells. Williamson designated
these spaces *infranodal canals.* While proving that they had
nothing to do with the attachment of lateral members,
he suggested that they might be concerned with secretion;
but their physiological significance is still a matter of speculation. The casts of infranodal canals are especially large and
conspicuous in the subgenus *Arthrodendron,* a form of Calamite
characterised by certain histological features to be referred to
later. Williamson[1] originally regarded the presence of infranodal canals as one of the distinguishing features of *Arthrodendron,* but they occur also in the casts of the commoner
type *Arthropitys.* As a rule we have only the cast of the
inner ends of the infranodal canals preserved as slight projections like those in fig. 83, *A*; but in one exceptionally interesting pith-cast described by Williamson, these casts of the
infranodal canals have been preserved as slender spoke-like
columns radiating from the upper ends of the ridges of the
infranodal region of a pith-cast.

This specimen, which was figured by Williamson[2] in two of
his papers, and by Lyell[3] in the fifth edition of his *Elementary*

[1] Williamson (71), p. 507.
[2] Williamson (71²), Pl. I. fig. 1; (78), Pl. XXI. fig. 31.
[3] Lyell (55), p. 368.

Geology, is historically interesting as being one of the first
important plants obtained by Williamson early in the fifties,
when he began his researches into the structure of Carboniferous plants. A joiner, who was employed by Williamson
to make a piece of machinery for grinding fossils, brought a
number of sandstone fragments as an offering to his employer,
whom he found to be interested in stones. The specimens
" were in the main the merest rubbish, but amongst them,"
writes Williamson, " I detected a fragment which was equally
elegant and remarkable... In later days, when the specimen so
oddly and accidentally obtained, came to be intelligently studied,
its history became clear enough, and the priceless fragment
is now one of the most precious gems in my cabinet[1]."

*Comparison of three types of structure met with in Calamitean
stems,*—Arthropitys, Arthrodendron, *and* Calamodendron.

The anatomical features which have so far been described
as characteristic of *Calamites* represent the common type met
with in the English Coal-Measures. The same type occurs also
in France, Germany and elsewhere. It is that form of stem
known as *Arthropitys,* a sub-genus of *Calamites.*

Arthropitys may be briefly diagnosed as follows,—confining
our attention to the structure of the stem: A ring of collateral
bundles surrounds a large hollow pith, each primary xylem
strand terminates internally in a more or less bluntly rounded
apex traversed by a longitudinally carinal canal. The principal
medullary rays consist of large-celled parenchyma, of which the
individual elements are usually tangentially elongated as seen
in transverse section, and four or five times longer than broad
as seen in a tangential longitudinal section. The secondary
xylem consists of scalariform and reticulately pitted tracheids ;
the interfascicular xylem may be formed completely across each
primary ray at an early stage in the growth of the stem[2], or
it may be developed more gradually so as to leave a tapering

[1] Williamson (96), p. 194.
[2] *Vide* specimens 15—17, etc. in the Williamson Collection.

principal ray of parenchyma between each primary xylem bundle. In the latter case the principal rays present the characteristic appearance shown in figs. 71, 74, *A*, 75 and 78, a type of stem which we may refer to as *Calamites (Arthropitys) communis*. In the former case the stem presents the appearance shown in fig. 83, *D* [1]. A third variety of *Arthropitys* stem is one which was originally named by Göppert *Arthropitys bistriata*; in this form the principal rays retain their individuality as bands of parenchyma throughout the whole thickness of the wood [2]. Such stems as those of figs. 73 and 74, *B*, may be young examples of *Arthropitys communis* or possibly of *A. bistriata*. The narrow secondary medullary rays of *Arthropitys* usually consist of a single row of cells which are three to five times higher than broad, as seen in tangential longitudinal section. Infranodal canals occur in some examples of *Arthropitys*.

In the subgenus *Arthrodendron*, a type of stem first recognised by Williamson and named by him *Calamopitys* [3], the principal medullary rays consist of *prosenchymatous cells* (*i.e.* elongated pointed elements) and not parenchyma. These elongated elements are not pitted like tracheids, and they are shorter and broader than the xylem elements. In some examples of this subgenus the primary rays are bridged across at an early stage by the formation of secondary interfascicular xylem, and in others they persist as bands of ray tissue, as in *Arthropitys*. Other characteristics of *Arthrodendron* are the abundance of reticulated instead of scalariform tracheids in the secondary wood, and the large size of the infranodal canals.

Fig. 83, *D* represents part of a transverse section of *Arthrodendron*; in this stem the rays have been occupied by interfascicular xylem at a very early stage of the secondary growth. The section from which fig. 83, *D* is drawn was described by Williamson in 1871; the complete section shows about 80 carinal canals and primary xylem groups. The prosenchymatous

---

[1] The stem of fig. 83 is an example of *Arthrodendron*, but the appearance of the secondary xylem agrees with that in some forms of *Arthropitys*.

[2] For figures of this type of stem *vide* Göppert (64); Cotta (50), Pl. xv. (specimens 13787 in the British Museum Collection); Mougeot (52), Pl. v.; Stur (87), pp. 27—31; Renault (93), Pls. xliv. and xlv. etc.

[3] Williamson (71), (71²), (87), fig. 5.

form of the principal medullary rays is seen in fig 83, *C*, and
the reticulate pitting on the radial wall of a tracheid is shown

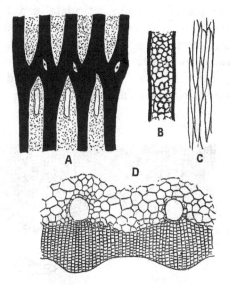

Fig. 83.   *Calamites (Arthrodendron).*
      A. Tangential section (diagrammatic) showing the course of the vas-
            cular strands, also leaf-traces and infranodal canals.
      B. Radial face of a tracheid.
      C. Prosenchymatous elements of a principal medullary ray.
      D. Transverse section of the wood.   (After Williamson.)   No. 36 in
            the Williamson Collection.

in fig. 83, *B*.   Fig. 83, *A* illustrates the large infranodal canals
as seen in a tangential section of a stem.   The same section
shows also the course of the vascular bundles characteristic of
Calamites as of *Equisetum,* and the position of outgoing leaf-
traces is represented by unshaded areas in the black vascular
strands.

    The subgenus *Arthrodendron* is very rarely met with, and
our information as to this type is far from complete[1].

    The third subgenus *Calamodendron* has not been discovered
in English rocks, and our knowledge of this type is derived from

---

[1] Williamson and Scott (94), p. 879.

French and German silicified specimens[1]. There is the same
large hollow pith surrounded by a ring of collateral bundles
with carinal canals, as in the two preceding subgenera. The
tracheids are scalariform and reticulate, and the secondary
medullary rays consist of rows of parenchymatous cells which
are longer than broad, as in *Arthropitys* and *Arthrodendron*.

The most characteristic feature of *Calamodendron* is the
occurrence of several rows of radially disposed thick-walled
prosenchymatous elements (fig. 84, *b*) on either flank of each

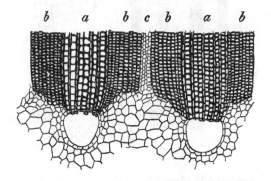

Fig. 84. *Calamites (Calamodendron) intermedium*, Ren.
Transverse section through two vascular bundles.
a, a, xylem tracheids, *b, b*, bands of prosenchyma, *c*, medullary ray. (After
Renault.)

wedge-shaped group of xylem. Each principal ray is thus
nearly filled up by bands of fibrous cells on the sides of adjacent
xylem groups, but the centre of each principal ray is occupied
by a narrow band of parenchyma (fig. 84, *c*). The relative
breadth of the xylem and prosenchymatous bands has been
made use of by Renault as a specific character in *Calamodendron* stems. Fig. 84 is copied from a drawing recently published
by this French author of a new species of *Calamodendron*,
C. intermedium[2]. In this case the bands of fibrous cells, *b*,
are slightly broader, as seen in a transverse section of the

[1] *Vide* Williamson (87[2]). In this paper Williamson compares the three
subgenera of Calamite stems. Renault and Zeiller (88), Pl. LXXV. Renault
(93), Pls. LVIII. and LIX.
[2] Renault (96), p. 125; (93), Pl. LIX. fig. 2.

stem, than the bands of xylem tracheids, *a*. The narrow band,
c, consists of four rows of the parenchymatous tissue of a medullary ray. At the inner end of each group of tracheids there is
a large carinal canal.

The question of the recognition of the pith-casts of stems
possessing the structure of any of the three subgenera of
Calamites is referred to in a later section of this chapter.

b. Leaves.

Leaves of Calamites and Calamitean foliage-shoots, including
an account of (*a*) Calamocladus (Asterophyllites) *and*
(*β*) Annularia.

Our knowledge of the structure and manner of occurrence
of Calamite leaves is very incomplete. There are numerous
foliage-shoots among the fossils of the Coal-Measures which are
no doubt Calamitean, but as they are nearly always found apart
from the main branches and stems, it is generally impossible
to do more than speak of them as probably the leaf-bearing
branches of a Calamite. The familiar fossils known as *Astero-*
phyllites, and in recent years often referred to the genus
Calamocladus, are no doubt Calamitean shoots; but they are
usually found as isolated fragments, and it is seldom that we
are able to refer them to definite forms of *Calamites*. Another
common Coal-Measure genus, *Annularia*, is also Calamitean,
and at least some of the species are no doubt leafy shoots of
Calamites. Although it is generally accepted that the fossils
referred to as *Asterophyllites* or *Calamocladus* are portions of
Calamites, and not distinct plants, it is convenient, and indeed
necessary, to retain such a term as *Calamocladus* as a means of
recording foliage-shoots, which may possess both a botanical and
a geological value.

Some of the Calamite casts, especially those referred to
the subgenus *Calamitina*, are occasionally found with leaves
attached to the nodes. In some stems the leaves are arranged
in a close verticil, and each leaf has a narrow linear form and
is traversed by a single median vein. Figures of Calamite

stems with verticils of long and narrow leaves may be found
in Lindley and Hutton[1], and in the writings of many other
authors[2]. In the specimen shown in fig. 85 the leaves are
preserved apart from the stem, but from their close association
with a Calamite cast, and from the proofs afforded by other
specimens, it is quite certain they formed part of a whorl of
leaves attached to the node of a true Calamite, and a stem

FIG. 85.　Linear leaves of a Calamite (*Calamitina*).　After Weiss,
slightly reduced.

having that particular type known as *Calamitina*[3] (figs. 99, 100).
It is probable that in some Calamites, and especially in younger
shoots, the leaves had the form of narrow sheaths split up into
linear segments. This question has already been referred to in
dealing with certain Palaeozoic fossils referred to *Equisetites*[4].

A few years ago the late Thomas Hick[5], of Manchester,
described the structure of some leaves which he believed to be
those of a Calamite. He found them attached to a slender
axis which possessed the characteristics of a young Calamite
branch. There can be little doubt that his specimens are true
Calamite leaves. The sketches of fig. 86 have been made from
the sections originally described by Hick. Fig. 86, 1 shows a
leaf in transverse section; on the outside there is a well-defined

[1] Lindley and Hutton (31), Pls. cxiv., cxc. etc. Most of the specimens
figured by these authors are in the Newcastle Natural History Museum. For
notes on the type-specimens of Lindley and Hutton, *vide* Howse (88) and
Kidston (90[2]).

[2] Weiss (88), Stur (87), *etc.*　　　　　　[3] *Vide*, p. 367.
[4] *Ante*, p. 260.　　　　　　　　　　　[5] Hick (95).

epidermal layer with a limiting cuticle.   Internal to this we
have radially elongated parenchymatous cells forming a loose or
spongy tissue, the cells being often separated by fairly large spaces

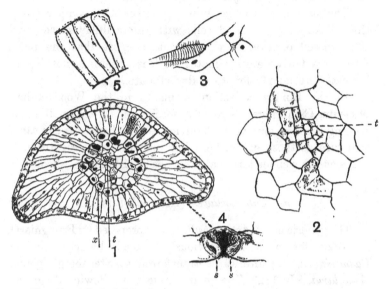

FIG. 86.   A leaf of *Calamites*.
  1.   Transverse section; *t*, vascular bundle; *x*, sheath of cells.   × 35.
  2.   Vascular bundle consisting of a few small tracheids, *t*.
  3.   A tracheid and a few parenchymatous cells, the latter with nuclei.
  4.   A stoma; *s*, *s*, guard-cells.
  5.   Pallisade cells and intercellular spaces.
    From a section in the Manchester Museum, Owens College.

(fig. 86, 5), especially in the region of the blunt lateral wings of
the leaf.   Some of these cells contain a single dark dot, which
in all probability is the mineralised nucleus.   These pallisade-
like cells probably contained chlorophyll and constituted the
assimilating tissue of the leaf.   In the centre there is a circular
strand of cells limited by a layer of larger cells with black
contents, enclosing an inner group of small-celled parenchyma
and traversed by a few spiral or scalariform tracheids constituting
the single median vein.   It is hardly possible to recognise any
phloem elements in the small vascular bundle; there appear to
be a few narrow tracheids surrounded by larger parenchymatous

elements (fig. 86, 2). At one point in the epidermis of
fig. 86, 1, there appears to be a stoma, but the details are
not very clearly shown (fig. 86, 4); the two cells, *s, s,* bordering
the small aperture are probably guard-cells.

The nature of the assimilating tissue, the comparatively
thick band of thin-walled cells with intercellular spaces, and
the exposed position of the stomata suggest that the plant
lived in a fairly damp climate; at least there is nothing to
indicate any adaptation to a dry climate.

In the Binney collection of plants in the Woodwardian
Museum, Cambridge, there is a species of a very small shoot
bearing three or four verticils of leaves which possess the same
structure as those of fig. 86. We may probably regard such
twigs as the slender terminal branches of Calamitean shoots.

a. *Calamocladus* (*Asterophyllites*).

The generic name *Asterophyllites* was proposed by Brongniart[1]
in 1822 for a fossil previously named by Schlotheim[2]
Casuarinites, and afterwards transferred to Sternberg's genus
Annularia. In 1828 Brongniart[3] gave the following diagnosis
of the fossils which he included under the genus *Asterophyllites* :—"Stems rarely simple, usually branched, with opposite
branches, which are always disposed in the same plane; leaves
flat, more or less linear, pointed, traversed by a simple median
vein, free to the base." Lindley and Hutton described examples
of Brongniart's genus as species of *Hippurites*[4], and other
authors adopted different names for specimens afterwards referred to *Asterophyllites.*

At a later date Ettingshausen[5] and other writers expressed
the view that the fossils which Brongniart regarded as a distinct
genus were the foliage-shoots of *Calamites,* and Ettingshausen
went so far as to include them in that genus. In view of the
generally expressed opinion as to the Calamitean nature of
Asterophyllites, Schimper[6] proposed the convenient generic

[1] Brongniart (22), p. 235.   [2] Schlotheim (20).
[3] Brongniart (28), p. 159.   [4] Lindley and Hutton (31), Pl. cxc.
[5] Ettingshausen (55).   [6] Schimper (69), p. 323.

name *Calamocladus* for "rami et ramuli foliosi" of *Calamites*.
Some recent authors have adopted this genus, but others prefer
to retain *Asterophyllites*. In a recent important monograph by
Grand'Eury[1] Calamitean foliage-shoots are included under the
two names, *Asterophyllites* and *Calamocladus*; the latter type
of foliage-shoots he associates with the stems of the subgenus
Calamodendron, and the former he connects with those Calamitean stems which belong to the subgenus *Arthropitys*.

It is an almost hopeless task to attempt to connect the
various forms of foliage-shoots with their respective stems, and
to determine what particular anatomical features characterised
the plants bearing these various forms of shoots. We may
adopt Schimper's generic name *Calamocladus* in the same
sense as *Asterophyllites*, but as including such other foliage-
shoots as we have reason to believe belonged to *Calamites*.
Those leaf-bearing branches which conform to the type known
as *Annularia* are however not included in *Calamocladus*, as we
cannot definitely assert that these foliage-shoots belong in all
cases to Calamitean stems. Grand'Eury's use of *Calamocladus*
in a more restricted sense is inadvisable as leading to confusion,
seeing that this name was originally defined in a more comprehensive manner as including Calamitean leaf-bearing branches
generally. We may define *Calamocladus* as follows:—

Branched or simple articulated branches bearing whorls of
uni-nerved linear leaves at the nodes; the leaves may be either
free to the base or fused basally into a cup-like sheath (*e.g.*
Grand'Eury's *Calamocladus*). The several acicular linear leaves
or segments which are given off from the nodes spread out
radially in an open manner in all directions; they may be
either almost at right angles to the axis or inclined at different
angles. Each segment is traversed by a single vein and terminates in an acuminate apex.

As a typical example of a Calamitean foliage-shoot the
species *Calamocladus equisetiformis* (Schloth.) may be briefly
described. The synonymy of the commoner species of fossil
plants is a constant source of confusion and difficulty; in order
to illustrate the necessity of careful comparison of specimens

[1] Grand'Eury (90).

and published illustrations, it may be helpful to quote a few
synonyms of the species more particularly dealt with. The
exhaustive lists drawn up by Kidston in his *Catalogue of
Palaeozoic plants in the British Museum* will be found
extremely useful by those concerned with a systematic study
of the older plants.

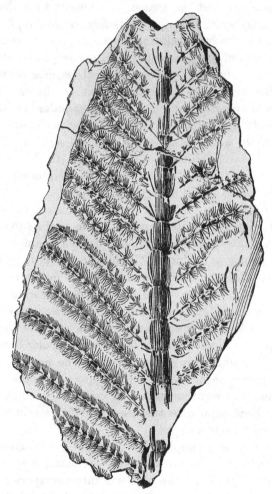

Fig. 87. *Calamocladus equisetiformis* (Schloth.).
From a specimen in the British Museum (McMurtrie Collection, no. v.
2963). *ca.* ⅓ nat. size.

Calamocladus equisetiformis (Schloth.).   Fig. 87.

1809.   *Phytolithus*, Martin[1].
1820.   *Casuarinites equisetiformis*, Schlotheim[2]
1825.   *Bornia equisetiformis*, Sternberg[3].
1828.   *Asterophyllites equisetiformis*, Brongniart[4].
1836.   *Hippurites longifolia*, Lindley and Hutton[5].
1855.   *Calamites equisetiformis*, Ettingshausen[6].
1869.   *Calamocladus equisetiformis*, Schimper[7].
1869.   *Annularia calamitoides*, Schimper[7].

The above synonyms do not exhaust the list[8], but they suffice
to illustrate the necessity of a careful comparison in drawing
up tables of species, in connection with geographical distribution
or for other purposes.

Calamocladus equisetiformis may be briefly defined as
follows :—A central axis possessing a hollow pith of Calamitean
character, divided externally into well-marked slightly constricted nodes and internodes; from the nodes long narrow
and free leaves are borne in whorls; from the axils of some of
the leaves lateral branches are given off inclined at a fairly
wide angle to the main axis, and bearing crowded verticils of
spreading acicular leaves.

The unusually good specimen, 38·5 cm. long, shown on a
much reduced scale in fig. 87, illustrates the characteristic habit
of this form of *Calamocladus*. It is from the Radstock coal-field
of Somersetshire, one of the best English localities for Coal-
Measure plants. An exceedingly good collection of Radstock
plants has recently been presented to the British Museum by
Mr J. McMurtrie; it includes many fine specimens of *Calamites*.
A small example—probably of this species—from Coalbrook
Dale, near Dudley, in Shropshire, and now in the British
Museum, illustrates very well the appearance of a young and

---

[1] Martin (09), Pl. xx. figs. 4 and 6.   [2] Schlotheim (20), p. 397.
[3] Sternberg (25), p. xxviii.   [4] Brongniart (28), p. 154.
[5] Lindley and Hutton (31), Pl. cxci.   [6] Ettingshausen (55), p. 28.
[7] Schimper (69), Pls. xxii. and xxvi. fig. 1.

[8] For other lists and synonyms, *vide* Zeiller (88), p. 368, and Kidston (86),
p. 38 and (93), p. 316, also Potonié (93), p. 162.

partially expanded Calamitean foliage-shoot. The central axis, 6·5 cm. in length, includes about 15 internodes, and terminates in a bud covered by several small leaves. Lateral branches are given off at a wide angle, and small unexpanded buds occur in the axils of several of the leaves.

As an example of the leaf-bearing branches which Grand'-Eury has recently described as *Calamocladus,* using the genus in a more restricted sense than is adopted in the present chapter, reference may be made to the fragment shown in fig. 68, *A*. The foliage-shoots of this type bore verticils of linear leaves, coherent basally in the form of a cup, at the ends of branches and not in a succession of whorls on each branch. The association of reproductive organs, in the form of long and narrow strobili, with *Calamocladus* is referred to in the sequel.

The specimens described by Grand'Eury are in the École des Mines Museum, Paris; some of the shoots which are well preserved bear a resemblance in habit of growth to the genus *Archaeocalamites.*

β. *Annularia.*

In 1820 this generic name was applied by Sternberg[1] to some specimens of branches bearing verticils of linear leaves. In 1828 Brongniart[2] thus defined the genus *Annularia*:— "Slender stem, articulated, with opposite branches arising above the leaves. Leaves verticillate, flat, frequently obtuse, traversed by a single vein, fused basally and of unequal length."

In the works of earlier writers we find frequent illustrations of specimens of *Annularia,* which are compared with Asters and other recent flowering plants. Lehmann[3] contributed a paper to the Royal Academy of Berlin in 1756, in which he referred to certain fossil plants as probable examples of flowers, among them being a specimen of *Annularia.* He refers to the occurrence of fossil ferns and other plants, and asks why we do

---

[1] Sternberg (20).   [2] Brongniart (28), p. 155.
[3] Lehmann (1756), p. 127. *Vide* also Volkmanns (1720), Pl. xv. p. 113.

not find flowers of the rose or tulip; his object being "not to acquire vain glory, but to give occasion for others to look into the matter more clearly."

The general habit of the fossils which are now included under *Annularia* agrees closely with that of *Calamocladus.* There is the same spreading form and a similar foliage in the two genera, but in *Annularia* the members of a whorl are always fused into a basal sheath, and the segments are not of equal length. We may thus summarise the characteristic features of the genus :—

Opposite branches are given off in one plane from the nodes of a main axis; the leaves are in the form of narrow sheaths divided into numerous and unequal linear or narrow lanceolate segments, each with a median vein. The segments in each whorl appear to be spread out in one plane very oblique to the axis of a branch, instead of spreading radially in all directions; the lateral segments are usually longer than the upper and lower members of a whorl. The vegetative branches possess the same type of structure as *Calamites.*

A comparison of *Annularia* and *Phyllotheca* has already been made in Chapter IX. (p. 282). Potonié[1] has recently given a detailed account of Annularian leaves; he compares them with those of *Equisetum,* and describes the occurrence on the lamina of each leaf-segment of a broad central band or midrib, with a groove, probably containing stomata, on either side. He shows that in well-preserved specimens of *Annularia,* it is possible to recognise certain minute surface-features, such as the presence of hairs and stomata, which enable one to detect a close resemblance between the leaves of Calamite stems and those of Annularian shoots.

It is not always easy to distinguish between *Annularia* and *Calamocladus* ; the collar-like basal sheath in the leaves of the former is a characteristic feature, but that cannot always be recognised. On the other hand, the leaves of *Calamocladus* may sometimes be flattened out on the surface of the rock and simulate the deeply cut sheaths of *Annularia.* It is difficult to decide how far the manner of occurrence of Annularian

[1] Potonié (93), pp. 169 *et seq.*, Pl. xxiv.

leaves in one plane, which is commonly insisted on as a generic character, is an original feature, or how far it is the result of compression in fossilisation. Probably the leaves of a living *Annularia* were spread out at right angles to the axis, as in the verticils' of such a plant as *Galium*.

Dawson[1] has described some fossils from the Devonian rocks of Canada as species of *Asterophyllites*; the figures bear a closer resemblance to the genus *Annularia*. The same author figures some irregularly whorled impressions as *Protannularia*, which appear to be identical with a fossil described by Nicholson[2] from the Skiddaw slates (Ordovician) of Cumberland as *Buthotrephis radiata*, but the specimens are too imperfect to admit of accurate determination.

Annularia stellata (Schloth.). Fig. 88.

1820.   *Casuarinites stellatus*, Schlotheim[3].
1826.   *Bornia stellata*, Sternberg[4].
1828.   *Annularia longifolia*, Brongniart[5].
1834.   *Asterophyllites equisetiformis*, Lindley and Hutton[6].
1868.   *Asterophyllites longifolius*, Binney[7].
1887.   *Annularia Geinitzi*, Stur[8].
1887.   *Annularia westphalica*, Stur.

This species was figured by Scheuchzer[9] in his *Herbarium Diluvianum*, and compared by him with a species of *Galium* (Bedstraw). Brongniart first made use of the generic name *Annularia* for this common Coal-Measure species, which may be defined as follows:—

Stem reaching a diameter of about 6—8 cm., with internodes 6—12 cm. in length, the surface either smooth or faintly ribbed. Primary branches given off in opposite pairs from the nodes, the lateral branches giving off smaller branches disposed in the same manner. The smaller branches bear verticils of leaves at each node; both leaves and ultimate branches being in one plane. The leaves are narrow, lanceolate-spathulate in form, broadest

[1] Dawson (71).

[2] Nicholson (69) Pl. xviii. *B.* Nicholson's specimens are in the Woodwardian Museum, Cambridge.   [3] Schlotheim (20), p. 397.

[4] Sternberg (26), p. xxviii.   [5] Brongniart (28), p. 156.

[6] Lindley and Hutton (31), Pl. cxxiv.   [7] Binney (68), Pl. vi. fig. 3.

[8] Stur (87), Pl. xvi b, and Pls. iv b and xiii.

[9] Scheuchzer (1723), p. 63, Pl. xiii. fig. 3.

about the middle, 1—5 cm. in length and 1—3 mm. broad, hairy
on the upper surface[1]; each leaf is traversed by a single vein.

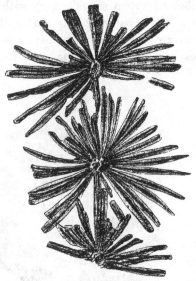

Fig. 88.   Branch of *Annularia stellata* (Schloth.).   ⅓ nat. size.
    From a specimen in the Collection of Mr R. Kidston.   Upper Coal-
    Measures, Radstock.

Each whorl contains 16—32 segments, which are connected
basally into a collar or narrow sheath; the lateral segments are
usually longer than the upper and lower.   The branches are
about 6—20 mm. broad, with finely ribbed internodes 3—7 cm.
long, bearing verticils of leaves; the ultimate branches arise in
pairs in the axils of the lateral segments of the verticils.

The strobili are of the *Calamostachys*[2] type and are borne
on the main branches or possibly on the stem; they have a long
and narrow form and are attached in verticils at the nodes.
Each strobilus consists of a central axis bearing alternate whorls
of linear lanceolate sterile bracts and sporangiophores, about
half as numerous as the sterile bracts; each sporangiophore
bears four ovoid sporangia.

The anatomical structure of a specimen referred to *Annularia stellata* has been described by Renault[3].   The cortex

[1] Potonié (93), p. 166.   [2] *Vide* pp. 351 *et seq.*
[3] Renault (96), p. 66; (93), Pl. xxviii.

consists of parenchyma traversed by lacunae and limited peripherally by a denser hypodermal tissue. In the stele Renault describes 14 xylem strands, each with a large carinal canal. The pith was apparently large and hollow. The same author describes an *Annularia* strobilus in which the lower sporangiophores bear macrosporangia, and the upper microsporangia.

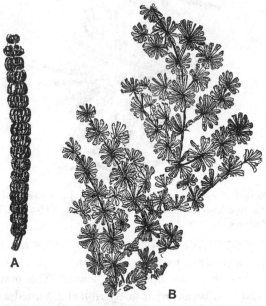

Fig. 89. *Annularia sphenophylloides* (Zenk.).

    A. Strobilus (*Stachannularia calathifera*, Weiss). ⅔ nat. size. *B*. Vegetative shoot. ¼ nat. size.

      From specimens in the Collection of Mr R. Kidston. Upper Coal-Measures, Radstock.

The references in the footnote should be consulted for figures of this species of *Annularia*; it is from the examination of such specimens as are referred to in the note that the above diagnosis has been compiled[1].

[1] One of the finest specimens of *Annularia stellata* is figured by Stur (87), Pl. xvi b; it is in the Leipzig Museum. *Vide* also Schenk (83), Pl. xxxix.; Germar (44), Pl. ix.; Renault and Zeiller (88), Pls. xlv. and xlvi. There are some well-preserved impressions of *A. stellata* in the British Museum from Radstock, Newcastle and elsewhere.

Annularia sphenophylloides (Zenk.).   Fig. 89.

1833.   *Galium sphenophylloides*, Zenker[1].
1865.   *Annularia brevifolia*, Heer[2], Strobilus.
1876.   *Calamostachys (Stachannularia) calathifera*, Weiss[3].

Principal branches 8—12 mm. wide, with internodes 8—10 cm. in length, giving off two opposite branches at the nodes; from the secondary branches arise smaller branches in opposite pairs.   The leaf-verticils and branches are all in one plane. Each verticil consists of 12—18 spathulate segments, 3—10 mm. long, cuneiform at the base and broader above, with an acuminate tip; the lateral segments are slightly longer than the upper and lower members of a whorl.

The small and crowded leaf-whorls give to this species a characteristic appearance, which readily distinguishes it from the larger-leaved forms such as *Annularia stellata*.   A fossil figured by Lhwyd[4] in 1699 as *Rubeola mineralis* is no doubt an example of *Annularia sphenophylloides*.

Annularian branches are occasionally found with cones given off from the axils of some of the leaf-whorls.   An interesting specimen, which is now in the Leipzig Museum, was described by Sterzel in 1882[5], showing cones attached to a vegetative shoot of *Annularia sphenophylloides*.   The long and narrow strobili—2·5 cm. long and about 6 mm. broad—appear very large in proportion to the size of the vegetative branches.   A fertile shoot consists of a central axis bearing whorls of bracts alternating with sporangiophores, to each of which are attached four sporangia.   The specimen in fig. 89, *A*, does not show the details clearly; each transverse constriction represents the attachment of a whorl of linear bracts; the whole cone appears to consist of a series of short broad segments.   The divisions in the lower half of each segment mark the position of the sterile bracts, while those of the upper half represent the out-

[1] Zenker (33), Pl. v. pp. 6—9.
[2] Heer (65), fig. 6, p. 9, and other authors.
[3] Weiss (76), p. 27, Pl. III. fig. 2.   [4] Lhwyd (1699), Pl. v. fig. 202.
[5] Sterzel (82).

lines of the upper sporangia of each whorl of sporangiophores, the lower sporangia being hidden by the ring of linear bracts[1]. On some portions of the specimen of fig. 89, *A*, it is possible to recognise the outlines of cells on the coaly surface-film; these probably belong to the sporangium wall. This type of cone is included under the genus *Calamostachys*, a name applied to Calamitean strobili with certain morphological characters, as described on p. 351.

c. Roots.

In 1871 Williamson[2] described some sections of what he considered to be a distinct variety of a Calamite stem. The chief peculiarity which he noticed lay in the absence of carinal canals, and in the solid pith. Some years later the same observer[3] came to the conclusion that the specimens were probably those of a plant generically distinct from *Calamites*; he accordingly proposed a new name *Astromyelon*. Subsequently Cash and Hick[4] gave an account of some examples of apparently another form of plant, to which they gave the name *Myriophylloides Williamsonis*; and Williamson[5] suggested the term *Helophyton* as a more suitable generic designation. It was, however, demonstrated by Spencer[6] that the plant described by Cash and Hick was identical with Williamson's *Astromyelon*. Williamson[7] then gave an account of several specimens of this type illustrating various stages in the growth and development of the *Astromyelon* 'stems,' which he compared with the rhizome of the recent genus *Marsilia*.

In 1885 Renault[8] published an account of *Astromyelon* in which he brought forward good evidence in favour of regarding it as a Calamitean root. The same author has recently given some excellent figures and a detailed description of certain specific types of these Calamite roots, and Williamson and Scott's memoir on the roots of *Calamites* has rendered our knowledge

[1] *Vide* Weiss (76), Pl. III. and Weiss (84), p. 178.
[2] Williamson (71), p. 487, Pls. xxv. and xxvi.
[3] *Ibid.* (78), p. 319, Pl. xix.   [4] Cash and Hick (81), p. 400.
[5] Williamson (81), *vide* also Spencer (81).   [6] Spencer (83), p. 459.
[7] Williamson (83), p. 459, Pls. xxvii.—xxx.   [8] Renault (85).

of *Astromyelon* almost complete. Some of the finest specimens, in which the organic connection between typical Calamite stems and *Astromyelon* roots is clearly demonstrated, are in the Natural History Museum, Paris. There are several sections also from English material which show the connection between root and stem very clearly.

Casts of the hollow pith of Calamite rhizomes or aerial branches are occasionally found in which slender appendages are given off either singly or in tufts from the nodal regions. Many examples of such casts have been figured by Lindley and Hutton[1], Binney[2], Grand'Eury[3], Weiss[4], Stur, and other writers[5].

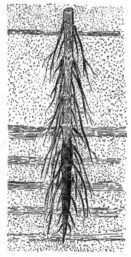

Fig. 90.   Pith-cast of a Calamite stem, with roots; embedded in sandstone and shale. (After Grand' Eury.)   Much reduced.

The large stem-cast of fig. 90 illustrates the manner of occurrence of long branched roots on the nodes of a Calamite growing in sandy or clay soil. The lower and more darkly

[1] Lindley and Hutton (31), Pls. LXXVIII. and LXXIX. (The specimens are figured in a reversed position.)

[2] Binney (68), p. 5, fig. 1.

[3] Grand'Eury (77), Pls. I. and II.; (87), Pls. XXVII., XXVIII.

[4] Weiss (84), Pls. II.—IV., VIII. and IX.

[5] Stur (87), Pls. III., VI., VII., *etc.*; Zeiller (86) Pl. LIV.

shaded portion of the specimen is covered by a layer of coal representing the carbonised wood and cortex, which has been moulded on to the sandstone pith-cast. In fig. 77 (p. 316) a fairly thick root is seen, in organic connection with one of the nodes, N 3, and on N 2 there is a scar of another root.

There are certain external characters by which one may often recognise a Calamitean root. There is no division into nodes and internodes as in stems, and as the pith of the root was usually solid the parallel ribs and grooves of stem-casts are not present. In smaller flattened roots there may sometimes be seen a central or excentric black line representing the stele, and the surface of the root presents a curious wrinkled or shagreen texture, probably due to the shrinkage of the loose lacunar cortex. The occasional excentric position of the stele is no doubt due to the displacement of the vascular cylinder as a result of the rapid decay of the cortical tissues. In the Bergakademie of Berlin there are some unusually good examples of Calamite casts bearing well-preserved root-impressions; these include the original specimens figured by Weiss[1].

No doubt some of the roots figured by various writers under the names *Pinnularia*[2] and *Hydatica*[3] belong to *Calamites*, but it is often impossible to identify detached specimens with any certainty.

The section figured diagrammatically in fig. 91 A shows the characteristic single series of large lacunae, l, in the middle cortical region. In the centre there is a wide solid pith surrounded by a ring of vascular tissue, x. The appearance of the middle cortex is very like that of the stem of a water-plant such as *Myriophyllum*, the Water Milfoil; it shows that the Calamite roots grew either in water or swampy ground. In fig. 91 B, the root characters are clearly seen; the centre of the stele is occupied by large parenchymatous cells which are rather longer than broad in longitudinal view; at the periphery there are four protoxylem groups px, alternating with four groups of phloem, ph, the latter being situated a little

---

[1] Weiss (76), (84).

[2] For references, *vide* Kidston (86), p. 58.

[3] Artis (25), Pl. v.

further from the centre of the stele. The structure is therefore
that of a typical tetrach root. In the example represented in

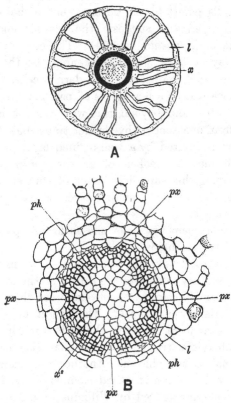

FIG. 91. *A*. Diagrammatic sketch of a transverse section of a young root of
Calamites. *x*, xylem; *l*, lacuna. After Hick.

   B. Central cylinder (stele) of root. *px*, protoxylem; *ph*, phloem; x^2,
secondary xylem; *l*, phloeoterma. × 75. After Williamson and Scott.

the figure secondary thickening has begun, and the cambial
cells internal to each phloem group have given rise to a few
radially disposed tracheids, x^2. Beyond the phloem there are
two layers of parenchyma representing, as regards position, a
pericycle and an endodermis. In the ordinary pericycle and
endodermis of the roots of most plants the cells of the two
layers are on alternate radii, but in the Calamite root, as in

Equisetum roots, the cells of these layers are placed on the
same radii, as seen in the neighbourhood of x^2 in the figure.
This correspondence of the radial walls of the endodermal and
pericyclic cells points to the development of both layers from
one mother-layer, and suggests the 'double endodermis' or
phloeoterma of *Equisetum* (p. 254). The cells in the outer of
these two layers have slight thickenings on the radial walls
recalling the usual character of endodermal cells. The phloeoterma is succeeded by a few layers of parenchyma, constituting
the inner cortex, and beyond this we have the large lacunae
separated from one another by slender trabeculae of cells. The
outer cortex is limited by a well-defined layer of thick-walled
cells, which may be spoken of as the *epidermoidal[1] layer*.
Roots possessing this superficial layer of thicker cells have no
doubt lost the original surface-layer which produced the
absorptive root-hairs.

The xylem elements have the form of spiral, reticulate
and scalariform tracheids.

In roots or rootlets smaller than that shown in fig. 91 *B*, the
primary xylem may extend to the centre of the stele, and form
a continuous axial strand; in such examples the structure may
be diarch, triarch or tetrach. The origin of the cambium
agrees with that in recent roots, the cells immediately external
to the protoxylem tracheids become meristematic, as also
those internal to the phloem. Another root-character is seen
in the endogenous origin of lateral members. Good examples of
branching roots are figured by Williamson[2] and by Williamson
and Scott[3].

Older roots[4] are usually found in a decorticated condition.
A transverse section of root in which secondary thickening has
been active for some time presents on a superficial view a close
resemblance to a stem of *Calamites*, but a careful comparison
at once reveals important points of difference. The specimen

[1] Williamson and Scott (95), p. 694.
[2] Williamson (83²), Pl. xxix. fig. 7.
[3] Williamson and Scott (95), Pls. xv.—xvii.
[4] For figures *vide* Williamson, *loc. cit.*, Williamson and Scott, and Renault
(85), (93).

diagrammatically sketched in fig. 92 illustrates very clearly the

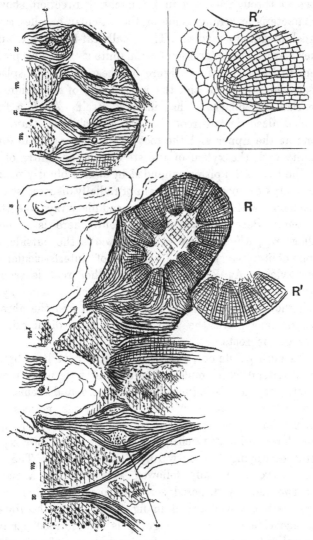

FIG. 92. Tangential section through a node of *Calamites*, showing a root in
   organic connection with the stem.

   R, R′. Root (*Astromyelon*) in transverse and oblique section. *x*, xylem;
      m, primary medullary ray; *t*, leaf-trace; *s*, Stigmarian appendage.

   R″, the inner portion of one of the xylem wedges of *R′* more highly magnified.
      Sketched from a section in the Cambridge Botanical Laboratory Collection.

origin of a root from the node of a Calamite stem. The section
has passed through a stem in a tangential direction, showing
the characteristic arrangement of the vascular bundles x, and
principal medullary rays m. The small leaf-traces, t, t, afford
another feature characteristic of a Calamite stem. The portion
of stem to the right of the figure has been slightly displaced,
and between this piece and the root R, one of the ubiquitous
Stigmarian appendages, s, has inserted itself. At R a fairly
thick and decorticated root is seen in oblique transverse
section; at the upper end the root tracheids are seen in direct
continuity with the xylem of the stem. In the centre of the
root is the large solid pith surrounded by twelve bluntly pointed
xylem groups, composed in the main of radially disposed scalariform elements with narrow secondary medullary rays like those
in a stem. Between each xylem group there is a broad
medullary ray, which tapers rapidly towards the outside, and
is soon obliterated by the formation of interfascicular secondary xylem. At R' a portion of another root is seen in
transverse section, and R'' the inner part of a single xylem
group is shown more clearly. The solid pith and the absence
of carinal canals are the two most obvious distinguishing
features of the roots.

As Renault points out, roots of *Calamites* have been figured
by some writers[1] as examples of stems, but it is usually comparatively easy to distinguish between roots and stems. On
examining the xylem groups more closely, one notices that
the apex of each is occupied by a triangular group of centripetally-developed primary tracheids, the narrow spiral protoxylem
elements occupying the outwardly directed apex. The protoxylem apex is usually followed externally by a ray of
one or two radially disposed series of parenchymatous cells.
This ray is not distinguished in fig. 92 R'' from the rows of
xylem tracheids. Each xylem group is thus formed partly of
centripetal xylem and in part of secondary centrifugal xylem;
the latter is associated with secondary medullary rays, as in
stems, and contains a broader ray (*fascicular ray* of Williamson

[1] *E.g.* Schenk (90) in Zittel's *Handbuch*, p. 237.

and Scott[1]) immediately opposite each protoxylem strand. In the roots of recent plants (*e.g. Cucurbita, Phaseolus,* &c.) a broad medullary ray is often found opposite the protoxylem, and such an arrangement is a perfectly normal structure in roots[2].

Renault has recently described several species of Calamite roots which he designates by specific names, some of them belonging to stems with the *Arthropitys* structure, and others to *Calamodendron.* Some of the roots figured by the French author have an axial strand of xylem with 7—15 projecting angles of protoxylem[3]. These he considers true roots, but the larger specimens with a wide pith he prefers to regard as stolons. In the latter he mentions the union of the primary centripetal with the secondary centrifugal wood as a distinguishing feature. It has been shown, however, that each group of secondary xylem includes a median ray of parenchyma, and that the whole structure is essentially that of a root, and not that of a modified stem or stolon. The organs described by Renault as true roots are probably rootlets, and as Williamson and Scott have demonstrated, there is every gradation between the smaller specimens with a solid xylem axis and those with a large central pith.

It is interesting to note that Renault's figures of *Calamodendron* roots show the closest resemblance to those of the subgenus *Arthropitys.*

d. Cones.

The occurrence of fossil plants in the form of isolated fragments is a constant source of difficulty, and is well illustrated by the numerous examples of strobili which cannot be connected with their parent stems. We are, however, usually able to recognise Calamitean cones if the impressions or petrified specimens are fairly well preserved, but it is seldom possible to correlate particular types of cones with the corresponding species of foliage-shoots or stems. Palaeobotanical

---

[1] Williamson and Scott, *loc. cit.* p. 689.

[2] de Bary (84), p. 474; van Tieghem (91), p. 720.

[3] Renault (96), pp. 118, 126; (93), Pl. LV.

literature contains numerous illustrations and descriptions of long and narrow strobili designated by different generic terms such as *Volkmannia, Brukmannia, Calamostachys, Macrostachya* and others; many of these have since been recognised as the cones of *Calamites,* while some species of *Volkmannia* have been identified with *Sphenophyllum* stems. Before further considering the general question of Calamite cones, a few examples may be described in detail as types of fructification which are known to have been borne by *Calamites.* The examples selected are species of the two provisional genera *Calamostachys* and *Palaeostachya.*

The usual form of a Calamite cone is illustrated in fig. 93, which represents a fertile shoot bearing a few narrow linear leaves of the *Calamocladus* type; in the axils of some of these are borne the long strobili.

FIG. 93. *Calamostachys* sp. A fertile Calamitean shoot. From a specimen in the Geological Survey Museum, Jermyn Street, London. From the Upper Coal-Measures of Monmouthshire (No. 5539).

Calamostachys Binneyana (Carr.). Figs. 94 and 95.

In 1867 Carruthers[1] gave an account of the structural features of the species of cones named by him *Volkmannia Ludwigi* and *V. Binneyi*, the generic term having been originally used by Sternberg[2] for some impressions of Carboniferous strobili. Brongniart[3] in 1849 referred to the various forms of *Volkmannia* as cones of Asterophyllitean branches, and the latter he regarded as the foliage-shoots of a Calamite stem. In 1868 Binney[4] published a description, with several illustrations, of the cones named by Carruthers *Volkmannia Binneyi*, and referred to them as the fructification of that type of Calamite stem spoken of in a previous section of this chapter (p. 311) as *Calamites (Arthropitys) communis* (Binney). This cone is now usually spoken of as *Calamostachys Binneyana*; the specific name *Binneyana* being suggested by Schimper[5] in 1869 as more euphonious than that proposed by Carruthers. In recent years our knowledge of both *C. Binneyana* and *C. Ludwigi* has been considerably extended. We shall confine our attention in the following account to the former species[6]. Some excellent figures of the latter species may be found in Weiss' Memoir[7] on Calamarieae.

One of the largest examples of *Calamostachys Binneyana* so far recorded has a length of 3—4 cm. and a maximum diameter of about 7·5 mm. The axis of the cone bears whorls of sterile leaves or bracts at equal distances; the linear bracts of each whorl are coherent basally as a disc or plate of tissue attached at right angles to the central axis of the cone. The periphery of each of these discs divides up into twelve linear segments, which curve upwards in a direction more or less parallel to the strobilus axis, and at right angles to the

[1] Carruthers (67), Pl. LXX.
[2] Sternberg (25), Pl. XLVIII. and LI.
[3] Brongniart (49), p. 51.
[4] Binney (68), p. 23, Pls. IV. and V.
[5] Schimper (69), p. 330.
[6] For figures and descriptions of this type of cone *vide* Williamson (73), (80), (89); Hick (93), (94) and Williamson and Scott (94).
[7] Weiss (84), Pls. XXII.—XXIV.

coherent portion of each whorl. The manner of occurrence of the whorls is shown in fig. 94, which has been sketched from a large section in the Williamson collection. The segments of the successive sterile verticils alternate with one another, so that in the surface-view of a cone the long and narrow free bracts appear spirally disposed. Midway between

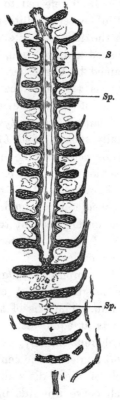

FIG. 94. *Calamostachys Binneyana* (Carr.) in longitudinal (radial and tangential) section.
Sp, sporangiophores; *S*, sporangia.
    (From specimen no. 1022 in the Williamson Collection, British Museum.)

these alternating sterile verticils there is a series of fertile appendages, also given off in regular whorls. Each fertile whorl consists of about half as many members as the segments

of a sterile whorl, and the members of the several fertile whorls
are superposed and not alternate. Each member has the form
of a stalk or sporangiophore given off at right angles from the
cone axis; this is expanded distally into a peltate disc bearing
four sporangia attached to its inner face. In fig. 94 we can only
see the basal portions of the sporangiophores, which are shown
in the upper part of the sketch as pointed projections, *Sp*, from
the cone axis. Each sporangiophore is traversed by a vascular
strand which sends off a branch to the base of a sporangium
(fig. 95, A, *t*).

The axis of the cone is occupied by a single stele, usually
triangular in section; the stele consists of a solid pith of
elongated cells surrounded by six vascular bundles, two at
each corner. A somewhat irregular gap marks the position
of the protoxylem of each strand, and portions of spiral or
annular tracheids may occasionally be seen in the cavity.
These cavities, which may be spoken of as the carinal canals,
disappear at the nodes, where there is a mass of short reticulately pitted tracheids, as in a Calamite stem. Vascular bundles
pass upwards in an oblique direction from the central stele
to supply the bracts, each of which is traversed by a single
strand of tracheids. The coherent portion, or disc, of each
sterile whorl consists of sclerenchymatous elements towards
the upper surface, and of parenchyma below. The pedicel of
the sporangiophores consists of fairly thick-walled cells traversed
by a single vascular strand, and the peltate distal portions are
made up of parenchymatous cells arranged in a palisade-like
form at right angles to the free surface of the sporangiophores.
The vascular strand of the pedicel forks into two halves just
below the peltate head, and these branches again bifurcate to
send a branch to each sporangium. The four sporangia of
each sporangiophore are attached by a narrow band of tissue
to the shield-shaped distal expansion (fig. 95, A).

In a tangential section of a cone, such as the lower portion
of fig. 94 and in fig. 95, B, the sporangiophores present the
appearance of narrow stalks (fig. 95, B, *a*) in the middle
of a cluster of sporangia, and the latter appear more or less
square in outline. The wall of a sporangium is made of a

single layer of cells (fig. 95, B) which present a characteristic
appearance in surface-view (fig. 95, C), the thin walls being

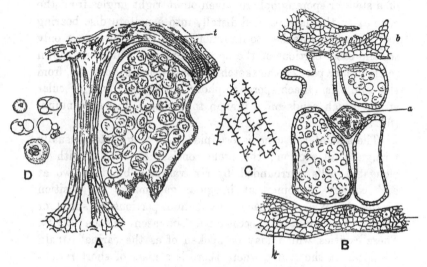

FIG. 95.   *Calamostachys Binneyana* (Carr.).

A. A sporangiophore and one sporangium. *t*, vascular bundle. × 45.

B. Tangential section showing portions of two sterile discs, *b*, *b*; a sporangiophore, *a*, with its four sporangia, in two of which are seen the spores;
t, vascular bundle. × 35.

C. Surface-view of cells of a sporangium wall. × 130.

D. Spores and remains of mother-cells. × 130.

(After Williamson and Scott.)

crossed at right angles by small vertical plates. In the
tangential section of the coherent sterile whorls (fig. 95, B,
b and *b*) the vascular strands are occasionally seen in
transverse section (fig. 95, B, *t*), as they pass outwards to the
several free bracts.

The spores in *Calamostachys Binneyana* are all of the same
size, and no macrospores have ever been seen. In well preserved specimens tetrads of spores may be seen, still enclosed by
the wall of the spore-mother-cell (fig. 95, A and D); and the torn
remnants of the mother-cell sometimes simulate in appearance
the elaters of an *Equisetum* spore. In surface-view a spore
often shows clearly the three-rayed marking, which is a characteristic feature of daughter-cells formed in a tetrad from a

mother-cell. The spores of a tetrad are in some cases of unequal size, some having developed more vigorously than others. This unequal growth and nourishment of spores is clearly shown in fig. 96, which represents a sporangium of a heteroporous Calamitean strobilus, *C. Casheana*. Williamson and Scott[1] have described striking examples of spores in different stages of abortion, and these authors draw attention to the importance of the phenomenon from the point of view of the origin of a heterosporous form of cone. The abortion of some of the members of a spore-tetrad and the consequent increased nutrition of the more favoured daughter-cells, might well be the starting-point of a process, which would ultimately lead to the production of well defined macrospores and microspores. The young microsporangia and macrosporangia of recent Vascular Cryptogams such as *Selaginella, Salvinia* and other heterosporous genera are identical in appearance[2]; it is not until the spore-producing tissue begins to differentiate into groups of spores, that the sporangia assume the form of macrosporangia and microsporangia. During the evolution of the various known types of pteridophytic plants heterospory gradually succeeded isospory, and this no doubt occurred several times and in different phyla of the plant kingdom. In the mature sporangia of some of the Calamitean strobili we have in the inequality of the spores in one sporangium an indication of the steps by which heterospory arose; and in the immature sporangia of some recent genera we are carried back to a stage still nearer the starting-point of the substitution of the heterosporous for the isosporous condition.

Calamostachys Casheana Will. Fig. 96.

To Williamson[3] again is largely due the information we possess as to the structure of this type of Calamitean strobilus. Its special interest lies in the occurrence of macrospores and microspores in the same cone.

---

[1] Williamson and Scott (94), p. 911, Pls. LXXXI. and LXXXII.

[2] *Vide* Heinricher (82); Bower (94), p. 495; Campbell (95), pp. 396, 503.

[3] Williamson (81), Pl. LIV.

The strobilus axis agrees in structure with that of *C. Binneyana*, but in *C. Casheana* a band of secondary xylem forms the peripheral portion of the triangular stele. Were any further proof needed of the now well-established fact that secondary growth in thickness is by no means unknown as an attribute of Vascular Cryptogams, the co-existence in the same cone of a cambium layer producing secondary wood and bark, and cryptogamic macrospores and microspores, affords conclusive evidence[1]. The dogma accepted by many writers for a considerable number of years that the power of secondary thickening is evidence against a cryptogamic affinity, has been responsible for no little confusion in palaeobotanical nomenclature.

On the axis of *Calamostachys Casheana* there are borne alternate whorls of fertile and sterile appendages similar to those in the homosporous *C. Binneyana*, but they are inclined more obliquely to the axis of the cone. Macrospores and microspores have been found in sporangia borne on the same sporangiophore.

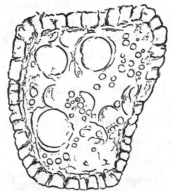

Fig. 96. *Calamostachys Casheana* Will.
A sporangium with macrospores and abortive spores.   × 65.
(After Williamson and Scott.)

The spore-tetrads in the macrosporangia occasionally include

---

[1] An excellent figure illustrating the co-existence of heterospory and secondary thickening is given by Williamson and Scott, *loc. cit.*, Pl. LXXXII. fig. 36.

aborted sister-cells like those noticed in *C. Binneyana*; this phenomenon is well illustrated by the unequally nourished spores in the sporangium of fig. 96, but no such starved spores have been found in the microsporangia. In this cone, then, heterospory has become firmly established, but the occurrence of undersized spores in a macrospore-tetrad leads us back to the probable lines of development of heterospory, which are seen in *C. Binneyana* at their starting-point.

In the two species of strobili which have been described, *Calamostachys Binneyana* and *C. Casheana*, the sporangiophores or sporophylls are given off at right angles to the axis, and midway between the sterile whorls. These are two of the most important distinguishing features of the Calamitean cones included under the generic term *Calamostachys*. In another form of cone, which also belongs to Calamitean stems, the sporangiophores arise in the axil of the sterile leaves, and are inclined obliquely to the axis of the cone. To this type the generic name *Palaeostachya* has been applied by the late Prof. Weiss[1] of Berlin. The portion of a cone shown in fig. 97 shows the arrangement of the sterile and fertile appendages characteristic of *Palaeostachya*.

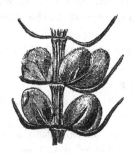

Fig. 97. *Palaeostachya pedunculata* Will. Part of a cone. × 3. (After Weiss.)

It is practically impossible to distinguish between cones of the *Calamostachys* and *Palaeostachya* type in the case of imperfectly preserved impressions; indeed we cannot assume that all long and narrow cones with spirally disposed verticillate bracts are Calamitean. We must have the additional

[1] Weiss (76), p. 103.

evidence of internal structure or of the direct association of
the cones with Calamitean foliage.

Palaeostachya vera sp. nov.   Fig. 98.

In 1869 Williamson[1] described a fragment of a strobilus
which showed certain anatomical features indicative of a close
relationship or even identity with *Calamites*. Some years later[2]
a much more perfect example was obtained from the Coal-
Measures of Lancashire, and the additional evidence which it
afforded definitely confirmed the earlier views of Williamson.
The cone was more fully described by Williamson in 1888,
as "the true fruit of *Calamites*." It is clearly a form of
Weiss' genus *Palaeostachya*; Williamson and Scott[3] refer to
it in their Memoir as *Calamites pedunculatus*. It is preferable,
however, to retain the generic designation *Palaeostachya* for
cones of this type. As the name *P. pedunculata* has previously been adopted by Weiss[4] for a cone figured by Williamson[5] in 1874, and afterwards referred to by that author in
writing as *P. pedunculata*, it is proposed to substitute the
specific name *vera*; this specific name being chosen with a
view to put on record the fact that it was this type of cone
that Williamson first proved to be the *true* fructification
of the Calamite.

The axis of *P. vera* is practically identical in structure with
a Calamitean twig. There is a hollow pith in the centre of the
stele surrounded by a ring of 16—20 collateral bundles, each
of which is accompanied by a carinal canal as in a vegetative
shoot. As the pedicel of the strobilus passes into the cone
proper it undergoes some modification in structure, but retains
the characteristic features of a Calamite. The diagrammatic
longitudinal section of fig. 98, which is copied from a drawing
by Williamson[6], shows the broadening of the vascular strands
at the nodes, and here and there a carinal canal is seen internal
to the wood.

[1] Williamson (71[2]).   [2] *Ibid.* (88[2]).
[3] Williamson and Scott (94), p. 900.   [4] Weiss (84), Pl. xxi. fig. 4.
[5] Williamson (74), Pl. v. fig. 32.   [6] Williamson (88[2]), Pl. ix. fig. 20.

The axis of the cone bears whorls of bracts at right angles
to the central column. Each whorl consists of about 30—40

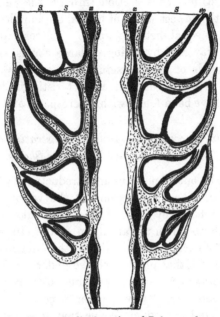

Fig. 98.   Diagrammatic longitudinal section of *Palaeostachya vera*, sp. nov.
S, S, S, sporangia; x, xylem; sp, sporangiophore. (After Williamson.)

segments coherent basally into a disc of prosenchymatous and
parenchymatous tissue. The free linear bracts curve sharply
upwards from the periphery of the disc, approximately parallel
to the axis of the cone. From each of these sterile whorls
there are given off 16—20 long and slender obliquely-inclined
sporangiophores, *sp*, which arise from the upper surface of the
disc close to the axis. Each sporangiophore no doubt bore four
sporangia, S, containing spores of one size,—about ·075 mm. in
diameter. The specimens of *Palaeostachya vera* so far obtained
do not show the actual manner of attachment of the sporangia,
but more complete examples of other species of *Palaeostachya*[1]
enable us to assume with certainty that the sporangiophores
terminated in a distal peltate expansion bearing four sporangia
on its inner face.

[1] *E.g.* Renault (82), Pl. xix. fig. 1; (96), Pl. xxix. figs. 1 and 4.

A transverse section of the axis of the cone in the region of
the sterile and fertile appendages shows the vascular bundles
arranged in pairs. In a section through the peduncle of the
cone, below the lowest whorl of bracts, the bundles of the stele
are situated at equal distances apart. The cortical tissue of the
peduncle is traversed by a ring of large canals[1] similar to the
vallecular canals of an Equisetum stem.

Isospory is not a constant characteristic of *Palaeostachya*;
some forms have been found with macrospores and microspores[2].

Other Calamitean cones, and examples illustrating the connection between Cones and Vegetative Shoots.

It would be out of place in an introduction to Palaeobotany
to attempt an exhaustive account of the various cones which
were probably borne by Calamitean plants, but there are a few
general points to which the attention of the student should be
directed. The examples dealt with in the foregoing description
illustrate the fact, that plants included under the comprehensive
genus *Calamites* bore cones possessing distinct morphological
features. There are, however, other types of strobili which
have been found in organic connection with *Calamites*; and
some of these must be taken into account in dealing with
Calamarian plants. The genera *Volkmannia, Brukmannia, Huttonia, Macrostachya,* in addition to *Calamostachys* and *Palaeostachya* and others, have been applied by different writers to
Calamitean cones. As Solms-Laubach[3] has suggested, it is
wiser to discard *Volkmannia* and *Brukmannia,* as they have
been made to do duty for cones of widely different forms.
It is better to adhere to the provisional generic names used by
Weiss, as they enable us to conveniently systematise the various
Calamarian strobili.

The following classification may be given of the better
known cones, some of which we are able to describe in
considerable detail, while others are still very imperfectly
known. We have good evidence that all these strobili were

---

[1] Williamson (88[2]), Pl. VIII. figs. 1 and 4.
[2] Renault (93), Pl. XXIX. fig. 7.   [3] Solms-Laubach (91), p. 325.

borne by vegetative shoots of the type of *Calamites, Calamocladus* or *Annularia*.

1. *Calamostachys*[1] (including *Paracalamostachys* and *Stachannularia*).

Cones long and narrow, consisting of a central axis bearing alternate whorls of sterile and fertile appendages, the latter having the form of sporangiophores attached at right angles to the axis midway between the sterile verticils, and bearing four sporangia on the inner face of a peltate distal expansion.

Calamostachys Binneyana Schimp., *C. Ludwigi* Carr., *C. Casheana* Will., may be referred to as examples of this type of cone; also some of the strobili described by different authors as species of *Volkmannia*[2], *Brukmannia*[3], &c.

Although one cannot make out the detailed structure of a Calamite cone in the absence of internal structure, it is often possible to recognise the essential features in specimens preserved in ironstone nodules, such as those from Coalbrook Dale in Shropshire, or by carefully examining the carbonised impressions on shale under a simple microscope.

Weiss applies the term *Paracalamostachys*[4] to cones of the *Calamostachys* form, but in which the manner of attachment cannot be made out. Such a cone as that of fig. 93 should probably be referred to this sub-type of *Calamostachys* in the absence of definite evidence as to the position of the sporangia.

Another term *Stachannularia*, originally used by Weiss as a genus[5], was afterwards[6] applied to cones of the same general type as *Calamostachys*, in which the sporangiophores have the form of thorn-like structures bearing on their upper side a lamellar expansion. There is however some doubt as to the correct interpretation of the features associated with cones included in *Stachannularia*, for an account of such forms

---

[1] Weiss (84), p. 161. Solms-Laubach, *loc. cit.* p. 326.

[2] *E.g. Volkmannia Ludwigi* Carr., also *Volkmannia elongata* Presl. [Solms (91), p. 332 and Weiss (76), p. 108].

[3] *E.g. Brukmannia Grand'Euryi* Ren. [Renault (76)].

[4] Weiss (84), p. 190.   [5] Weiss (76), p. 1.

[6] Weiss (84), p. 161.

reference must be made to the writings of Weiss, Renault[1], Solms-Laubach[2] and others[3].

Calamostachys cones have been found in organic union with branches bearing leaves of the *Annularia* type, also with *Calamocladus* foliage, and the branches bearing such cones have been found in actual connection with Calamitean stems. The association of cones and vegetative stems and branches is shown in tabular form on p. 363.

2. *Palaeostachya*[4].

In this genus the general habit agrees with that of *Calamostachys*, and in imperfectly preserved specimens it may be impossible to discriminate between *Calamostachys* and *Palaeostachya*. The latter form is characterised by the attachment of the sporangiophores in the axil of the sterile bracts, or immediately above them, as shown in figs. 97 and 98.

EXAMPLES. *Palaeostachya vera* sp. nov., *P. pedunculata* Will. afford examples of this form of strobilus. The genus *Palaeostachya* includes several species previously described under the genus *Volkmannia*[5].

Strobili of this generic type are known in organic association with Annularian branches, as well as with *Calamocladus* and *Calamites*.

3. *Macrostachya*.

This generic name was originally applied by Schimper[6] to certain forms of Calamitean stems, of the type afterwards referred to the sub-genus *Calamitina* by Weiss, bearing long and thick cones. The name is, however, more appropriately restricted to strobili, which differ from the two preceding genera in their greater length (14—16 cm.) and in the more crowded and imbricating whorls of bracts. The internodes of the cones are very short, and each whorl of bracts consists of about 20 coherent members separated at the periphery of the disc

---

[1] Renault (82), p. 139; (76). [2] Solms-Laubach (91), p. 330.
[3] Schenk (88), p. 132; (83), p. 232. [4] Weiss (84), p. 161.
[5] *E.g. Volkmannia gracilis* Sternb. [Renault (76), Pl. II.].
[6] Schimper (69), p. 332. *Vide* also Renault and Zeiller (88), p. 420.

into short pointed teeth. The internal structure of *Macrostachya*
has not been satisfactorily determined. An account by Renault[1]
of a petrified specimen does not present a very clear idea as to
the structural features of this form of Calamitean strobilus.

THE ASSOCIATION OF CALAMITEAN VEGETATIVE SHOOTS AND CONES.

Strobilus	Foliage-shoot	Stem
Calamostachys (*Stachannularia*) *ramosa* Weiss[2]	*Annularia ramosa* Weiss	*Calamites ramosus* Artis
C. (*Stachannularia*) *calathifera* Weiss[3]	*A. sphenophylloides* Zenk.	Stem bearing verticils of long and narrow leaves[4]. Probably a young *Calamites*
C. (*Stachannularia*) *tuberculata* (Stern.)	*A. stellata* (Schloth.)[5] (*A. longifolia* Brongn.)	*Calamites* sp.[6]
C. Solmsi[7] Weiss	*Calamocladus* sp.	*Calamites* (*Calamitina*) sp.
C. longifolia (Stern.)[8]	*Calamocladus* sp.	
Palaeostachya pedunculata Will.[9]	*Calamocladus*	
P. arborescens (Stern.)[10]		*Calamites* (*Stylocalamites*) *arborescens* (Stern.)
Macrostachya[11]	*Calamocladus equisetiformis* (Schloth.)	*Calamites* (*Calamitina*) sp.

The generic name *Huttonia*, suggested by Sternberg[12] in
1837, is applied to cones which closely resemble *Macrostachya*
in habit, but differ—so far as our scanty knowledge enables us
to judge—in the arrangement of the members. The student

[1] Renault (82), p. 120, Pl. XIX.; (93), Pl. XXIX. figs. 8—14; (96), p. 77.
[2] Weiss (84), p. 98, Pl. XX. etc.   [3] Sterzel (82).
[4] Renault and Zeiller (88), Pl. XLVI. fig. 7.
[5] Kidston (86), p. 47; (93), p. 319. *Vide* also Renault (93), Pl. XXVIII. Renault and Zeiller (88), Pl. XLV.
[7] Solms-Laubach (91), p. 339. Weiss (84), p. 159.
[8] Weiss (84), Pl. XX. fig. 6.   [9] Weiss (84), Pl. XX. fig. 7; Pl. XXI. fig. 4.
[10] *Ibid.* Pls. XIV. and XV. *Cf.* also Stur (87), Pls. VI. and VII b, and Lesquereux (84), Pl. XC. fig. 1.
[11] Grand'Eury (90), pp. 205, 208. Renault and Zeiller (88), Pl. LI.
[12] *Vide* Unger (50), p. 63.

must refer to Weiss[1], Solms-Laubach[2] and other writers[3] for a further account of these types, and of another rare and little-known form of cone, called by Weiss *Cingularia*[4].

Macrostachyan cones have been found attached to stems of *Calamites* which are included in the sub-genus *Calamitina* (p. 367). The larger size of *Macrostachya* as a distinguishing feature is not always a safe test; some cones which belong to *Palaeostachya* [*e.g. P. arborescens* Sternb.] and *Calamostachys* (*e.g. C. Solmsi*) are much thicker and larger than the majority of species of these two genera.

It would appear from the examples selected to illustrate the connection between strobili and vegetative shoots, that the *Annularia* type of branch usually bears cones which conform to the genus *Calamostachys* (*Stachannularia*); while the Asterophyllitean branches—*Calamocladus*—are associated with *Palaeostachya* and *Macrostachya*. But this rule is not constant, and we are not in a position to speak of cones of a particular type as necessarily characteristic of definite types of Calamitean shoots.

Although it is admitted by the great majority of Palaeobotanists that the Calamites were all true Vascular Cryptogams, the older view that some members of the Calamarieae are gymnospermous has not been given up by Renault[5]. This observer has recently described some seeds which he believes were borne by Calamitean stems; he admits, however, that no undoubted female cones of *Calamodendron* have so far been found. In view of the unsatisfactory evidence on which Renault's opinion is based, we need not further discuss the questions which he raises.

[The following specimens in the Williamson Cabinet in the British Museum, may be found useful in illustration of the structure of *Calamites*.

Stems. (i. *Arthropitys.*) *Young twigs and small branches* 1, 2, 6, 10, 14, 19, 116*, 1002, 1007, 1020.

[1] Weiss, *loc. cit.*
[2] Solms-Laubach (91), p. 328.   [3] Schenk (83), p. 234.
[4] Weiss (76), p. 88; (84), p. 162. Solms-Laubach, *loc. cit.* p. 334, fig. 47.
[5] Renault (96), p. 132.

III. Pith-casts of Calamites.

A. *Calamitina.*   B. *Stylocalamites.*   C. *Eucalamites.*

Palaeobotanical literature contains a large number of species
of *Calamites* founded on pith-casts alone. Many of these so-
called species are of little or no value botanically, but while we
may admit the futility of attempting to recognise specific types
in the same sense as in the determination of recent plants, it is
necessary to pay attention to such characters as are likely to
prove of value for descriptive and comparative purposes. From
the nature of the specimens it is clear that many of the
differences may be such as are likely to be met with in
different branches of the same species, while in others the
pith-casts of distinct species or genera may be almost identical.

The most striking differences observable in Calamite casts
are in the character of the internodes, the infranodal canals,
the number and disposition of branch-scars, and other surface
features. Occasionally it is possible to recognise certain anatomical characters in the coaly layer which often surrounds a
shale- or sandstone-cast, and the surface of a well preserved cast
may give a clue to the nature of the wood in the faint outlines
of cells which can sometimes be detected on the cast itself[1].
The breadth of the carbonaceous envelope on a cast has been
frequently relied on by some writers as an important character.
It has been suggested[2] that we may arrive at the original
thickness of the wood of a stem by measuring the coaly layer

---

[1] Zeiller (88), Pl. LIV. fig. 4.   [2] Stur (87), p. 17.

and multiplying the breadth by 27; the explanation being that a zone of wood 27 mm. in thickness is reduced in the process of carbonisation to a layer 1 mm. thick.

The breadth of the coal on the same form of cast may vary considerably; on this account, and for various other reasons, such a character can have but little value. Our knowledge of anatomy may often help us to interpret certain features of internal casts and to appreciate apparently unimportant details. One occasionally notices that a Calamite pith-cast has large infranodal canals, and in some specimens each internodal ridge may be traversed by a narrow median line or small groove; large infranodal canal casts suggest the type of stem referred to the subgenus *Arthrodendron*, and the median line on the ridges may be due to bands of hard tissue in each principal medullary ray.

In attempting to identify pith-casts the student must keep in view the probable differences presented by the branching rhizome, the main aerial branches and the finer shoots of the same individual. The long internodal ridges of some casts may be mistaken for the parallel veins of such a leaf as *Cordaites*, a Palaeozoic Gymnosperm, if there are no nodes visible on the specimen. The fossil figured by Lindley and Hutton[1] as *Poacites*, and regarded by them as a Monocotyledon, is no doubt a portion of a Calamite with very long internodes. An interesting example of incorrect determination has recently been pointed out by Nathorst[2] in the case of certain casts from Bear Island, originally described by Heer as examples of *Calamites*; the vertical rows of leaf-trace casts on a *Knorria* were mistaken for the ribs of a Calamite stem. The specimens in the Stockholm Museum fully bear out Nathorst's interpretation. The undulating course of internodal ridges and grooves is not in itself a character of specific value. If a Calamite stem were bent slightly, the wood and medullary-ray tissues on the concave side might adapt themselves to the shortening of the stem by becoming more or less folded, and a

[1] Lindley and Hutton (31), Pl. cxlii B. The original specimen is in the University College Collection, London.

[2] Nathorst (94), p. 56, Pl. xv. figs. 1 and 2.

cast of such a stem would show undulating ridges and grooves on one side and straight ones on the other[1].

A convenient classification of Calamite casts was proposed by Weiss in 1884, founded chiefly on the number and manner of occurrence of branch-scars—or rather branch-depressions—on the surface of pith-casts. Weiss[2] recognised the imperfection of his proposed grouping, and Zeiller[3] has also expressed reasonable doubts as to the scientific value of such group-characters. Weiss instituted three subgenera—*Calamitina*, *Eucalamites* and *Stylocalamites*, which are made use of as convenient terms in descriptive treatment of Calamite casts. The following account of a few of the more typical casts may serve to illustrate the methods employed in the description of such specimens; the synonymy given for the different species is not intended to be complete, but it is added with a view to drawing attention to the necessity for careful comparison in systematic work.

A. *Calamitina.*

This sub-genus of *Calamites*, as instituted by Weiss[4], includes Calamitean stems or branches, which are characterised by the periodic occurrence of branch-whorls usually represented by fairly large oval or circular scars just above a nodal line (figs. 99, 100 and 101). The branch-scars may form a row of contiguous discs, or a whorl may consist of a smaller number of branches which are not in contact basally. A form described by Weiss as *C. pauciramis*, Weiss[5], has only one branch in each whorl, as represented by a single large oval scar on some of the nodes of the cast. A stem of this form is by no means a typical *Calamitina*, but it serves to show the existence of forms connecting Weiss' sub-genera *Calamitina* and *Eucalamites*. The number of internodes and nodes between the branch whorls varies in different specimens, and is indeed not constant in the same plant. Each nodal line bears numerous elliptical scars which mark the points of attachment of leaves; each

---

[1] Seward (88).   [2] Weiss (84), p. 54.   [3] Zeiller (88), p. 329.
[4] Weiss (76), p. 117; (84), p. 55.   [5] Weiss (84), p. 93, Pl. xi. fig. 1.

branch-whorl is situated immediately above a node, and in some forms this nodal line pursues a somewhat irregular course across the stem, following the outlines of the several branch-scars[1]. The surface of the internodes is either perfectly smooth or it is more frequently traversed by short longitudinal ridges or grooves probably representing fissures in the bark of the living stem; these are indicated by lines in fig. 99 and by elongated elliptical ridges in fig. 101. On young stems the leaves are occasionally found in place, as for example in an example figured by Weiss[2] (*C. Göpperti*), or we may have leaf-verticils still in place in much older and thicker branches[3] (cf. fig. 85, p. 330).

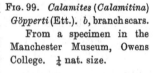

FIG. 99. *Calamites (Calamitina) Göpperti* (Ett.). *b*, branch scars. From a specimen in the Manchester Museum, Owens College. ¼ nat. size.

It occasionally happens that the bark of *Calamitina* stems has been preserved as a detached shell[4] reminding one of the sheets of Birch bark often met with in forests, the separation being no doubt due in the fossil as in the recent trees to the manner of occurrence of the cork-cambium.

In a few cases branches have been preserved still attached to a stem or branch of higher order; examples of such specimens are figured by Lindley and Hutton[5], Stur[6], and others. Grand'Eury[7] has given an idealised drawing of a typical *Calamitina* bearing a whorl of branches with the foliage and habit of *Asterophyllites equisetiformis*. The specimen on which this drawing is based

[1] *Vide* Weiss (84), Pl. xxv. fig. 2; Pl. xvi *a*, etc.
[2] Weiss (76), Pl. xvii. fig. 1.   [3] Weiss (84), Pl. i.
[4] Grand'Eury (90), p. 208, and (77), Pl. v.
[5] Lindley and Hutton (31), Pl. cxc.   [6] Stur (87), Pl. v. fig. 1.
[7] Grand'Eury (77), Pl. iv.

is in the Natural History Museum, Paris; it shows Asterophyllitean branches in organic connection with a Calamitean
stem, but it is not quite clear if the stem is a true *Calamitina*.
A large drawing of this interesting specimen is given by Stur[1]
in his monograph on *Calamites*, also a smaller sketch by
Renault[2] in his *Cours de botanique fossile*. Similar branches
of the *Asterophyllites* type attached to an undoubted *Calamitina* are figured also by Lindley and Hutton. There is, in
short, good evidence that stems of this sub-genus bore branches
with Asterophyllitean shoots.

The wood of stems of the Calamitina group of *Calamites*,
in some instances at least, was of the *Arthropitys* type; this
has been shown to be the case in some French specimens from
the Commentry coal-field[3] and in others described by Stur[4].
The pith-casts of *Calamitina* are characterised by comparatively
short internodes separated by deep nodal constrictions, as shown
in fig. 100. From Permian specimens from Neu Paka in
Bohemia, described by Stur[5], we learn that there were the usual
Calamite diaphragms bridging across the wide pith-cavity at
each node. Such a cast as that shown in fig. 100 is often referred to as *Calamites approximatus* Brongn.; the length of the
internodes and the periodic occurrence of branch-scars in the
form of circular or oval depressions along a nodal line enable us
to recognise the *Calamitina* casts. Weiss[6] points out that in
pith-casts of this form the branch-scars occur on the nodal
constriction, and not immediately above the node as is the case
on the surface of a typical *Calamitina*. This distinction is
however of little or no value; the point of attachment of a
branch may be above the nodal line, while on the pith-cast
of the same stem the point of origin of the vascular bundles
of the branch is on the nodal constriction[7].

The specimen shown in fig. 100 illustrates the appearance of
a *Calamitina* cast. There is a verticil of branch-scars on the

---

[1] Stur (87), Pl. xvii.   [2] Renault (82), Pl. xvii. fig. 2.
[3] Renault and Zeiller (88), Pt. ii. p. 434, Pls. lii. and liii.
[4] Stur (87), p. 37, fig. 17. *Vide* also Grand'Eury (90), p. 208.
[5] Stur, *loc. cit.*   [6] Weiss (84), p. 61.
[7] *Vide* Grand'Eury (77), Pl. v. fig. 5.

lowest nodal constriction; on the right of the pith-cast the broad
band of wood is faintly indicated by the smooth surface of the

Fɪɢ. 100.   *Calamites (Calamitina) approximatus* Brongn. Lower Coal-Measures
        of Ayrshire.
    x, impression of the wood.
            (From a specimen in the collection of Mr R. Kidston.)

rock (*x*). Other examples demonstrating the existence of a broad
woody cylinder in *Calamitina* stems have been figured by Weiss[1]
and other writers, and some good examples may be seen in the
British Museum.

We have so far noticed the connection of certain forms of
pith-casts (*e.g. Calamites approximatus*), and Asterophyllitean
shoots with stems of the sub-genus *Calamitina*.

As regards the strobili our knowledge is far from satisfactory.
Stur[2] figures some fertile branches bearing long and narrow

---

[1] Weiss, *loc. cit.* Pl. xxi. fig. 5.   [2] Stur (87), Pl. xi. fig. 1.

strobili, either *Palaeostachya* or *Calamostachys*, in close association with *Calamitina* stems, and Renault and Zeiller[1] give
illustrations of the association of *Calamitina* stems with large
strobili of the *Macrostachya* form.

Before Weiss proposed the term *Calamitina*, various
authors had figured this form of Calamite under a distinct
generic name (*e.g. Hippurites* of Lindley and Hutton[2],
Cyclocladia[3], *Macrostachya*[4], &c.). Stems of this type have also
been described by more recent writers under different names,
and considerable confusion has been caused by the use of
numerous generic designations for forms of *Calamitina*. Some
small fragments of *Calamitina* stems were described by Salter[5]
in 1863 as portions of a new species of the Crustacean
Eurypterus (*E. mammatus*). In 1869 Grand'Eury proposed
the generic name *Calamophyllites*[6] for stems bearing verticils
of *Asterophyllites* shoots; his description of such stems agrees
with Weiss' *Calamitina*, but as Grand'Eury's name is used in
a narrower sense as implying a connection with *Asterophyllites*,
it is more convenient to adopt Weiss' term in spite of the
priority of *Calamophyllites*. In the *Fossil flora of Commentry*
we find some flattened stems of the *Calamitina* type described
under different generic names, as *Arthropitys approximatus*[7]
and as *Macrostachya*[8].

The determination of distinct species of the sub-genus
Calamitina is rendered almost hopeless by the variation in the
different branches of the same individual, and by the difficulty
of connecting surface-impressions with casts of the pith-cavity.

A typical example of the *Calamitina* type of *Calamites* was
figured by Sternberg[9] in 1821 as *Calamites varians*. This has
been adopted by Weiss[10] as a comprehensive species including

[1] Renault and Zeiller (88), Pt. ii. Pl. li. p. 423.

[2] Lindley and Hutton (31), Pl. cxiv. and Pl. cxc. The original specimens are in the Natural History Museum, Newcastle-on-Tyne.

[3] *Ibid.* Pl. cxxx. and Feistmantel (75), Pl. i. fig. 8.

[4] Lesquereux (79), Pl. xiii. fig. 14.

[5] Salter (63), figs. 6 and 7. *Vide* also Carruthers in Woodward, H. (72), p. 168.

[6] Grand'Eury (69); *vide* also (77) and (90).

[7] Renault and Zeiller (88), Pt. ii. Pls. lii. and ciii.   [8] *Ibid.* Pl. li.

[9] Sternberg (21), Pl. xii.   [10] Weiss (84), p. 61.

several different 'forms' of stems, which differ from Sternberg's
fossil in such points as the number of nodes between the
branch-whorls and the number of branches in each whorl. The
result of this system of nomenclature is the separation of portions
of one specific type under different form-names. It must be
clearly recognised that accurate specific diagnoses are practically
impossible when we have to deal with fragments of plants,
some of which are mere pith-casts, while others show the surface
features. The specimen represented in fig. 99 agrees with a
stem described by Ettingshausen[1] in 1855 as *Calamites Göpperti,*
and as a matter of convenience a member of the *Calamitina*
group showing such characters may be referred to as *Calamites
(Calamitina) Göpperti* (Ett.). The following list, which includes
a few synonyms of this form, may suffice to illustrate the
difficulties connected with accurate systematic determinations.

Calamites (Calamitina) Göpperti (Ett.).   Fig. 99.

1855.   *Calamites Göpperti,* Ettingshausen[2]
1869.   *Calamites (Calamophyllites) Göpperti,* Grand'Eury[3]
1874.   *Cyclocladia major,* Feistmantel[4].
1874.   *Calamites verticillatus,* Williamson[5].
1876.   *Calamitina Göpperti,* Weiss[6].
1884.   *Calamites (Calamitina) varians abbreviatus,* Weiss[7].
1884.   *Calamites (Calamitina) varians inconstans,* Weiss[8].
1887.   *Calamites Sachsei,* Stur[9].
1888.   *Calamophyllites Göpperti,* Zeiller[10].

This species is characterised by the smooth bark, which may
be traversed by a few irregular longitudinal fissures; most of
the nodes bear a series of small leaf-scars, and at fairly regular
intervals a node is immediately succeeded by a circle of
contiguous branch-scars, 8—12 in a whorl. The pith-cast of
this type of stem has short ribbed internodes separated by

[1] Ettingshausen (55), Pl. I. fig. 4.   [2] Ettingshausen, *loc. cit.*
[3] Grand'Eury (69), p. 709.   [4] Feistmantel (75), Pl. I. fig. 8.
[5] Williamson (74), Pl. VII. fig. 45.   [6] Weiss (76), Pl. XVII. figs. 1 and 2.
[7] Weiss (84), Pl. XVI a. figs. 10 and 11.   [8] *Ibid.* Pl. XXV. fig. 2.
[9] Stur (87), Pl. II. etc.   [10] Zeiller (88), p. 363, Pl. LVII. fig. 1.

rather deep nodal constrictions; the branch-whorls being represented by a series of pits on the nodal constrictions recurring
at corresponding intervals to the whorls of branch-scars on the
surface of the stem. Leaves narrow and linear in form, like
those on Asterophyllitean branches, are occasionally associated
with this type of stem.

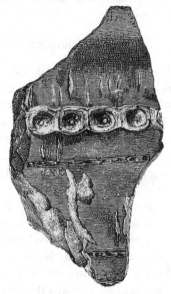

Fig. 101. *Calamites (Calamitina)* sp. From a specimen in the British Museum.
(After Carruthers.) Slightly reduced.

The fragment of a *Calamitina* stem shown in fig. 101 is the
counterpart of a specimen originally figured by Steinhauer[1] in
1818 as a species of *Phytolithus*. This may be specifically
identical with *C. Göpperti*; but it is better to speak of so small
a specimen as merely one of the *Calamitina* stems, to be compared with *Calamites (Calamitina) Göpperti*. The specimen
measures 14·5 cm. in length and 7 cm. in breadth.

The form of pith-cast represented in fig. 100 is no doubt that
of one of the *Calamitina* species, but as it is seldom possible to
determine the connection between such casts and the particular

---

[1] Steinhauer (18), Pl. VI. fig. 1.

species of stems to which they belong, they are often referred
to as *Calamites (Calamitina) approximatus* (Brongn.). The
specimen of which fig. 100 is a photograph was originally
described and figured by Mr Kidston[1] from the lower Coal-
Measures of Ayrshire. Both *Calamites (Calamitina) Göpperti*
(Ett.) and *C. (Calamitina) approximatus* (Brongn.) are recorded
from the Transition, Middle and Lower Coal-Measures[2].

B. *Stylocalamites.*

In the members of this sub-genus the branch-scars are
either irregular in their occurrence or absent. In some Calamites the branch-scars are very few and far between, and other
species appear to have been almost without branches; pith-casts
of such stems may be referred to the sub-genus *Stylocalamites*[3].

An exceedingly common Calamitean cast, *C. Suckowi* Brongn.
(fig. 82) affords a good illustration of this type of stem. In
the specimen shown in fig. 82 we have a cast of a rhizome,
which is rather exceptional in showing three branches in
connection with one another. The appearance of the fossil
suggests a rhizome, rather than an aerial shoot, bearing lateral
branches; the narrowing of the branches and the rapid
decrease in the length of the internodes towards the point of
attachment being features associated with rhizomes rather than
with aerial branches.

Calamites (Stylocalamites) Suckowi, Brongn. Fig. 82.

1818. *Phytolithus sulcatus*, Steinhauer[4].
1825. *Calamites decoratus*, Artis[5]
1828. *Calamites Suckowi*, Brongniart[6].
1833. *Calamites cannaeformis*, Lindley and Hutton[7].

For more complete lists of synonyms of this species

[1] Kidston (93), p. 311, Pl. II.
[2] Kidston (94), p. 248.
[3] Weiss (84), p. 119.
[4] Steinhauer (18), Pl. v. figs. 1 and 2.
[5] Artis (25), Pl. XXIV.
[6] Brongniart (28²), Pls. XV. and XVI.
[7] Lindley and Hutton (31), Pl. LXXIX.

reference should be made to Kidston[1], Zeiller[2], and other
authors.

Casts of *Calamites Suckowi* are characterised by flat or
slightly convex internodal ridges separated by shallow depressions, the ridges are rounded at the upper end of each internode,
and usually bear circular casts of infranodal canals. There are
some unusually large examples of casts of this species in the
British Museum from the Radstock Coal-Measures; one of these
has a length of 81 cm., and a diameter of 27 cm. Specimens
are not infrequently found with verticils of slender roots in
close proximity to the nodes of the cast; figures of such root-
bearing casts have been given by Lindley and Hutton[3], Weiss[4],
and other authors.

Renault[5] has drawn attention to the thinness of the layer
of wood which is often associated with large casts of *C. Suckowi*;
he concludes that the stems must have possessed little or no
secondary wood. In a more recent work by Grand'Eury[6]
Calamites Suckowi is spoken of as having had wood of the
Calamodendron type, but as wood of this kind has not been
found in England, it is suggested that the plant may not have
assumed an arborescent habit until late in the Coal-Measure
period. During the Lower and Middle Coal-Measures, at which
horizon it commonly occurs in England, it may have been
herbaceous. This suggestion has little to commend it; the
close agreement between *C. Suckowi* from English and French
localities points to a plant of the same form, and we have no
satisfactory evidence as to any difference in stem-structure in
the two cases.

Stur has figured a specimen of a Calamite cast, which he
compares with *C. Suckowi*, surrounded by a band of silicified
wood apparently of the *Arthropitys* type. From this and other
facts it would appear probable that some of the English stems
with the *Arthropitys* structure possessed casts referable to *Calamites (Stylocalamites) Suckowi*.

We are not in a position to speak with confidence as to the

[1] Kidston (93), p. 314; (86), p. 24.   [2] Zeiller (88), p. 333.
[3] Lindley and Hutton (31), Pl. LXXIX.   [4] Weiss (84), Pl. IV. fig. 1.
[5] Renault and Zeiller (88), p. 385.   [6] Grand'Eury (90), p. 214.

strobili of *C. Suckowi*, but Stur adduces evidence in support
of a connection between this species of Calamite and certain
Asterophyllitean branches (*Calamocladus equisetiformis*) bearing Calamostachyan cones. He does not appear to have found
the foliage-shoots and stems in organic contact, but draws
this conclusion from the association of the fertile branches and
stems in the same rocks[1]. This species is abundant in the
Lower, Middle and Upper Coal-Measures; it has also been
recorded from the Millstone Grit[2].

C. *Eucalamites.*

In this sub-genus branch-scars occur on every node; the
scars never form a contiguous whorl as in *Calamitina*, but
there may be from 3 to 10 on each node. The scars of
successive nodes often alternate in position, and thus form
more or less regular vertical series as shown in fig. 102.
The most obvious feature as regards the arrangement of the
branch-scars is their spiral disposition on the surface of the
pith-cast. The internodes are fairly uniform in length, and
there is no periodic recurrence of narrower internodes as in
Calamitina. From an examination of specimens of *Eucalamites*
in which the pith-cast is covered with a coaly layer representing
the carbonised remains of the wood and cortex, it would appear
that the surface of the stems was practically smooth. The
coaly investment on *Eucalamites* casts varies considerably in
thickness[3]; it is very unsafe to make use of the thickness of
this layer as a test of the breadth of the wood in Calamitean
stems. The branch-scars as seen in a surface-view of a stem
are situated a little above the nodal lines, while depressions on
the pith-casts occur in the slight nodal constriction or immediately above it. Small leaf-scars have been described as occurring
on the nodes between the branch-scars in specimens showing the
surface features[4].

The species long known as *Calamites cruciatus* Sternb. is
usually taken as the type of the sub-genus *Eucalamites*.

[1] Stur (87), p. 160, Pl. ix. fig. 2.　　　[2] Kidston (94), p. 249.
[3] Grand'Eury (89), p. 1087.　　　　　[4] Zeiller (88), p. 355.

Weiss[1] has subdivided this species into several 'forms,' which he
bases on the number of branch-scars on each node and on other
characters; a more extended subdivision of *C. cruciatus* has

Fig. 102.   *Calamites (Eucalamites) cruciatus*, Sternb.
From a specimen in the Barnsley Museum, Yorkshire.   ½ nat. size.

[1] Weiss (84), p. 96.

recently been made by Sterzel[1], who admits the impossibility
of separating the specific forms by means of the data at our
disposal, but for purposes of geological correlation he prefers
to express slight differences by means of definite 'forms' or
varieties. The more comprehensive use of the specific name
cruciatus as adopted by Zeiller in his *Flore de Valenciennes*[2] is,
I believe, the better method to adopt. The specimen shown in
fig. 102 affords a good example of a typical *Calamites cruciatus*,
it was found in the Middle Coal-Measures near Barnsley,
Yorkshire.

Calamites (Eucalamites) cruciatus (Sternb.).　Fig. 102.

1826. *Calamites cruciatus*, Sternberg[3].

1828. *Calamites cruciatus*, Brongniart[4].

1831. *Calamites alternans*, Germar and Kaulfuss[5].

1837. *Calamites approximatus*, Lindley and Hutton[6].

1877. *Calamodendrofloyos cruciatus*, Grand'Eury[7].

1878. *Calamodendron cruciatum*, Zeiller[8].

1884. *Calamites (Eucalamites) cruciatus ternarius*, Weiss[9]

1884.　　　　　 ,,　　　　　 ,,　　　　　 ,,　　*quaternarius*, Weiss[9].

1884.　　　　　 ,,　　　　　 ,,　　　　　 ,,　　*genarius*, Weiss[9].

1884.　　　　　 ,,　　　　　 ,,　　　　*multiramis*, Weiss[9].

1888. *Calamites (Calamodendron) cruciatus*, Zeiller[10].

This species occurs in the Upper, Middle and Lower Coal-
Measures[11]. The casts of the *cruciatus* type have been found
associated with wood possessing the structural features of the
sub-genus *Calamodendron*[12], but our knowledge of the structure
of the stem, and of the fertile branches of *C. cruciatus* is very
imperfect. A restoration of *Calamites (Eucalamites) cruciatus*
is given by Stur[13] in his classic work on the Calamites, but
he does not make quite clear the supposed connection with

[1] Sterzel (93), p. 66.　　　　　　　　[2] Zeiller, *loc. cit.* p. 353.

[3] Sternberg (25), Pl. XLIX. fig. 5.　　　[4] Brongniart (28[2]), p. 128, Pl. XIX.

[5] Germar and Kaulfuss (31), p. 221, Pl. XLV. fig. 1.

[6] Lindley and Hutton (31), Pl. CCXVI.　　　[7] Grand'Eury (77), p. 293.

[8] Zeiller (80), Pl. CLXXIV. (expl. plates) fig. 3.

[9] Weiss (84), pp. 112, 113, 114.　　　　　[10] Zeiller (88), p. 353.

[11] Kidston (94), p. 249.　　　　[12] Grand'Eury (90), p. 216 (expl. plates).

[13] Stur (87), p. 68.

the stems and the fertile shoots of the Asterophyllites type[1] which he describes. Another member of the *Eucalamites* group, which is better known as regards its foliage-shoots, is *Calamites ramosus*, a species first described by Artis[2] in 1825. Stems of this species have been found in connection with the branches and leaves of the *Annularia*[3] type, bearing *Calamostachys*[3] cones. In all probability pith-casts included in the sub-genus *Eucalamites* belonged to stems with foliage-shoots and probably also with cones of more than one form.

In the above account of a few common pith-casts it has been pointed out that there is occasionally satisfactory evidence for connecting certain casts with wood of a particular structure, and with sterile and fertile foliage-shoots of a definite type. It is, however, impossible in many cases to recognise with any certainty the leaf-bearing branches and strobili of the different casts of *Calamites*; it is equally impossible to determine what type of pith-cast or what type of foliage-shoots belongs to petrified stem-fragments in which it is possible to investigate the microscopical features. The scattered and piece-meal nature of the material on which our general knowledge of Calamitean plants is based, necessitates a system of nomenclature which is artificial and clumsy; but the apparent absurdity of attaching different names to fragments, which we believe to be portions of the same genus, is of convenience from the point of view of the geologist and the systematist. As our material increases it will be possible to further simplify the nomenclature for Calamarian plants, but it is unwise to allow our desire for a simpler terminology to lead us into proposals which are based rather on suppositions than on established fact. If it were possible to discriminate between pith-casts of stems having the different anatomical characters designated by the three sub-genera, *Arthropitys*, *Arthrodendron* and *Calamodendron*, the genus *Calamites* might be used in a much narrower and probably more natural sense than that which we have adopted. The tests made use of by some

---

[1] Stur (87), Pl. x.   [2] Artis (25), Pl. ii.   [3] Weiss, *loc. cit.* Pls. v. vi. and x.

authors for separating pith-casts of *Calamodendron* and *Arthropitys* stems do not appear to be satisfactory; we want some term to apply to all Calamitean casts irrespective of the anatomical features of the stems, or of the precise nature of the foliage-branches. As used in the present chapter, *Calamites* stands for plants differing in certain features but possessing common structural characters, which must be defined in a broad sense so as to include types which may be worthy of generic rank, but which for convenience sake are included in a comprehensive generic name. The attempts to associate certain forms of foliage with *Arthropitys* on the one hand and with *Calamodendron* on the other, cannot be said to be entirely satisfactory; we still lack data for a trustworthy diagnosis of distinct Calamarian genera which shall include external characters as well as histological features. If we restricted the genus *Calamites* to stems with an *Arthropitys* structure and an Asterophyllitean foliage, we should be driven into unavoidable error. Within certain limits it is possible to distinguish generically or even specifically between petrified branches, and we already possess material enough for fairly complete diagnoses founded on internal structure; but it is not possible to make a parallel classification for pith-casts and foliage-shoots. For this reason, and especially bearing in mind the importance of naming isolated foliage-shoots and stem-casts for geological purposes, I believe it is better to admit the artificially wide application of the name *Calamites*, and to express more accurate knowledge, where possible, by the addition of a subgeneric term. In dealing with distinctions exhibited by Calamitean stems it may be advisable to make use of specific names, but we must keep before us the probability of the pith-cast and petrified stem-fragment of the same plant receiving different specific names. If the structural type is designated by a special sub-genus, this will tend to minimise the anomaly of using more than one binominal designation for what may be the same individual.

The following summary may serve to bring together the different generic and subgeneric terms which have been used in the foregoing account of *Calamites*.

CALAMITES.

Subgenera having reference to the method of branching as seen in casts or impressions of the stem-surface or in pith-casts.	Subgenera founded on anatomical characters in stems and branches.	Genus proposed for roots of *Calamites* before their real nature was recognised. The name refers to anatomical characters.
Calamitina, *Eucalamites,* *Stylocalamites.*	*Arthropitys,* *Calamodendron,* *Arthrodendron* (new sub-genus substituted for *Calamopytus*).	*Astromyelon.*
Genera of which some species, if not all, are the leaf-bearing branches of *Calamites.*	Generic names applied to strobili belonging to *Calamites.*	Genus including impressions of Calamite roots.
Calamocladus (including *Asterophyllites*), *Annularia.*	*Calamostachys,* *Palaeostachya,* *Macrostachya,* etc.	*Pinnularia.*

IV. Conclusion.

A brief sketch of the main features of *Calamites* suffices to bring out the many points of agreement between the arborescent Calamite plants and the recent Equisetums. The slight variation in morphological character among the present-day Horsetails, contrasts with the greater range as regards structural features among the types included in *Calamites.* The Horsetails probably represent one of several lines of development which tend to converge in the Palaeozoic period; the Calamite itself would appear to mark the culminating point of a certain phylum of which we have one degenerate but closely allied descendant in the genus *Equisetum.* We shall, however, be in a better position to consider the general question of plant-evolution after we have made ourselves familiar with other types of Palaeozoic plants. Grand'Eury's[1] striking descriptions

[1] Grand'Eury (77), (90).

of forests of Calamites in the Coal-Measures of central France, enable us to form some idea of the habit of growth of these plants with their stout branching rhizomes and erect aerial shoots.

By piecing together the evidence derived from different sources we may form some idea of the appearance of a living Calamite. A stout branching rhizome ascended obliquely or spread horizontally through sand or clay, with numerous whorls or tufts of roots penetrating into swampy soil. From the underground rhizome strong erect branches grew up as columnar stems to a height of fifty feet or more; in the lower and thicker portions the bark was fissured and somewhat rugged, but smoother nearer the summit. Looking up the stem we should see old and partially obliterated scars marking the position of a ring of lateral branches, and at a higher level tiers of branches given off at regular or gradually decreasing intervals, bearing on their upper portions graceful green branchlets with whorls of narrow linear leaves. On the younger parts of the main shoot rings of long and narrow leaves were borne at short intervals, several leaf-circles succeeding one another in the intervals between each radiating series of branches. On some of the leaf-bearing branchlets long and slender cones would be found here and there taking the place of the ordinary leafy twigs. Passing to the apical region of the stem the lateral branches given off at a less and less angle would appear more crowded, and at the actual tips there would be a crowded succession of leaf-segments forming a series of overlapping circles of narrow sheaths with thin slender teeth bending over the apex of the tree.

Thus we may feebly attempt to picture to ourselves one of the many types of Calamite trees in a Palaeozoic forest, growing in a swampy marsh or on gently sloping ground on the shores of an inland sea, into which running water carried its burden of sand and mud, and broken twigs of Calamites and other trees which contributed to the Coal Period sediments. The large proportions of a Calamite tree are strikingly illustrated by some of the broad and long pith-casts occasionally seen in Museums; in the Breslau Collection there is a cast of a stem belonging to the sub-genus *Calamitina*, which measures about 2 m. in length and 23 cm. in breadth, with 36 nodes. In the Natural History Museum, Paris, there is a cast nearly 2 metres long and more than 20 cm. wide, which is referred to the sub-genus *Calamodendron*.

V. *Archaeocalamites*.

In the Upper Devonian and Culm rocks casts of a well-defined Calamitean plant are characteristic fossils; stems, leaf-bearing branches, roots and cones have been described by several authors, and the genus *Archaeocalamites* has been instituted for their reception. Although this genus agrees in certain respects with *Calamites*, and as recent work has shown this agreement extends to internal structure, it has been the custom to regard the Lower Carboniferous and Devonian plants as generically distinct. The surface features of the stem-casts, the form of the leaves, and apparently the cones, possess certain distinctive characters which would seem to justify the retention of a separate generic designation.

We may briefly summarise the characteristics of the genus as follows:—

Pith-casts articulated, with very slightly constricted nodes; the internodes traversed by longitudinal ribs slightly elevated or almost flat, separated by shallow grooves. The ribs and grooves are continuous from one internode to another, and do not usually show the characteristic alternation of *Calamites*[1] Along the nodal line there are occasionally found short longitudinal depressions, probably marking the points of origin of outgoing bundles. Branches were given off from the nodes without any regular order; a pith-cast may have branch-scars on many of the nodes, or there may be no trace of branches on casts consisting of several nodes. The leaves[2] are in whorls; in some cases they occur as free, linear, lanceolate leaves, or on younger branches they are long, filiform and repeatedly forked.

---

[1] *Vide* Stur (75), *etc.* for remarks on the course of the vascular strands.

[2] For good figures of the leaves *vide* Stur (75), Rothpletz (80), Ettings-hausen (66), Solms (96).

The structure of the wood agrees with that of some forms of *Arthropitys*. The strobili consist of an articulated axis bearing whorls of sporangiophores, and each sporangiophore has four sporangia. Our knowledge of the fertile shoots is, however, very imperfect.

Renault[1] has recently described the structure of the wood in some small silicified stems of *Archaeocalamites* from Autun. A large hollow pith is surrounded by a cylinder of wood consisting of wedge-shaped groups of xylem tracheids associated with secondary medullary rays; at the apex of each primary xylem group there is a carinal canal. The primary medullary rays appear to have been bridged across by bands of xylem at an early stage of secondary thickening, as in the Calamite of fig. 83, *D*.

Our knowledge of the cones of *Archaeocalamites* is far from satisfactory. Renault[2] has recently described a small fertile branch bearing a succession of verticils of sporangiophores; each sporangiophore stands at right angles to the axis of the cone and bears four sporangia, as in *Calamostachys*. It is not clear how far there is better evidence than that afforded by the association of the specimen with pith-casts of stems, for referring this cone to *Archaeocalamites*, but the association of vegetative and fertile shoots certainly suggests an organic connection. The cone described by the French author agrees with *Equisetum* in the absence of sterile bracts between the whorls of sporangiophores. It is an interesting fact that such a distinctly Equisetaceous strobilus is known to have existed in Lower Carboniferous rocks.

Stur[3] has also described *Archaeocalamites* at considerable length; he gives several good figures of stem-casts and foliage-shoots bearing long and often forked narrow leaves. The same writer describes specimens of imperfectly preserved cones in which portions of whorls of forked filiform leaves are given off

---

[1] Renault (96), p. 80; (93), Pls. xlii. and xliii. Since the above was written an account of the internal structure of *Archaeocalamites* has been published by Solms-Laubach (97); he describes the wood as being of the *Arthropitys* type.

[2] Renault, *loc. cit.* Pl. xlii. figs. 6 and 7.

[3] Stur (75), p. 2, Pls. ii.—v.

from the base of the strobilus[1]. Kidston[2] published an important memoir on the cones of *Archaeocalamites* in 1883, in which he advanced good evidence in support of the view that certain strobili, which were originally described as Monocotyledonous inflorescences, under the generic name *Pothocites*[3], are the fertile shoots of this Calamarian genus. Kidston's conclusions are based on the occurrence on the *Pothocites* cones, of leaves like those of *Archaeocalamites*, on the non-alternation of the sporangiophores of successive whorls, and on the close resemblance between his specimens and those described by Stur. Good specimens of the cones, formerly known as *Pothocites*, may be seen in the Botanical Museum in the Royal Gardens, Edinburgh; as they are in the form of casts without internal structure it is difficult to form a clear conception as to their morphological features.

The fossils included under *Archaeocalamites* have been referred by different authors to various genera, and considerable confusion has arisen in both generic and specific nomenclature. The following synonomy of the best known species, *A. scrobiculatus* (Schloth.) illustrates the unfortunate use of several terms for the same plant.

Fig. 103. *Archaeocalamites scrobiculatus* (Schloth.). From a specimen in the Woodwardian Museum, Cambridge. From the Carboniferous limestone of Northumberland. ¼ nat. size.

---

[1] An examination of the specimens in the Museum of the Austrian Geology Survey did not enable me to satisfactorily verify the features of the cone as described by Stur; the impressions are far from clear.

[2] Kidston (83[2]).

[3] *Vide* Paterson (41); Lyell (67), vol. I. p. 149 *etc.*

386   CALAMITES.   [CH.

Archaeocalamites scrobiculatus (Schloth.).   Fig. 103.

1720.   *Lithoxylon,* Volkmann[1].
1820.   *Calamites scrobiculatus,* Schlotheim[2].
1825.   *Bornia scrobiculata,* Sternberg[3].
1828.   *Calamites radiatus,* Brongniart[4].
1841.   *Pothocites Grantoni,* Paterson[5]
1852.   *Calamites transitionis,* Göppert[6].
——   *Stigmatocanna Volkmanniana, ibid.*
——   *Anarthrocanna tuberculata, ibid.*
——   *Calamites variolatus, ibid.*
——   *C. obliquus, ibid.*
——   *C. tenuissimus, ibid.*
——   *Asterophyllites elegans, ibid.*
1866.   *Calamites laticulatus,* Ettingshausen[7].
——   *Equisetites Göpperti, ibid.*
——   *Sphenophyllum furcatum, ibid.*
1873.   *Asterophyllites spaniophyllus,* Feistmantel[8].
1880.   *Asterocalamites scrobiculatus,* Zeiller[9]

For other lists of synonyms reference may be made to Binney[10], Stur[11], Kidston[12] and other authors.

Some of the best specimens of this species are to be seen in the Museums of Breslau and Vienna, which contain the original examples described by Göppert and Stur. An examination of the original specimens, figured by Göppert under various names, enables one to refer them with confidence to the single species, *Archaeocalamites scrobiculatus.* The generic name *Archaeocalamites,* which has been employed by some authors, was suggested by Schimper[13] in 1862, as a subgenus of *Calamites,* on account of the occurrence of a deeply divided leaf-sheath, attached to the node of a pith-cast, which seemed to differ from

[1] Volkmann (1720), p. 93, Pl. vII. fig. 2.
[2] Schlotheim (20), p. 402, Pl. xxII. fig. 4.   [3] Sternberg (25).
[4] Brongniart (28²), p. 122, Pl. xxvI. figs. 1 and 2.
[5] Paterson (41), Pl. III.   [6] Göppert (52), Pls. III., v., vI., vIII., xxxvIII.
[7] Ettingshausen (66), Pls. I.—Iv.   [8] Feistmantel (73), Pl. xIv. fig. 5.
[9] Zeiller (80), p. 17.   [10] Binney (68), p. 7.
[11] Stur (75), p. 3.   [12] Kidston (86), p. 35.
[13] Schimper and Koechlin-Schlumberger (62), Pl. I. The original specimens of Schimper's figures are in the Strassburg Museum.

the usual type of Calamitean leaf. The specimens described by Schimper are in the Strassburg Museum; the leaf-sheath which he figures is not very accurately represented.

The example given in fig. 103 shews very clearly the continuous course of the ribs and grooves of the pith-cast. Each rib is traversed by a narrow median groove which would seem to represent the projecting edge of some hard tissue in the middle of each principal medullary ray of the stem. The specimen was found in a Carboniferous limestone quarry, Northumberland; there is a similar cast from the same locality in the Museum of the Geological Survey.

Affinities of Archaeocalamites.

This genus agrees very closely with *Calamites* both in the anatomical structure of the stem and in the verticillate disposition of the leaves. The strobili appear to be Equisetaceous in character, and there is no satisfactory evidence of the existence of whorls of sterile bracts in the cone, such as occur in *Calamostachys* and in other Calamitean strobili. The continuous course of the vascular bundles of the stem from one internode to the next is the most striking feature in the ordinary specimens of the genus; but it sometimes happens that the grooves on a pith-cast shew the same alternation at the node as in *Calamites*. This is the case in a specimen in the Göppert collection in the Breslau Museum, and Feistmantel[1] has called attention to such an alternation in specimens from Rothwaltersdorf. In the true *Calamites*, on the other hand, the usual nodal alternation of the vascular strands is by no means a constant character[2]. Stur[3], Rothpletz[4], and other authors have pointed out the resemblance of *Archaeocalamites* to *Sphenophyllum*. The deeply divided leaves of some Sphenophyllums and those of *Archaeocalamites* are very similar in form; and the course of the vascular strands in *Sphenophyllum* may be compared with that in *Archaeocalamites*. But the striking difference in the structure

---

[1] Feistmantel (73), p. 491, Pl. xxiv. figs. 3 and 4.

[2] *Vide* specimens 20 A, 20 B, 24 in the Williamson Collection.

[3] Stur (75), p. 17.   [4] Rothpletz (80), p. 8.

of the stele forms a wide gap between the two genera. We have evidence that the Calamites and Sphenophyllums were probably descended from a common ancestral stock, and it may be that in *Archaeocalamites*, some of the *Sphenophyllum* characters have been retained ; but there is no close affinity between the two plants.

On the whole, considering the age of *Archaeocalamites* and the few characters with which we are acquainted, it is probable that this genus is very closely related to the typical *Calamites*, and may be regarded as a type which is in the direct line of development of the more modern Calamite and the living *Equisetum*. Weiss[1] includes *Archaeocalamites* as one of his subgenera with *Calamitina* and others, and it is quite possible that the genus has not more claim to stand alone than other forms at present included in the comprehensive genus *Calamites*.

The student will find detailed descriptions of this genus in the works which have been referred to in the preceding pages.

---

[1] Weiss (84), p. 56.

CHAPTER XI.

II. SPHENOPHYLLALES.

Sphenophyllum.

THE genus *Sphenophyllum* is placed in a special class, as representing a type which cannot be legitimately included in any of the existing groups of Vascular Cryptogams. Although this Palaeozoic genus possesses points of contact with various living plants, it is generally admitted by palaeobotanists that it constitutes a somewhat isolated type among the Pteridophytes of the Coal-Measures. Our knowledge of the anatomy of both vegetative shoots and strobili is now fairly complete, and the facts that we possess are in favour of excluding the genus from any of the three main divisions of the Pteridophyta.

In Scheuchzer's *Herbarium Diluvianum* there is a careful drawing of some fragments of slender twigs, from an English locality, bearing verticils of cuneiform leaves, which the author compares with the common *Galium*[1]. As regards superficial external resemblance, the *Galium* of our hedgerows agrees very closely with what must have been the appearance of fresh green shoots of *Sphenophyllum*.

A twig of the same species of *Sphenophyllum* is figured by Schlotheim[2] in the first part of his work on fossil plants; he regards it as probably a fragment of some species of Palm. Sternberg[3] was the first to institute a generic name for this genus of plants, and specimens were described by him in 1825

[1] Scheuchzer (1723), p. 19, Pl. IV. fig. 1.
[2] Schlotheim (04), Pl. II. fig. 24, p. 57.   [3] Sternberg (25), p. 32.

as a species of the genus *Rotularia*. The name *Sphenophyllites*
was proposed by Brongniart[1] in 1822 as a substitute for
Schlotheim's genus, and in a later work[2] the French author
instituted the genus *Sphenophyllum*. Dawson[3] was the first to
make any reference to the anatomy of this genus; but it is from
the examination of the much more perfect material from
St. Etienne, Autun, and other continental localities, the North
of England and Pettycur in Scotland, by Renault, Williamson,
Zeiller and Scott, that our more complete knowledge has been
acquired.

The affinity of *Sphenophyllum* has always been a matter of
speculation; it has been compared with Dicotyledons, Palms,
Conifers (*Ginkgo* and *Phyllocladus*), and various Pteridophytes,
such as *Ophioglossum, Tmesipteris, Marsilia, Salvinia, Equisetum*
and the Lycopodiaceae[4].

We may define the genus *Sphenophyllum* as follows:—

Stem comparatively slender (1·5—15 mm. ?), articulated,
usually somewhat tumid at the nodes; the surface of the
internodes is marked by more or less distinct ribs and grooves
which do not alternate at the nodes, but follow a straight course
from one internode to the next. A single branch is occasionally
given off from a node. Adventitious roots are very rarely seen,
their surface does not show the ridges and grooves of the foliage-
shoots.

The leaves are borne in verticils at the nodes, those in the
same whorl being usually of the same size, but in some forms
two of the leaves are distinctly smaller than the others. Each
verticil contains normally 6, 9, 12, 18 or more leaves, which are
separate to the base and not fused into a sheath; the number
of leaves in a verticil is not always a multiple of six. They vary
in form from cuneiform with a narrow tapered base, and a lamina
traversed by several forked veins, to narrow uninerved leaves
and leaves with a lamina dissected into dichotomously branched

[1] Brongniart (22), Pl. xiii. fig. 8, p. 234.
[2] *Ibid.* (28), p. 68.
[3] Dawson (66), p. 153, Pl. xii.
[4] For reference *vide* an excellent monograph by Coemans and Kickx (64), also Potonié (94).

linear segments. The leaves of successive whorls are super-posed.

The strobili are long and narrow in form, having a length in some cases of 12 cm., and a diameter of 12 mm.; they occur as shortly stalked lateral branches, or terminate long leaf-bearing shoots. The axis of the cone bears whorls of numerous linear lanceolate bracts fused basally into a coherent funnel-shaped disc, bearing on its upper surface sporangiophores and sporangia.

The strobili are usually isosporous, but possibly heterosporous in some forms.

The stem is monostelic, with a triarch or hexarch triangular strand of centripetally developed primary xylem, consisting of reticulate, scalariform and spiral tracheae; the protoxylem elements being situated at the blunt corners of the xylem-strand. Foliar bundles are given off, either singly or in pairs, from each angle of the central primary strand. The secondary xylem consists of radially disposed reticulate or scalariform tracheae, developed from a cambium-layer. The phloem is made up of thin-walled elements, including sieve-tubes and parenchyma. Both xylem and phloem include secondary medullary rays of parenchymatous cells. The cortex consists in part of fairly thick-walled elements; in older stems the greater part of the cortical region is cut off by the development of deep-seated layers of periderm.

The roots are apparently diarch in structure, with a lacunar and smooth cortex.

The branch of *Sphenophyllum emarginatum* Brongn. given in fig. 109 shows the characteristic appearance of the genus as represented by this well-known species which Brongniart figured in 1822. The Indian species shown in fig. 111 illus-trates the occurrence of unequal leaves in the same whorl, and in fig. 110, *B*, we have a form of verticil in which the leaves are deeply divided into filiform segments. A larger-leaved form is represented by *S. Thoni*, Mahr. (fig. 110, *A*), a species occasionally met with in Permian rocks.

No specimens of *Sphenophyllum* have so far been found attached to a thick stem; they always occur as slender shoots, which sometimes reach a considerable length. One of the longest examples known is in the collection of the Austrian Geological Survey; the axis is 4 mm. in breadth and 85 cm. long, bearing a slender branch 61 cm. in length. The manner of occurrence of the specimen as a curved slender stem on the surface of the rock suggests a weak plant, which must have depended for support on some external aid, either water or another plant. The anatomical structure and other features do not favour the suggestion of some writers that *Sphenophyllum* was a water-plant[1], but there would seem to be no serious obstacle in the way of regarding it as possibly a slender plant which flung itself on the branches and stems of stronger forest trees for support.

I. The anatomy of Sphenophyllum.

The following account of the structural features of the stem and root is based on the work of Renault[2], Williamson[3] and Williamson and Scott[4]. We may first consider such characters as have been recognised in different examples of the genus, and then notice briefly the distinguishing peculiarities of two well-marked specific types.

a. Stems.

i. Primary structure.

In a transverse section of a young *Sphenophyllum* stem such as that diagrammatically sketched in fig. 105, *A*, we find in the centre the xylem portion of a single stele with a characteristic triangular form. The primary xylem consists mainly of fairly large tracheae with numerous pits on their walls; towards the end of each arm the tracheids become scalariform, and at the apex there is a group of narrower spiral protoxylem elements. In the British species there is a single protoxylem group at the apex of each arm, but Renault has described some French stems in which the stele appears to be hexarch, having two protoxylem groups at the end of each of the three rays of the stele. The primary xylem strand of *Sphenophyllum* has therefore a

[1] *e.g.* Newberry (91).   [2] Renault (73), (76[2]), (96).
[3] Williamson (74), (78).   [4] Williamson and Scott (94), p. 919.

root-like structure, the tracheids having been developed centripetally from the three initial protoxylem groups. This type of
structure is typical of roots, but it also occurs in the stems of
some recent Vascular Cryptogams.

As a rule the tissue next the xylem has not been petrified,
but in exceptionally well-preserved examples it is seen to
consist of a band of thin-walled elements, of which those in
contact with the xylem may be spoken of as phloem, and those
beyond as the pericycle. Succeeding this band of delicate
tissue there is a broader band of thicker-walled and somewhat
elongated elements, constituting the cortex. The specimen
drawn in fig. 105, *A*, shows very prominent grooves in the cortex
opposite the middle of each bay of the primary wood. It is
these grooves that give to the ordinary casts of *Sphenophyllum*
branches the appearance of longitudinal lines traversing each
internode. In a longitudinal section of a stem, the cortical
tissue (fig. 104, *c*) is found to be broader in the nodal regions,

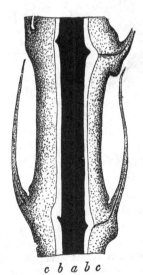

c b a b c

Fɪɢ. 104.   Diagrammatic longitudinal section of *Sphenophyllum*.
            c, outer cortex; *b*, space next the stele, originally occupied by
                    phloem *etc.*; *a*, xylem strand. (After Renault[1].)   × 7.

---

[1] Specimen 929 in the Williamson Cabinet is a longitudinal section of the
French *Sphenophyllum*, as described by Renault (76[2]).

thus giving rise to the tumid nodes referred to in the diagnosis. The increased breadth at the nodes does not mean that the xylem is broader in these regions, as it is in Calamite stems. Small strands of vascular tissue are given off from the three edges of the triangular stele (fig. 105 *A*) at each node;

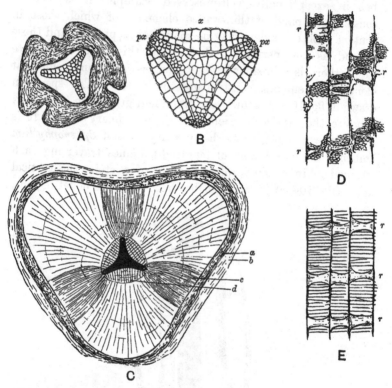

FIG. 105. *Sphenophyllum.*

    A. Transverse section of young stem.

    B. Transverse section of the wood of a young stem; *px*, protoxylem; *x*, secondary xylem. (*A* and *B*. *Sphenophyllum plurifoliatum*.) × 20.

    C. Transverse section of an old stem; (*S. insigne*); *a*, phloem; *b*, periderm; *c*, fascicular secondary xylem; *d*, interfascicular secondary xylem. × 9. (No. 914 in the Williamson Collection.)

    D. Longitudinal section of the reticulate tracheae and medullary rays; *r, r, r*, of *S. plurifoliatum.* × 36.

    E. Similar section of *S. insigne.* × 75. (*D* and *E* after Williamson and Scott.)

these branch in passing through the cortex on their way to
the verticils of leaves. The space b in the diagrammatic
section of fig. 104 was originally occupied by the phloem and
inner cortex. In some species of *Sphenophyllum* the apex of
each arm of the xylem strand, as seen in transverse section, is
occupied by a longitudinal canal surrounded by spiral tracheids,
as in the primary xylem of the old stem shown in fig. 105, C.

ii. Secondary structure.

With the exception of very young twigs the petrified *Sphenophyllum* stems usually show a greater or less development of
secondary wood. In the xylem-strand of fig. 105, B, the broad
concave bays of the primary wood have been filled in by the
development of two rows of large secondary tracheids, x, but
opposite the protoxylem groups, px, there are no signs of
cambial activity. In the unusually large stem represented by
a rough sketch in fig. 105, C, the triangular primary xylem
lies in the centre of a thick mass of secondary vascular tissue.
The secondary and primary wood together have a diameter of
about 5 mm.

After the bays between each protoxylem corner have been
filled in, the formation of secondary wood proceeds uniformly
along the stem radii, but the rows of tracheids and medullary
rays which are developed opposite the corners of the primary
strand, c, differ in certain characters from the broader masses
of wood opposite the bays. For convenience, the secondary
wood, c, opposite the protoxylem groups has been spoken of as
fascicular wood, and the rest, d, as *interfascicular wood*.

The secondary xylem consists either of tracheae with
numerous bordered pits on their radial walls (fig. 105, D), or of
tracheae with broad and bordered scalariform pits (fig. 105, E).
The suggestion of concentric rings of growth in the wood in
fig. 105, C, is rather deceptive ; there are no well-marked regular
rings in *Sphenophyllum* stems, but irregular bands of smaller
elements occasionally interrupt the uniformity of the secondary
xylem. In some stems the medullary rays have the form of
rows of parenchymatous cells, which in tangential longitudinal

section are found to consist frequently of a single row of radially
disposed elements; this type of medullary rays occurs in the
species *Sphenophyllum insigne*, in which the tracheae are scalariform. Three medullary rays, *r*, are seen on the radial face of
the scalariform tracheids in fig. 105, *E*, which represents a radial
section of this species. In other species, *e.g. S. plurifoliatum*, the
medullary rays have a peculiar and characteristic structure; in
a transverse section of the stem they appear as groups of a
few parenchymatous cells in the spaces between the truncated
angles of the large tracheae (fig. 106). In longitudinal section
these medullary-ray elements resemble thick bars stretching
radially across the face of the tracheae (fig. 105, *D*, *r*); the
apparent septa or bars are however thin-walled cells connecting
the different groups of medullary-ray cells, as seen in a transverse
section. These radial connecting cells are occasionally seen as
short rays in transverse sections of stems.

The cambium and phloem elements are occasionally preserved in good specimens of older stems; the former consist
of tabular flatted thin-walled cells, and the latter in some cases
include large sieve-tubes and narrower parenchymatous elements.

The sections shown in fig. 107, *E* and *F*, illustrate the
preservation of cambial and phloem tissue. In the transverse
section of fig. 107, *F*, the secondary xylem with the medullary
rays, *r*, is succeeded by a few tabular cambium cells, and external
to these there are thin-walled elements of unequal size representing the phloem. In fig. 107, *E*, the scalariform tracheids
are succeeded by narrow thin-walled cells, and the larger
elements with transverse and oblique septa are no doubt
sieve-tubes.

In the large stem of fig. 105, *C*, the xylem is succeeded by
a band of tissue, *a*, which is no doubt phloem, and external to
this there is a considerable development of periderm (*b*). The
periderm in *Sphenophyllum* stems had a deep-seated origin,
the phellogen or cork-cambium occasionally being formed in
the secondary phloem-parenchyma, and in other cases in the
pericycle, as in the stems of some living dicotyledons. Williamson and Scott[1] describe stems in which a succession of phellogens

[1] Williamson and Scott (94), p. 926.

were formed at different levels, thus producing a scaly type of
bark, such as we find in the Pine or the Plane tree.

Before describing the structure of the strobili of *Sphenophyllum*, we may briefly point out the distinguishing features of
two specific types of the genus recently described by Williamson
and Scott. One of these species, *S. insigne*, was originally
described by Williamson as an *Asterophyllites*; the numerous
narrow linear leaves in each verticil led to the inclusion of the
specimens in the latter genus. The material on which this
species is founded is from the volcanic beds of Pettycur,
Burntisland, on the coast of the Firth of Forth.

1. *Sphenophyllum insigne* (Will.).   Figs. 105, *C* and *E*, and
107, *E* and *F*.

1891 *Asterophyllites insignis*, Williamson[1]

An *intercellular space* occurs at each angle of the three-
rayed primary xylem strand, and spiral tracheae are abundant.
The tracheae of the secondary wood have *scalariform markings*
on the radial walls. *Regular medullary rays* extend through
the secondary wood. The phloem contains large sieve-tubes.

This species occurs in the Calciferous sandstone rocks of
Burntisland, and has lately been recorded from Germany. It
characterises a lower horizon than *S. plurifoliatum* (Will. and
Scott).

2. *Sphenophyllum plurifoliatum* (Williamson and Scott)[2].
Figs. 105, *A*, *B*, and *D*, and 106.

1891. *Asterophyllites sphenophylloides.* Will.[3]

The specific name *plurifoliatum* was proposed by Williamson
and Scott for a type of stem originally described by Williamson[4]
as an *Asterophyllites*, from the Coal-Measures of Oldham, Lancashire. This form of stem has not so far been connected
with any of the older species founded on external characters,

---

[1] Williamson (91), p. 13.   [2] Williamson and Scott (94), p. 920.
[3] Williamson (91), p. 12.   [4] *Ibid.* (74).

but it evidently bore foliage in which the leaves were deeply
divided, as in *Sphenophyllum trichomatosum* (fig. 110, *B*).

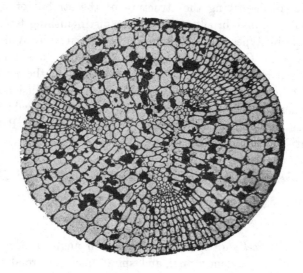

FIG. 106.   *Sphenophyllum plurifoliatum*, Will. and Scott.
From a photograph by Mr Highly from a section in the Williamson Collection (no. 899).   × 27.

In this species there are *no canals* at the angles of the
primary xylem, and there are fewer spiral tracheae than in
S. insigne. The tracheae of the secondary wood have *numerous
small pits* on the radial walls, and the medullary rays are
chiefly composed of parenchymatous cells, which appear in
transverse section as *groups of cells* between the truncated
angles of the tracheae. The characters are fairly well seen in
the xylem portion of a stele shown in fig. 106. The fascicular
wood includes some rows of parenchymatous medullary-ray
cells in addition to the characteristic groups, as seen in the
figure. A slightly oblique transverse section of a stem is often
convenient in the interpretation of histological features; one of
the sections of *S. plurifoliatum* in the Williamson collection
(no. 893), which has been cut somewhat obliquely, shows very
clearly the differences in pitting exhibited by the different
xylem elements.

b.   Roots.

Our knowledge of the anatomy of *Sphenophyllum* roots is very limited. Renault has described a somewhat imperfect example of a silicified root from St. Etienne and Autun. The drawing in fig. 107, *B*, which is copied from one of Renault's figures, shows a cylindrical mass of xylem with a small band of narrower elements occupying the centre, and surrounded by rows of larger secondary tracheae. The central bipolar band is described as the diarch primary xylem, around which the secondary pitted elements have been developed.

It is probable that the specimen described by Renault is a root of *Sphenophyllum*, but my impression gained from an examination of the section was that the diarch primary strand is not quite so clear as in the published figures. Until we possess better material we cannot attempt any very satisfactory description of the anatomical features of the roots of this genus.

A section of a *Sphenophyllum* stem has been figured by Felix[1], in which a lateral member is being given off; this may possibly represent the origin of an adventitious root, but the preservation is not sufficiently distinct to render this certain.

c.   Leaves.

Renault[2] has described some silicified leaves of *Sphenophyllum* from Autun in which the laminae consist of thin-walled loose parenchyma, traversed by small groups of tracheids constituting the simple or forked veins. The epidermis is made up of a single layer of cells, with here and there indistinct indications of stomata. A more perfect stoma has, however, been described by Solms-Laubach from the epidermis of a bract in a strobilus (fig. 107, *A*).

---

[1] Felix (86), Pl. vi. fig. 2.
[2] For figures *vide* Renault (82), Pl. xvi. fig. 1, (76²) Pls. vii. and ix.

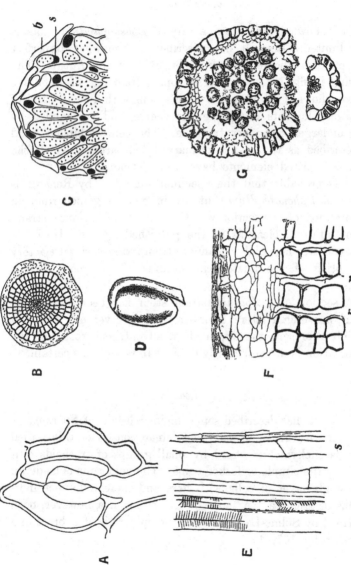

FIG. 107. A. Stoma in a bract of *Sphenophyllostachys*. B. Root of *Sphenophyllum*. C. *Sphenophyllostachys Römeri*, Solms. *s*, sporangiophore, *b*, bract. D. Sporangium. E and F. Sections through the cambium, phloem and secondary xylem of *Sphenophyllum insigne* (Will.). *s*, sieve-tube. G. Sporangium and pedicel. A, C, D. After Solms-Laubach. B. After Renault. E—G. After Williamson and Scott. E. F. × 100. G. × 115.

d. Cones.

The history of the recognition of the cones of *Sphenophyllum* has already been briefly alluded to in chapter V., p. 100. The main points in the structure of the cones of this genus were known for several years, before the fact was established that they belonged to *Sphenophyllum* stems. In 1871 Williamson[1] published an account of an imperfect fossil strobilus from the Lower Coal-Measures of Oldham, Lancashire, under the name of *Volkmannia Dawsoni*. The generic term *Volkmannia* has been used by different writers for cones varying considerably in structural features; in the case of Williamson's fossil, Weiss[2] substituted the name *Bowmanites*, a genus instituted by Binney[3] for a strobilus apparently of the same type as *Volkmannia Dawsoni*. In 1891 Williamson[4] described some additional specimens of *Bowmanites Dawsoni*, and, as in his earlier paper, he compared the strobilus with *Asterophyllites* and *Sphenophyllum*, but it was still a matter of speculation as to what was the form of the vegetative branches. Soon after the more complete account of the English cones was published, Zeiller[5] recognised a close agreement between some French and Belgian specimens of *Sphenophyllum* strobili and the strobilus described by Williamson. A closer comparison thoroughly established the connection between *Bowmanites Dawsoni* and *Sphenophyllum*; and there is little doubt that this strobilus belongs to the stem known as *Sphenophyllum cuneifolium* (Sternb.)—a well-known species of the genus.

The most important morphological features of the strobilus of *Sphenophyllum* may best be illustrated by a detailed account of one specific type, and by a brief reference to other forms which are characterised by certain differences in the number and attachment of the sporangia. When we know that a given strobilus must have grown on a *Sphenophyllum* stem, the obvious name to assign to it would seem to be that of the plant which bore it; but there are advantages in making use of special generic terms for detached cones, which cannot be referred with certainty to a

[1] Williamson (71²).   [2] Weiss (84), p. 200.   [3] Binney (71).
[4] Williamson (91²).   [5] Zeiller (93).

particular species of stem. The genus *Calamostachys* affords
an example of a name which is intended to denote that a cone
so called belongs to a Calamarian plant; similarly such a name
as *Sphenophyllostachys* may be used for Sphenophylloid cones
which cannot be connected with certainty to particular species
of *Sphenophyllum*. It has been suggested that the genus
Bowmanites, first used for a cone which was afterwards recognised as belonging to a *Sphenophyllum*, should be employed
instead of the sesquipedalian term *Sphenophyllostachys*. The
latter is used here as being in accordance with a generally
accepted and convenient system of nomenclature, and as a
name which at once denotes the fact that the fossil is not only
a cone but that it belongs to a *Sphenophyllum*.

Sphenophyllostachys Dawsoni (Will.).   Figs. 107, *A* and *G*, 108.

Probably the strobilus of *Sphenophyllum cuneifolium* (Sternb.).

FIG. 108. Diagrammatic longitudinal section of a *Sphenophyllum* strobilus.
The upper figure represents a portion of a whorl of bracts. (The smaller
figure, after Zeiller.)

The cone consists of a central axis bearing a number of
verticils of bracts coherent in their lower portions in the form

of a widely open funnel-shaped disc, which splits up peripherally
into 14—20 linear-lanceolate segments. The free segments of
each verticil have an obliquely ascending or almost vertical position, and extend upwards for a distance of about six internodes.
The smaller drawing in fig. 108 shows the appearance in side
view of the narrow bracts of a single whorl. A transverse
section of a strobilus would include, therefore, sections of
several concentric series of ascending bracts. The verticils of
Sphenophyllostachys Dawsoni are probably superposed, but this
point has not been definitely settled. From the upper surface
of the coherent basal portion of each verticil, there are given off
twice as many sporangiophores as there are free segments, and
these are attached close to the line of junction of the axis of the
cone and the funnel-shaped disc. Each sporangiophore has the
form of a slender stalk which bends inwards at its distal end
and bears a single sporangium (*cf.* fig. 107, *D*). The sporangiophores given off from the same verticil of bracts vary in length.
All the sporangiophores are attached to the coherent bracts at
the same distance from the axis of the cone ; but as the sporangia
between each verticil of bracts are arranged in two or three
concentric series, it follows that the length of the sporangiophores varies considerably. The diagrammatic longitudinal
section of a strobilus in fig. 108 shows three concentric series
of sporangia between successive bract-verticils. A similar
diagram was published by Williamson in 1892[1], and afterwards copied by Potonié[2], but in Williamson's restoration the
sporangiophores of the three series of sporangia are erroneously
represented as arising from different points on the surface of
the bracts. There is little doubt, as regards the strobilus of *S.
cuneifolium*, that the sporangiophores were given off in a single
series close to the axils of the bracts, as is partially shown in
fig. 108.

The central part of the axis of the cone is occupied by a
single triangular stele like that of the stem, except that each
ray of the xylem strand has a comparatively broad blunt
termination, and is not tapered to a narrow arm as in fig. 105,

[1] Williamson (92).   Potonié (94), fig. 1.

A and *B*.　The wood consists of pitted tracheae, with two groups
of protoxylem elements at each of the truncated angles of the
solid strand of xylem.　From the angles of the stele branches
of vascular tissue pass out through the cortex to supply the
sterile and fertile segments of each verticil.　One of the
transverse sections of the *Sphenophyllum* cone in the British
Museum Collection (no. 1898 *E*) affords a good example of the
misleading appearance occasionally presented by an intruded
'rootlet' of *Stigmaria*; the vascular tissue of the cone has
disappeared, and a Stigmarian appendage with its vascular
bundle occupies the position of the stelar tissues.

The bracts consist of parenchymatous tissue limited externally by an epidermis containing stomata.　A single stoma
with subsidiary cells is represented in fig. 107, *A*.　The
sporangiophores are composed internally of thin-walled cells
with stronger cells towards the surface.　The longer sporangiophores in a series may be more or less coherent for part of
their length to the upper surface of the verticil of bracts.　In
fig. 108 the slender sporangiophores do not appear to come
off always from the same portion of the bracts, but this is due
to some of them lying on the surface of the latter during part
of their course to support the external circle of sporangia.
The hook-like distal end of a sporangiophore, towards the point
of attachment of the sporangium, is characterised by the larger
size and greater prominence of the surface cells; these larger
cells, which pass over the upper surface of a sporangium base,
probably constitute a kind of *annulus* which determines the
dehiscence of the sporangial wall[1].

Fig. 107, *G*, represents a sporangiophore and its sporangium
cut through transversely just below the point of attachment of
the latter to the end of the hook-like termination of the former.
The spores are characterised by an irregularly reticulate
thickening of the outer coat or exospore, as seen in the figure.

One of the chief points of interest suggested by a *Sphenophyllum* cone is the exact morphological nature of the sporangiophores.　Are they branches borne in the axils of bracts, or

---

[1] For a more complete account of this strobilus *vide* Zeiller (93), and
Williamson (91[2]), *etc.*

may we regard each sporangiophore as a modified leaf, which has become coherent with the whorls of sterile leaves? Or is a sporangiophore merely a stalk of a sporangium; or a ventral lobe of a leaf, of which the sterile bracts represent the dorsal lobes? Although it is impossible without the evidence of development to decide with certainty between these alternatives, it would seem most probable that a sporangiophore may be looked upon as a ventral lobe of a leaf, the sterile lobes forming the bracts or members of the sterile whorls of the cone. This question is discussed by Zeiller[1] and Williamson and Scott[2], also more recently by Scott[3] in his memoir on *Cheirostrobus*.

Sphenophyllostachys Römeri (Solms-Laubach)[4].
Fig. 107, *C* and *D*.

In another type of *Sphenophyllum* strobilus, recently described by Solms-Laubach, the incurved end of each sporangiophore bore two sporangia. In most respects this species, which has not been found in connection with a vegetative shoot, agrees with *Sphenophyllostachys Dawsoni*.

In fig. 107, *C*, which is copied from one of Solms-Laubach's drawings[5], we have an oblique transverse section of part of a strobilus, including portions of two series of sporangia borne on one verticil of bracts, and at the right-hand edge the section has passed through the sporangia belonging to another whorl of bracts. There were probably three concentric series of sporangia attached to each verticil of bracts, as in the case of fig. 108. The unshaded area, *b* (fig. 107, *C*), represents the bracts of two successive sterile whorls in transverse section. The shaded areas are the sporangia, with their sporangiophores, *s*. The relative position of the sporangia and sporangiophores suggests that each pedicil bore two sporangia at its tip, instead of one, as in the strobilus of *Sphenophyllum cuneifolium* (Sternb.).

[1] Zeiller (93), p. 37. [2] Williamson and Scott (94), p. 943.
[3] Scott (97), p. 24. [4] Solms-Laubach (95[4]).
[5] *Ibid.* Pl. x. fig. 6.

A further variation in the structure of the strobili is illustrated by some specimens of *S. trichomatosum* Stur, described by Kidston[1], from the Coal-Measures of Barnsley. Each whorl of bracts bears a single series of oval sporangia which appear to be sessile on the basal portion of the whorl. It is possible that delicate sporangiophores may have been present, but in the imperfect examples in Kidston's collection[2] the sporangia present the appearance of being seated directly on the surface of the bracts. As the specimens do not show any internal structure, it would be unwise to lay too much stress on the apparent absence of the characteristic sporangiophores. In any case, Kidston's cones afford an illustration of the occurrence of a single series of sporangia in each whorl, instead of the pluriseriate manner of occurrence in some other species.

The statement is occasionally met with that some *Sphenophyllum* cones possessed two kinds of spores, but we are still in want of satisfactory evidence that this was really the case. Renault has described an imperfect specimen, which he considers points to the heterosporous nature of a *Sphenophyllum* cone, but Zeiller and Williamson and Scott have expressed doubts as to the correctness of Renault's conclusions. While admitting the possibility of undoubted heterosporous strobili being discovered, we are not in a position to refer to *Sphenophyllum* as having borne strobili containing two kinds of spores[3].

[The following are some of the specimens in the Williamson Cabinet which illustrate the structure of *Sphenophyllum* :--

S. plurifoliatum.   874, 882, 884, 893, 894, 897, 899, 901, 903, 908, 1893.
S. insigne.   910, 914, 919, 921, 922, 924, 926, 1420, 1898.
Sphenophyllostachys.   1049A—1049c, 1898.]

---

[1] Kidston (90).
[2] I am indebted to my friend Mr Kidston for an opportunity of examining these specimens.
[3] *Vide* Renault (77), (96), p. 158. Zeiller (93), p. 34. Williamson and Scott (94), p. 942.

II. Types of vegetative branches of Sphenophyllum.

1. *Sphenophyllum emarginatum* (Brongniart).   Fig. 109.

1822.   *Sphenophyllites emarginatus,* Brongniart[1].
1828.   *Sphenophyllum emarginatum,* Brongniart[2].
1828.   *Sphenophyllum truncatum,* Brongniart[2].
1828.   *Rotularia marsileaefolia,* Bischoff[3].
1862.   *Sphenophyllum osnabrugense,* Römer[4].

This species of *Sphenophyllum* bears verticils of six or eight wedge-shaped leaves varying in breadth and in the extent of dissection of the laminae; they are truncated distally, and terminate in a margin characterised by blunt or obtusely-rounded teeth, each of which receives a single vein. The larger leaves are usually more or less deeply divided by a median slit. The narrow base of each leaf receives a single vein which branches repeatedly in a dichotomous manner in the substance

Fig. 109.   *Sphenophyllum emarginatum* (Brongniart).
From a specimen in the Collection of Mr. R. Kidston, Upper Coal-Measures, Radstock.   ⅔ nat. size.

[1] Brongniart (22), p. 234, Pl. ii. fig. 8.   [2] Brongniart (28), p. 68.
[3] Bischoff (28), Pl. xiii. fig. 1.   [4] Römer, F. (62), p. 21, Pl. v. fig. 2.

of the lamina. Several drawings have been given by Sterzel[1] in a memoir on Permian plants, showing the variation in leaf-form in *Sphenophyllum emarginatum*, but as Kidston[2] and Zeiller[3] have pointed out Sterzel's specimens probably belong to *S. cuneifolium* (Sternb.).

Branches are given off singly from the nodes, and the cones are borne at the tips of branches or branchlets. The cone of *S. emarginatum* agrees very closely with that of *S. cuneifolium*, and is of the same type as that shown in fig. 108. The small branch of *S. emarginatum* represented in fig. 109 does not show clearly the detailed characters of the species, as the leaf-margins are not well preserved.

In one of the largest specimens of this species which I have seen, in the Leipzig Museum, the main stem has internodes of about 3·9 cm. in length, from which a lateral branch with much shorter internodes is given off from a node.

It is important to notice the close resemblance, as pointed out by Zeiller, between some of the narrower-leaved forms of *S. emarginatum* and *S. cuneifolium* (Sternb.)[4]; but in the latter species the margins of the leaves have sharp, and not blunt teeth.

The cone described and figured by Weiss[5] as *Bowmanites germanicus*, since investigated by Solms-Laubach[6], must be referred to this species. Geinitz[7] figured a cone in 1855 as that of *S. emarginatum*, but his determination of the species is a little doubtful. Good figures of the true cone of *S. emarginatum* have been given by Zeiller[8] in his *Flore de Valenciennes*, as well as in his important memoir on the fructification of *Sphenophyllum*.

2. *Sphenophyllum trichomatosum* Stur. Fig. 110, *B*.

The finely-divided leaves of the single whorl shown in fig. 110, *B* (from the Middle Coal-Measures of Barnsley, Yorkshire),

[1] Sterzel (86), pp. 26, 27, etc.　　　[2] Kidston (93), p. 333.

[3] Zeiller (88), p. 414.

[4] *Ibid.* p. 411.　　　　　　　　　[5] Weiss (84), p. 201, Pl. xxi. fig. 12.

[6] Solms-Laubach (95⁴), p. 232.　　　[7] Geinitz (55), Pl. xx. fig. 7.

[8] Zeiller (88), Pl. lxiv. figs. 3—5, and (93), p. 24, Pl. ii. fig. 4.

afford an example of a form of *Sphenophyllum* which is represented by such species as *S. tenerrimum* Ett.[1], *S. trichomatosum*
Stur[2], and *S. myriophyllum*[3] Crép. Probably the specimen
should be referred to *S. trichomatosum*, but it is almost
impossible to speak with certainty as to the specific value
of an isolated leaf-whorl of this form. It has long been known
that the leaves of *Sphenophyllum* may vary considerably, as
regards the size of the segments, on the same plant; and the
occurrence of such finely-divided leaves has lent support to
an opinion which was formerly held by some writers, that
Asterophyllites and *Sphenophyllum* could not be regarded as well-
defined separate genera. This heterophylly of *Sphenophyllum*
has thus been responsible for certain mistaken opinions both as
to the relation of the genus to *Calamocladus*[4] (*Asterophyllites*),
and as regards the view that the finely-divided laminae belonged
to submerged leaf-whorls, while the broader segments were those
of floating or subaerial whorls.

There is a very close resemblance between some of the
deeply-cut and linear segments of a *Sphenophyllum* and the
leaves of *Calamocladus*, but in the former genus the linear
segments are found to be connected basally into a narrow
common sheath. The assertion[5] that the deeply-cut leaves occur
on the lower portions of stems is not supported by the facts.
Kidston[6] has pointed out that the cones are often borne on
branches with such leaves, and the same author refers to a
figure by Germar, in which entire and much-divided leaves
occur mixed together in the same individual specimen. M. Zeiller
recently pointed out to me a similar irregular association of
broader and narrower leaf-segments on the same shoots in some
large specimens in the École des Mines, Paris. Cones of
Sphenophyllum tenerrimum have been figured by Stur[7] and
others; they are characterised by their small size and by the
dissection of the slender free portions of the narrow bracts[8].

[1] Stur (75), p. 108.

[2] Stur (87), Pl. xv. and Kidston (90), p. 59, Pl. i.

[3] Zeiller (88), Pl. lxii. figs. 2—4.

[4] Stur (87); Williamson (74); Seward (89), *etc.*

[5] Renault (82), p. 84, and Newberry (91).   [6] Kidston (90), p. 62.

[7] Stur (75), p. 114, Pl. vii.   [8] Zeiller (93), p. 32.

3. *Sphenophyllum Thoni* Mahr.   Fig. 110, *A*.

Another type of *Sphenophyllum* is illustrated by *S. Thoni*
Mahr as shown in fig. 110, *A*.   This species was first described

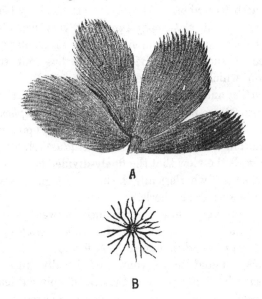

A

B

FIG. 110.   *A.*   *Sphenophyllum Thoni*, Mahr.   (After Zeiller.)
          B.   *Sphenophyllum trichomatosum*, Stur.   From a specimen in the
               Woodwardian Museum; from the Coal-Measures of Barnsley,
               Yorks.   *A* and *B* ¾ nat. size.

by Mahr[1] from the Coal-Measures of Ilmenau, and has since
been figured by Zeiller and other authors.   Each whorl consists
of six large obcuneiform leaves with the broad margin somewhat
irregularly fringed.   The unusually good specimen of which
fig. 110, *A*, represents a single verticil was originally described
and figured by Zeiller in 1880[2]; it is now in the École des
Mines Museum, Paris.

The leaf-forms illustrated by figs. 109 and 110 are some of
the more extreme types of *Sphenophyllum* leaves; but these
are more or less connected by a series of intermediate forms.

[1] Mahr (68), Pl. VIII.   [2] Zeiller (80), p. 34, Pl. CLXI. fig. 9.

For a more complete systematic account of the different species the student should consult such works as those by Coemans and Kickx[1], Zeiller, Schimper, and others.

4. *Sphenophyllum speciosum* (Royle). Fig. 111.

1834. *Trizygia speciosa*, Royle[2]

The species shown in fig. 111 has been usually described as a separate genus *Trizygia*, a name instituted by Royle in 1834 for some Indian fossils from the Lower Gondwana rocks of India[3]. Zeiller[4] has lately pointed out the advisability of including this Asiatic type in the genus *Sphenophyllum*. The slender stem bears verticils of cuneate leaves in three pairs at each node, the anterior pair being smaller than the two lateral pairs. The characteristic *Sphenophyllum* venation is clearly seen in the enlarged leaf, fig. 111, *B*.

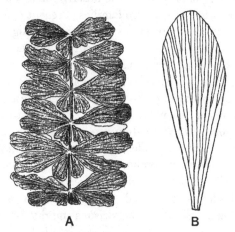

A   B

FIG. 111. *Sphenophyllum speciosum* (Royle).
A. Nat. size.   *B*. enlarged leaf.
From the Raniganj Coal-field, India. (After Feistmantel.)

The inequality of the members of a single whorl, which characterises this Indian plant, is sometimes met with in

[1] Coemans and Kickx (64); Zeiller (80), (88); Schimper (69).

[2] Royle (39), p. 431.

[3] For other figures of this plant, *vide* Feistmantel (81), Pls. XI. A and XII. A.

[4] Zeiller (91).

European species. A specimen of *Sphenophyllum oblongifolium*,
which Prof. Zeiller showed me in illustration of this point, was
practically indistinguishable from *Trizygia*[1].

In some of the earlier descriptions of the Indian species the
generic name *Sphenophyllum*[2] was used by McClelland and
others, but the supposed difference in the leaf-whorls was made
the ground of reverting to the distinct generic term *Trizygia*.
Now that a similar type of leaf-whorl is known to occur in
Sphenophyllum, it is better to adopt that genus rather than to
allow the question of locality to unduly influence the choice
of a separate generic name for an Indian plant.

III. Affinities, range and habit of Sphenophyllum.

It has been pointed out in the description of *Sphenophyllum*,
that the most widely separated families of recent plants have been
selected by different authors as the nearest living allies of this
Palaeozoic genus. It is now generally admitted that *Sphenophyllum* is a generic type apart; it cannot be classed in any
family or subclass of recent or fossil plants, without considerably
extending or modifying the recognised characteristics of existing divisions of the plant-kingdom. The anatomical characters of the *Sphenophyllum* stem are such as one finds in some
recent genera of the Lycopodinae, especially *Psilotum*. If the
stele of *Psilotum* were composed internally of a solid strand of
xylem, we should have a close correspondence between the
centripetally-developed wood of this genus and that of
Sphenophyllum. Similar comparisons might be drawn with
other existing genera, but the more detailed consideration of
the affinities of the Palaeozoic plant will be more easily dealt
with after other members of the Pteridophytes have been
described. The recent discovery of an entirely new type of Carboniferous strobilus in rocks of Calciferous sandstone age on the
shores of the Firth of Forth has thrown new light on the
position of *Sphenophyllum*. *Cheirostrobus Pettycurensis*, the
new cone which Scott has described in an able memoir, affords

[1] *Vide* also Zeiller (92²), p. 75.   [2] Feistmantel (81), p. 69.

certain points of contact with *Sphenophyllum* on the one hand
and with *Calamites* on the other. This important question will
be dealt with after we have given an account of *Cheirostrobus*[1].
To put the matter shortly, *Sphenophyllum* agrees with some
Lycopodinous plants in its anatomical features; with the
Equisetales it is connected by the verticillate disposition of the
leaves, and some of the forms of *Sphenophyllum* strobili present
features which also point to Equisetinous affinities.

In his Presidential address to the Botanical Section at the
British Association Meeting of 1896 Scott[2] thus refers to the
Sphenophyllums :—" We may hazard the guess that this interesting group may have been derived from some unknown
form lying at the root of both Calamites and Lycopods. The
existence of the Sphenophyllae certainly suggests the probability of a common origin for these two series." The result of
the subsequent investigation of the new cone *Cheirostrobus*
amply justifies this opinion as to the position of *Sphenophyllum*.

It is probable that *Sphenophyllum* lived during the
Devonian period, but the unsatisfactory specimens on which
Dawson has founded a species of this age, *S. antiquum*[3], can
hardly be said to afford positive evidence of the Pre-Carboni-
ferous existence of the genus. From the Culm rocks and
other strata older than the Coal-Measures, we have such species
as *S. insigne* (Will.), *Sphenophyllostachys Römeri* (Solms-
Laubach), and *Sphenophyllum tenerrimum*, Ett.[4] while *S. emarginatum*[5], Brongn. occurs in the Upper Coal-Measures and in the
Transition rocks. *S. cuneifolium*[5] (Sternb.) has been recorded
from the Transition, Middle and Lower Coal-Measures. *Sphenophyllum oblongifolium*, Germ.[6], is recorded from Lower Permian
rocks, as is also *S. Thoni*[7], Mahr.

The comparison which has naturally been drawn between
Sphenophyllum with its slender stems bearing occasionally
dimorphic leaves, and water-plants is not, I believe, supported
by the facts of anatomy or external characters. The entire
and finely-dissected leaves do not exhibit that regularity of

[1] Scott (97). [2] Scott (96²), p. 15.
[3] Dawson (61), p. 10, fig. 7. [4] Kidston (94), p. 250.
[5] Kidston (94), p. 250. [6] Sterzel (93), p. 143. [7] Zeiller (94), p. 172.

relative disposition which is characteristic of aquatic plants; the two forms of leaves may occur indiscriminately on the same branch.   The well-developed and thick xylem is not in accordance with the anatomical features usually associated with water-plants.   It is true that in some living dicotyledons of the family Leguminosae, which inhabit swampy places, the secondary xylem is represented by a thick mass of unlignified and thin-walled parenchyma, as in the genus *Aeschynomene*[1], from which the material of 'pith'-helmets is obtained; but the wood of *Sphenophyllum* was obviously thick-walled and thoroughly lignified.

It is not improbable that the long and slender stems of this plant may have grown like small lianas in the Coal-Measure forests, supporting themselves to a large extent on the stouter branches of Calamites and other trees.   The anatomical structure of a *Sphenophyllum* stem would seem to be in accord with the requirements of a climbing plant.   It has been shewn[2] that in recent climbing plants the tracheae and sieve-tubes are characterised by their large diameter, a fact which may be correlated with the small diameter of climbing stems and the need for rapid transport of food material.   In *Sphenophyllum* the tracheae of the xylem have a wide bore, and in *S. insigne* the phloem contains unusually wide sieve-tubes. The central position of the stele is another feature which is not inconsistent with a climbing habit.   Schwendener and others[3] have demonstrated that in climbing organs, as in underground stems and roots, there is a tendency towards a centripetal concentration of mechanical or strengthening tissue.   The axial xylem strand of *Sphenophyllum* would afford an efficient resistance to the tension or pulling force which climbing stems encounter.

[1] De Bary (84), p. 499.   [2] Westermaier and Ambronn (81).
[3] Schwendener (74), p. 124.   Haberlandt (96), p. 165.

LIST OF WORKS REFERRED TO IN THE TEXT

Adamson, S. A. (88) Notes on a recent discovery of *Stigmaria ficoides* at Clayton, Yorkshire. *Quart. Journ. Geol. Soc.* vol. XLIV. p. 375, 1888.

Agassiz, S. (88) Three Cruises of the U. S. Coast and Geodetic Survey Steamer "Blake" in the Gulf of Mexico, in the Caribbean Sea, and along the Atlantic Coast of the United States, from 1877 to 1880. *London,* 1888.

Andrae, K. J. (53) Fossile Flora Siebenbürgens und des Banates. *Abhand. k. k. geol. Reichsanst.* Bd. II. Abth. III. No. 4, 1853.

Andrussow, N. (87) Eine fossile *Acetabularia* als Gestein-bildender Organismus. *Ann. k. k. nat. hist. Hofmuseum, Wien.* Bd. II. Heft II. p. 77, 1887.

d'Archiac, E. J. A. (43) Description géologique du département de l'Aisne. *Mém. Soc. Géol. France,* vol. V. pt. II. p. 129, 1843.

Artis, F. T. (25) Antediluvian Phytology. *London,* 1825.

Baker, J. G. (87) Handbook of the Fern-Allies. *London,* 1887.

Balfour, J. H. (72) Palaeontological Botany. *Edinburgh,* 1872.

Barber, C. A. (89) The Structure of *Pachytheca. Annals Bot.* vol. III. p. 141, 1889.

—— (90) The Structure of *Pachytheca. Ibid.* vol. V. p. 145, 1890.

—— (92) *Nematophycus Storriei,* nov. sp. *Ibid.* vol. VI. p. 329, 1892.

Barrois, C. (88) Note sur l'existence du genre *Oldhamia* dans les Pyrénées. *Ann. Soc. Géol. Nord,* vol. XV. p. 154, 1888.

Barton, R. (1751) Lectures in Natural Philosophy. *Dublin,* 1751.

Bary, A. de. (84) Comparative anatomy of the vegetative organs of the Phanerogams and Ferns. *Oxford,* 1884.

—— (87) Lectures on Bacteria. *Oxford* 1887.

Bates, H. W. (63) The Naturalist on the River Amazons. *London,* 1863. 2 vols.

Bateson, W. B. (88) Suggestion that certain fossils known as *Bilobites* may be regarded as casts of *Balanoglossus. Proc. Camb. Phil. Soc.* vol. VI. p. 298, 1888.

Batters, E. A. (90) *Vide* Holmes, E. M.

—— (92) On *Conchocelis*, a new genus of perforating Algae. *Phyco-logical Memoirs. London*, p. 25, 1892.

Benecke, E. W. (76) Geognostisch-paläontologische Beiträge. Vol. II. Heft III., 1876.

Bennett, J. J. and **Brown, R.** (38). Plantae Javanicae. *London*, 1838—52.

Bentham, G. (70) Presidential Address. *Proc. Linn. Soc.* p. lxxiv. 1869—70.

Bertrand, C. E. (93) Les Bogheads à Algues. *Bull. Soc. Belg. Géol. Paléont. Hydrol.* vol. VII. p. 45, 1893.

—— (96) Nouvelles remarques sur le Kerosene Shale de la Nouvelle-Galles du Sud. *Soc. d'hist. nat. Autun*, 1896.

Bertrand, C. E. and **Renault, B.** (92) *Pila bibractensis* et le Boghead d'Autun. *Soc. d'hist. nat. Autun.* 1892.

—— (94) *Reinschia Australis* et premières remarques sur le Kerosene Shale de la Nouvelle-Galles du Sud. *Soc. d'hist. nat. Autun*, 1894.

Binney, E. W. (68) Observations on the Structure of Fossil plants found in the Carboniferous Strata. Pt. I. *Calamites* and *Calamodendron. Palaeont. Soc. London*, 1868.

—— (71) *Ibid.* Pt. II. *Lepidostrobus* and some allied cones.

Bischoff, G. W. (28) Die cryptogamischen Gewächse. *Nürnberg*, 1828.

Bornemann, J. G. (56) Über organische Reste der Lettenkohlen-Gruppe Thüringens. *Leipzig*, 1856.

—— (86) Geologische Algenstudien. *Jahrb. K. Preuss. Geol. Landesanst. Bergakad.*, 1886.

—— (87) Die Versteinerungen des Cambrischen Schichtensystems der Insel Sardinien. *Nova Acta Acad. Caes. Leop. Car.* vol. LI., 1887.

—— (91) *Ibid.* vol. LVI., 1891.

Bornet, E. and **Flahault.** (89) Note sur deux nouveaux genres d'algues perforantes. *Journ. Bot.* vol. II. p. 161, 1888.

—— (89²) Sur quelques plantes vivant dans le test calcaire des mollusques. *Bull. Soc. Bot. France*, vol. XXXVI. p. clxvii., 1889.

Bower, F. O. (94) Studies in the morphology of Spore-producing members. Equisetineae and Lycopodineae. *Phil. Trans. R. Soc.* vol. CLXXXV. B. p. 473, 1894.

Brongniart, A. (22) Sur la classification et la distribution des végétaux fossiles en général. *Mém. Mus. Hist. Nat. Paris*, vol. VIII. p. 233, 1822.

—— (28) Prodrome d'une histoire des végétaux fossiles. *Paris*, 1828.

—— (28²) Histoire des végétaux fossiles. *Paris*, 1828.

—— (39) Observations sur la structure intérieure du *Sigillaria elegans. Arch. Mus. d'hist. nat.* vol. I. p. 405, 1839.

—— (49) Tableau des genres de végétaux fossiles. *Extrait du Dictionnaire d'histoire naturelle*, vol. XIII. p. 49. *Paris*, 1849.

Brown, A. (94) On the structure and affinities of the genus *Solenopora*. *Geol. Mag.* [4] vol. I. p. 145, 1894.

Brown, R. (11) Some observations on the parts of fructification in Mosses, with characters and descriptions of two new genera of that order. *Trans. Linn. Soc.* vol. x. p. 312, 1811.

—— (38) *Vide* Bennett, J. J.

Bryce, J. (72) The Geology of Arran. *Glasgow* and *London*, 1872.

Buckland, W. (37) Geology and Mineralogy considered with reference to Natural Theology. *London*, 1858.

Buckman, J. (50) On some fossil plants from the Lower Lias. *Quart. Journ. Geol. Soc.* vol. VI. p. 413, 1850.

Bunbury, C. J. F. (51) On some fossil plants from the Jurassic Strata of the Yorkshire Coast. *Quart. Journ. Geol. Soc.* vol. VII. p. 179, 1851.

—— (61) Notes on a collection of fossil plants from Nágpur, Central India. *Ibid.* vol. XVII. p. 325, 1861.

—— (83) Botanical Fragments. *London*, 1883.

Bütschli, O. (83) Bronn's Klassen und Ordnungen des Thier-Reichs, vol. I. *Leipzig* and *Heidelberg*, 1883—87.

Campbell, D. (95) The structure and development of the Mosses and Ferns. *London*, 1895.

Carpenter, W. B., Parker, W. K. and **Jones, T. R.** (62) Introduction to the study of the Foraminifera. (*Ray Society*) *London*, 1862.

Carruthers, W. (67) On the structure of the fruit of *Calamites*. *Journ. Bot.* Vol. v. 1867.

—— (70) On the structure of a Fern stem from the Lower Eocene of Herne Bay. *Quart. Journ. Geol. Soc.* vol. XXVI. p. 349, 1870.

—— (71) On some supposed vegetable fossils. *Ibid.* vol. XXVII. p 443, 1871.

—— (72) On the history, histological structure, and affinities of *Nematophycus Logani*, Carr. (*Prototaxites Logani*, Dawson), an alga of Devonian age. *Micros. Journ.* vol. VIII. p. 160, 1872.

—— (72²) *Vide* Woodward, H., p. 168.

—— (76) Address before the Geologists' Association, 1876—77. *Proc. Geol. Assoc.* vol. v. p. 17, 1876.

Cash, W. and **Hick, T.** (78) A contribution to the flora of the Lower Coal-Measures of the Parish of Halifax, Yorkshire. *Proc. Yorks. Geol. Polyt. Soc.* vol. VII. p. 73, 1878—81.

—— (78²) On fossil fungi from the Lower Coal-Measures of Halifax. *Proc. Yorks. Geol. Polyt. Soc.* vol. VII. p. 115, 1878.

—— (81) A contribution to the flora of the Lower Coal-Measures of the Parish of Halifax, Yorkshire. Pt. III. *Proc. Yorks. Geol. Polyt. Soc.* vol. VII. p. 400, 1881.

Castracane, F. (76) Die Diatomeen in der Kohlenperiode. *Pringsheim's Jahrb.* vol. x. p. 1, 1876.

Cayeux, L. (92) Sur la présence de nombreuses Diatomées dans les gaizes crétacées du bassin de Paris. *Compt. Rend.* vol. CXIV. p. 375, 1892.

—— (97) Contribution à l'étude micrographique des terrains sédimentaires. *Mém. Soc. Géol. Nord,* vol. IV. ii. *Lille,* 1897.

"Challenger." Reports on the Scientific results of the Voyage of H.M.S. "Challenger" during the years 1873—76, under the command of Captain Nares and Captain Thomson. Prepared under the Superintendence of Sir C. Wyville Thomson and John Murray. *London,* 1880—95.

—— (85) Narrative, vols. I. and II. 1882—85.

—— (91) Deep Sea Deposits. J. Murray and A. F. Renard, 1891.

Christison, R. (76) Notice of Fossil trees recently discovered in Craigleith Quarry near Edinburgh. *Phil. Trans. Edinburgh,* vol. XXVII. p. 203, 1876.

Church, A. H. (95) The structure of the Thallus of *Neomeris dumetosa,* Lam. *Annals Bot.* vol. IX. p. 581, 1895.

Coemans, E. and **Kickx, J. J.** (64) Monographie des *Sphenophyllum* d'Europe. *Bull. Acad. Roy. Belg.* vol. XVIII. [2] p. 160, 1864.

Cohn, F. (62) Über die Algen des Carlsbader Sprudels und deren Antheil an der Bildung des Sprudel-Sinters. *Abh. Schles. Ges. vaterland. Cultur,* 1862, p. 65.

Cole, G. (94) On some examples of cone-in-cone structure. *Mineral. Journ.* vol. X. 1894.

Comstock, T. H. (88) An introduction to Entomology. *Ithaca,* 1888.

Conwentz, H. (78) Über ein tertiäres Vorkommen Cypressenartiger Hölzer bei Calistoga in Californien. *Neues Jahrb.* p. 800, 1878.

—— (80) Die fossilen Hölzer von Karlsdorf am Zobten. 1880.

—— (90) Monographie der baltischen Bernsteinbäume. *Danzig,* 1890.

—— (96) On English Amber and Amber generally. *Nat. Science,* vol. IX. p. 99, 1896.

Corda, A. J. (45) Beiträge zur Flora der Vorwelt. *Prag,* 1845. (New edition, unaltered, 1867.)

Cormack, B. G. (93) On a cambial development in *Equisetum. Annals Bot.* vol. VII. p. 63, 1893.

Costa, Mendes da. (1758) An account of the impressions of plants on the slates of coal. *Phil. Trans. R. S. London,* vol. L. p. 228, 1758.

Cotta, C. B. (50) Die Dendrolithen in Beziehung auf ihren inneren Bau. *Dresden,* 1850. (Original edition, 1832.)

Cramer, C. (87) Über die verticillaten Siphoneen, besonders *Neomeris* und *Cymopolia. Denksch. Schweiz. Nat. Ges.* vol. XXV. 1887.

—— (90) Über die verticillaten Siphoneen, besonders *Neomeris* und *Bornetella. Denksch. Schweiz. Nat. Ges.* vol. XXXII. 1890.

Credner, H. (87) Elemente der Geologie. *Leipzig,* 1887.

Crépin, F. (81) L'emploi de la Photographie pour la reproduction des empreintes végétales. *Bull. Soc. Bot. Belg.* vol. XX. pt. II. 1881.

Damon, R. (88) A supplement to the geology of Weymouth and the Isle of Portland. *Weymouth,* 1888.

Darwin, C. (82) The Origin of Species. (Edit. 6.) *London,* 1882.

—— (87) The life and letters of Charles Darwin. Edited by Francis Darwin. *London,* 1887. (Edit. 2.)

—— (90) A Naturalist's Voyage. *London,* 1890.

Dawson, J. W. (59) On fossil plants from the Devonian rocks of Canada. *Quart. Journ. Geol. Soc.* vol. XV. p. 477, 1859.

—— (66) On the conditions of the deposition of coal. *Ibid.* vol. XXII. p. 95, 1866.

—— (71) The fossil plants of the Devonian and Upper Silurian formations of Canada. *Geol. Surv. Canada,* 1871.

—— (81) Notes on new Erian (Devonian) plants. *Quart. Journ. Geol. Soc.* vol. XXXVII. p. 299, 1881.

—— (82) Notes on *Prototaxites* and *Pachytheca* discovered by Dr Hicks in the Denbighshire Grits of Corwen, North Wales. *Ibid.* vol. XXXVIII. p. 103, 1882.

—— (88) The geological history of plants. *London,* 1888.

—— (90) On burrows and tracks of invertebrate animals in Palaeozoic rocks. *Quart. Journ. Geol. Soc.* vol. XLVI. p. 595, 1890.

Deecke, W. (83) Über einige neue Siphoneen. *Neues Jahrb.* Jahrg. I. p. 1, 1883.

Defrance. (26) Dictionnaire des Sciences naturelles. Article *Polytrype,* vol. XLII. p. 453, 1826.

Delgado, J. F. N. (86) Étude sur les Bilobites. *Secc. Trav. Géol. Portugal. Lisbon,* 1886.

Dixon, H. H. and **Joly, J.** (97) Coccoliths in our coastal waters. *Nature,* vol. LVI. p. 468, 1897.

Dixon, H. N. and **Jameson, H. G.** (96) The Student's Handbook of British Mosses. *London,* 1896.

Duncan, P. M. (76) On some unicellular algae parasitic within Silurian and Tertiary Corals, with a notice of their presence in *Calceolina sandalina* and other fossils. *Quart. Journ. Geol. Soc.* vol. XXXII. p. 205, 1876.

—— (76²) On some Thallophytes parasitic within recent Madreporaria. *Proc. R. Soc.* vol. XXV. p. 238, 1876.

Dunker, W. (46) Monographie der norddeutschen Wealdenbildung. *Braunschweig,* 1846.

Duval-Jouve, J. (64) Histoire naturelle des Equisetum de France. *Paris,* 1864.

Ehrenberg, C. G. (36) Über das Massenverhältniss der jetzt lebenden Kieselfusorien und über ein neues Infusorien-Conglomerat als Polirschiefer von Jastrabo in Ungarn. *Abh. k. Akad. Wiss. Berlin,* 1836, p. 109.

—— (54) Mikrogeologie. *Leipzig,* 1854.

420 LIST OF WORKS

Ellis, J. (1755) An essay towards a Natural History of the Corallines. *London*, 1755.

Engelhardt, H. (87) Über *Rosselinia conjugata* (Beck.) eine neue Pilzart aus der Braunkohlen-formation Sachsens. *Ges. Isis. Dresden, Abh.* 4, p. 33, 1887.

Etheridge, R. (81) *Vide* Hicks, H.

Etheridge, R. junr. (92) On the occurrence of microscopic fungi, allied to the genus *Palaeachyla*, Duncan, in the Permo-Carboniferous rocks of New South Wales and Queensland. *Rec. Geol. Surv. N.S.W.* Pt. III. p. 95, 1892.

—— (95) On the occurrence of a plant in the Newcastle or Upper Coal-Measures possessing characters both of the genera *Phyllotheca* Brongn. and *Cingularia* Weiss. *Ibid.* vol. IV. Pt. IV. p. 148, 1895.

Ettingshausen, C. von. (55) Die Steinkohlenflora von Radnitz in Böhmen. *Abhand. k. k. geol. Reichsanst. Wien*, vol. II. p. 1, 1855.

—— (66) Die fossile Flora des Mährisch-Schlesischen Dachschiefers. *Ibid.* vol. XXV. p. 77, 1866.

Ettingshausen, C. von and Gardner, J. S. (79) A monograph of the British Eocene Flora. *Palaeontographical Society. London*, 1879—1882.

Feilden, H. W. (96) Note on the glacial Geology of Arctic Europe and its islands. Pt. I. Kolguev Island. *Quart. Journ. Geol. Soc.* vol. LII. p. 52, 1896. Appendix by A. C. Seward, p. 61.

Feistmantel, O. (73) Das Kohlenvorkommen bei Rothwaltersdorf in der Grafschaft Glatz und dessen organische Einschlüsse. *Zeit. Deutsch. Geol. Ges.* vol. XXV. p. 463, 1873.

—— (75) Die Versteinerungen der böhmischen Kohlenablagerungen. *Palaeontograph.* vol. XXIII. p. 1, 1875—76.

—— (81) The Fossil Flora of the Gondwana System. *Mem. Geol. Surv. India*, (Ser. XII.) vol. III. 1881. (The Flora of the Damuda-Panchet Divisions, pp. 1—77. Pls. I. A—XVI. A *bis*, 1880.)

—— (90) Geological and palaeontological relations of the Coal and plant-bearing beds of Palaeozoic and Mesozoic age in Eastern Australia and Tasmania. *Mem. Geol. Surv. N. S. Wales. Palaeontology*, No. 3, 1890.

Felix, J. (86) Untersuchungen über den inneren Bau westfälischer Carbon-Pflanzen. *Abh. Geol. Landesanst.* 1886, p. 153.

—— (94) Studien über fossile Pilze. *Zeit. deutsch. geol. Ges.* Heft I. p. 269, 1894

—— (96) Untersuchungen über den inneren Bau westfälischer Carbonpflanzen. *Földtani Közlöny*, vol. XXVI. p. 165, 1896.

Fischer, A. (92) Die Pilze. *Rabenhorst's Kryptogamen Flora*, vol. I. Abt. IV. *Leipzig*, 1892.

Fliche, P. and Bleicher, —. Étude sur la flore de l'oolithe inférieure aux environs de Nancy. *Bull. Soc. Sci. Nancy*, vol. v. [2] p. 54, 1881.

Forbes, E. (56) On the Tertiary fluviomarine formation of the Isle of Wight. *Mem. Geol. Surv.* 1856.

Forbes, H. O. (85) A Naturalist's Wanderings in the Eastern Archipelago. *London*, 1885.

Früh, J. (90) Zur Kenntniss der gesteinbildenden Algen. *Abh. Schweiz. pal. Ges.* vol. XVIII. 1890.

Fuchs, T. (94) Über eine fossile *Halimeda* aus dem Eocänen Sandstein von Greifenstein. *Sitzbericht. Abh. Wien*, vol. CIII. Abt. I. p. 300, 1894.

—— (95) Studien über Fucoiden und Hieroglyphen. *Denksch. Akad. Wien*, vol. LXII. p. 369, 1895.

Gardner, J. S. (79) *Vide* Ettingshausen.

—— (84) Fossil plants. *Proc. Geol. Assoc.* vol. VIII. p. 299, 1884.

—— (86) On Mesozoic Angiosperms. *Geol. Mag.* vol. III. p. 193, 1886.

—— (87) On the leaf-beds and gravels of Ardtun, Carsaig, &c., in Mull. *Quart. Journ. Geol. Soc.* vol. XLIII. p. 270, 1887.

Geikie, A. (93) A Text-book of Geology. *London*, 1893. (Edit. iii.)

Geinitz, H. B. (55) Die Versteinerungen der Steinkohlenformation in Sachsen. *Leipzig*, 1855.

Germar, E. F. (44) Die Versteinerungen des Steinkohlengebirges von Wettin und Löbejün in Saalkreis, bildlich dargestellt und beschrieben. *Halle*, 1844—53.

Germar, E. F. and **Kaulfuss, F.** (31) Über einige merkwürdige Pflanzen-Abdrücke. *Nova Acta Leop. Car.* vol. XV. 1831.

Giesenhagen, K. (92) Über hygrophile Farne. *Flora*, vol. LXXVI. p. 157, 1892.

Gomont, M. (88) Recherches sur les enveloppes cellulaires des nostocacées filamenteuses. *Bull. Soc. Bot. France*, vol. XXXV. p. 204, 1888.

—— (92) Monographie des Oscillariées. *Ann. Sci. Nat.* vol. XV. [7] p. 263, 1892.

Göppert, H. R. (36) Über den Zustand in welchem sich die fossilen Pflanzen befinden, und über den Versteinerungsprocess insbesondere. *Poggendorf's Annalen*, vol. XXXVIII. p. 561, 1836. (Also the *Edinburgh New Philosophical Journal*, vol. XXIII. p. 73, 1837.)

—— (36²) Die fossilen Farnkräuter. *Nova Acta Leop. Car.* vol. XVII. (Supplt.) 1836.

—— (45) Tchihatcheff's Voyage Scientifique dans l'Altaï oriental, pp. 379—390. *Paris*, 1845.

—— (52) Über die Flora des Übergangsgebirges. *Nova Acta Ac. Caes. Leop. Car.* vol. XXII. (Supplt.) 1852.

—— (57) Über den versteinten Wald von Radowenz bei Adersbach in Böhmen und über den Versteinerungsprocess überhaupt. *Jahrb. k. k. Geol. Reichs. Wien*, vol. VIII. p. 725, 1857.

Göppert, H. R. (60) Über die fossile Flora der Silurischen, der Devonischen und unteren Kohlenformation. *Nova Acta Ac. Caes. Leop. Car.* vol. XXVII. p. 427, 1860.

—— (64) Die fossile Flora der Permischen Formation. *Palaeontograph.* vol. XII. 1864—65.

Göppert, H. R. and **Berendt, G. C.** (45) Der Bernstein und die in ihm befindlichen Pflanzenreste der Vorwelt. *Berlin*, 1845.

Göppert, H. R. and **Menge, A.** (83) Die Flora des Bernsteins. Vol. I. *Danzig*, 1883.

Gottsche, C. M. (86) Über die im Bernstein eingeschlossenen Lebermoose. *Bot. Central.* vol. XXV. p. 95, 1886.

Grand'Eury, C. (69) Observations sur les Calamites et les Asterophyllites. *Compt. Rend.* vol. LXVIII. p. 705, 1869.

—— (77) Flora Carbonifère du dépt. de la Loire et du centre de la France. *Mém. Ac. Sci. Paris*, vol. XXIV. 1877.

—— (82) Mémoire sur la formation de la houille. *Ann. Mines*, vol. I. 1882.

—— (87) Formation des couches de houille et du terrain houiller. *Mém. Soc. Géol. France*, vol. IV. [3] 1887.

—— (89) Calamariées. Arthropitus et Calamodendron. *Compt. Rend.* vol. CVIII. p. 1086, 1889.

—— (90) Géologie et paléontologie du bassin houiller du Gard. *St. Etienne*, 1890.

Greville, R. K. (47) Notice of a new species of *Dawsonia*. *Ann. Mag. Nat. Hist.* vol. XIX. p. 226, 1847.

Guillemard, F. H. H. (86) The Cruise of the "Marchesa" to Kamschatka and New Guinea. *London*, 1886.

Gümbel, C. W. (71) Die sogenannten Nulliporen. Pt. ii. *Abh. k. Münchener Akad. Wiss.* vol. XI. p. 231, 1871.

Gutbier, A. von. (49) Die Versteinerungen des Rothliegendes in Sachsen. *Dresden* and *Leipzig*, 1849.

Haberlandt, G. (96) Physiologische Pflanzenanatomie (*edit.* 2). *Leipzig*, 1896.

Hall, J. (47) and (52) Natural History of New York. Palaeontology. Vol. I. *Albany*, 1847, vol. II. *Albany*, 1852.

Harker, A. (95) Petrology for students. *Cambridge*, 1895.

Harper, R. A. (95) Die Entwickelung des Peritheciums bei *Sphaerotheca Castagnei*. *Bericht. deutsch. Bot. Ges.* vol. XIII. p. 475, 1895.

Harrison, J. B. (91) *Vide* Jukes-Browne.

Hartig, R. (78) Die Zersetzungserscheinungen des Holzes der Nadelholzbäume und der Eiche in förstlicher botanischer und chemischer Richtung bearbeitet. *Berlin*, 1878.

—— (94) Text-book of the diseases of trees. (Translation by W. Somerville.) *London*, 1894.

Harvey, W. H. (58) Phycologia Australica. *London*, 1858.

Hauck, F. (85) Die Meeresalgen Deutschlands und Oesterreichs. *Rabenhorst's Kryptogamen Flora,* vol. II. *Leipzig,* 1885.

Heer, O. (53) Beschreibung der angeführten Pflanzen und Insekten. *Neue Denksch. Schweiz. Ges. gesammt. Wiss.* Vol. XIII. 5, p. 117, 1853. (Appendix to a Memoir by A. Escher v. d. Linth.)

—— (55) Flora Tertiaria Helvetiae. *Winterthur,* 1855.

—— (65) Die Urwelt der Schweiz. *Zürich,* 1865.

—— (68) Flora fossilis Arctica. Die fossile Flora der Polarlände. Vol. I. 1868. Vol. II. 1871. Vol. III. 1875. Vol. IV. 1877. Vol. V. 1878. Vols. VI. and VII., *Flor. foss. Grön.* Die fossile Flora Grönlands. 1882—83.

—— (76) Flora fossilis Helvetiae. *Zürich,* 1876.

—— (77) Beiträge zur Jura-Flora Ost-Sibiriens und des Amurlandes. *Flora fossilis Arctica,* vol. V. *Zürich,* 1877.

—— (82) Nachträge zur Jura-Flora Sibiriens. (*Flor. foss. Grönlandica,* Th. i.) *Flora foss. Arct.* vol. VI. *Zürich,* 1882.

Heim, A. (78) Untersuchungen über den Mechanismus der Gebirgsbildung. *Basel,* 1878.

Heinricher, E. (82) Die näheren Vorgänge bei der Sporenbildung der *Salvina natans* verglichen mit den der übrigen Rhizocarpeen. *Sitzber. kais. Akad. Wiss. Wien,* vol. LXXXVII. Abt. I. p. 494, 1882.

Hensen, V. (92) Ergebnisse der in dem Atlantischen Ocean Plankton-Expedition der Humboldt Stiftung. *Kiel* and *Leipzig,* vol. I. 1892.

Herzer, H. (93) A new fungus from the Coal-Measures. *American Geologist,* vol. XI. p. 365, 1893.

Hick, T. (78) and (81) *Vide* Cash, W.

—— (93) *Calamostachys Binneyana,* Schimp. *Proc. Yorks. Geol. Polytech. Soc,* vol. XII. pt. IV. p. 279, 1893.

—— (93²) The fruit-spike of *Calamites. Nat. Science,* vol. II. p. 354, 1893.

—— (94) On the primary structure of the stem of *Calamites. Proc. Lit. and Phil. Soc.* [4] vol. VIII. p. 158. *Manchester,* 1894.

—— (95) On the structure of the leaves of *Calamites. Ibid.* vol. IX. [4] p. 179, 1895.

—— (96) On a sporangiferous spike, from the Middle Coal-Measures, near Rochdale. *Ibid.* vol. X. [4] p. 73, 1896.

Hicks, H. (81) On the discovery of some remains of plants at the base of the Denbighshire grits, near Corwen, N. Wales; with an appendix by R. Etheridge. *Quart. Journ. Geol. Soc.* vol. XXXVII. p. 482, 1881.

Hinde, G. J. and **Fox, H.** (95) On a well-marked horizon of Radiolarian rocks in the Lower Culm-Measures of Devon, Cornwall and West Somerset. *Quart. Journ. Geol. Soc.* vol. LI. p. 609, 1895.

Hirschwald, J. (73) Über Umwandlung von verstürzter Holzzimmerung in Braunkohle im alten Mann der Grube Dorothea bei Clausthal. *Zeit. deutsch. geol. Ges.* vol. XXV. p. 364, 1873.

Holmes, E. M. and **Batters, E. A.** (90) A revised list of British marine algae. *Annals Bot.* vol. v. p. 63, 1890—91.

Holmes, W. H. (80) Fossil forests of the Volcanic Tertiary formations of the Yellowstone National Park. *Bull. U. S. Geol. Surv.* vol. v. no. 2, 1880.

Hooker, J. D. (44) The Botany of the Antarctic voyage of H.M. Discovery Ships "Erebus" and "Terror" in the years 1839—43. *London,* 1844.

—— (53) *Vide* Strickland.

—— (81) Presidential Address before Section E of the York meeting of the British Association. *Annual Report,* p. 727, 1881.

—— (89) On *Pachytheca. Annals Bot.* vol. III. p. 135, 1889.

—— (91) Himalayan Journals. *London,* 1891. (*Minerva Library Edition.*)

Hooker, W. J. (20) Musci Exotici. Vol. II. *London,* 1820.

—— (61) A second century of ferns. *London,* 1861.

Howse, R. (88) A catalogue of fossil plants from the Hutton Collection. *Newcastle,* 1888.

Hughes, T. McKenny (79) On the relation of the appearance and duration of the various forms of life upon the earth to the breaks in the continuity of the sedimentary strata. *Proc. Phil. Soc. Cambridge,* vol. III. pt. VI. p. 247, 1879.

—— (84) On some tracks of terrestrial and freshwater animals. *Quart. Journ. Geol. Soc.* vol. XL. p. 178, 1884.

Humboldt, A. von. (48) Cosmos. (Translated by E. C. Otté.) *London,* 1848.

Huxley, T. H. (93) Science and Hebrew tradition. *Collected Essays,* vol. IV. *London,* 1893.

Jäger, G. F. (27) Über die Pflanzenversteinerungen welche in dem Bausandstein von Stuttgart vorkommen. *Stuttgart,* 1827.

James, J. F. (93) Studies in the problematic organisms. No. 2. The genus *Fucoides. Journ. Cincinnati Soc. Nat. Hist.* 1893.

—— (93²) Notes on fossil fungi. *U. S. Dpt. Agriculture,* vol. VII. no. 3, p. 268, 1893.

—— (93³) Fossil fungi. Translated from the French of R. Ferry, with remarks. *Journ. Cincinnati Soc. Nat. Hist.* 1893, p. 94.

Judeich, J. F. and **Nitsche, F.** (95) Lehrbuch der mitteleuropäischen Forstinsektenkunde. *Vienna,* 1895.

Jukes-Browne, A. J. (86) The Student's Handbook of Historical Geology. *London,* 1886.

Jukes-Browne, A. J. and **Harrison, J. B.** (91) The Geology of Barbados. *Quart. Journ. Geol. Soc.* vol. XLVII. p. 197, 1891.

Kayser, E. (95) Text-book of Comparative Geology. (Translated and edited by P. Lake.) *London,* 1895.

Kent, S. (93) The Great Barrier Reef of Australia. *London,* 1893.

Kickx, J. J. (64) *Vide* Coemans, E.

Kidston, R. (83) Report on fossil plants collected by the Geological Survey of Scotland in Eskdale and Liddesdale. *Trans. R. Soc. Edinburgh*, vol. XXX. p. 531, 1883.

—— (83[2]) On the affinities of the genus *Pothocites*, Paterson ; with the description of a specimen from Glencartholm, Eskdale. *Trans. Bot. Soc. Edinburgh*, vol. XVI. p. 28.

—— (86) Catalogue of the Palaeozoic plants in the department of geology and palaeontology, British Museum. *London*, 1886.

—— (88) *Vide* Young.

—— (90) On the fructification of *Sphenophyllum trichomatosum*, Stur, from the Yorkshire Coal-field. *Proc. R. Soc. Phys. Edinburgh*, vol. XI. p. 56, 1890—91.

—— (90[2]) Note on the Palaeozoic species mentioned in Lindley and Hutton's fossil Flora. *Proc. R. Soc. Phys. Edinburgh*, vol. X. p. 345, 1890—91.

—— (92) On the occurrence of the genus *Equisetum* (*E. Hemingwayi*, Kidston) in the Yorkshire Coal-Measures. *Annals Mag. Nat. Hist.* vol. IX. [6] p. 138, 1892.

—— (93) On the fossil plants of the Kilmarnock, Galston, and Kilwinning Coal-fields, Ayrshire. *Trans. R. Soc. Edinburgh*, vol. XXXVII. pt. II. p. 307, 1893.

—— (94) On the various divisions of British Carboniferous rocks as determined by their fossil flora. *Proc. R. Soc. Phys. Soc. Edinburgh*, vol. XII. p. 183, 1894.

Kinahan, J. R. (58) The genus *Oldhamia*. *Trans. R. Irish Acad.* vol. XXIII. p. 547, 1858.

Kippis, A. (78) A narrative of the voyage round the world, performed by Captain James Cook. *London*, 1878.

Kitton, F. (81) *Vide* Shrubsole, W. H.

Kjellman, F. R. (83) The Algae of the Arctic Sea. *Kongl. Svensk Vetenskaps Akad. Hand.* vol. XX. no. 5, 1883.

Knowlton, F. H. (89) Fossil wood and lignite of the Potomac formation. *Bull. U. S. Geol. Surv.* no. 56, 1889.

—— (89[2]) Description of a problematic organism from the Devonian at the falls of the Ohio. *Amer. Journ. Science*, vol. XXXVII. [3] p. 202, 1889.

—— (94) Fossil plants as aids to geology. *Journ. Geol.* vol. II. no. 4, p. 365, 1894.

—— (96) The nomenclature question. *Bot. Gazette*, vol. XXI. p. 82, 1896.

Koechlin-Schlumberger, J. *Vide* Schimper, W. P.

Kölliker, A. (59) On the frequent occurrence of vegetable parasites in the hard structures of animals. *Annals Mag. Nat. Hist.* vol. IV. [3] p. 300, 1859.

Kölliker, A. (59²) Über das ausgebreitete Vorkommen von pflanzlichen Parasiten in den Hartgebilden niederer Thiere. *Zeitsch. wiss. Zool.* vol. x. p. 215, 1859.

König, C. (29) *Vide* Murchison (29) p. 298.

Kuntze, O. (80) Über Geysirs und nebenan entstehende verkieselte Bäume. *Ausland*, 1880, p. 361.

Lake, P. (95) The Denbighshire series of South Denbighshire. *Quart. Journ. Geol. Soc.* vol. LI. p. 9, 1895.

Lamarck, de. (16) Histoire naturelle des animaux sans vertèbres. *Paris*, vol. II. 1816.

Lamouroux, J. (21) Exposition méthodique des genres de l'ordre des Polypiers. *Paris*, 1821.

Lapworth, C. (81) Appendix to W. Keeping's Geology of Central Wales. *Quart. Journ. Geol. Soc.* vol. XXXVII. p. 171, 1881.

Leckenby, J. (64) On the sandstones and shales of the Oolites of Scarborough, with descriptions of some new species of fossil plants. *Quart. Journ. Geol. Soc.* vol. XX. p. 74, 1864.

Lehmann. (1756) Dissertation sur les fleurs de l'Aster montanus. *Hist. Acad. R. Sci. et bell. lett. Berlin*, 1756.

Lesquereux, L. (66) Geological Survey of Illinois. Vol. II. *New York*, 1866.

—— (70) *Ibid.* vol. IV. 1870.

—— (79) Atlas to the Coal Flora of Pennsylvania and of the Carboniferous formation throughout the United States. *Second Geol. Surv. Penn. Report of Progress P. Harrisburg*, 1879.

—— (80) Description of the Coal Flora of the Carboniferous formation in Pennsylvania and throughout the United States. *Ibid.* 1880 and 1884.

—— (84) *Ibid.* vol. III. 1884.

—— (87) A species of Fungus recently discovered in the shale of the Darlington Coal bed at Canneton, Pennsylvania. *Proc. Amer. Phil. Soc.* vol. XVII. p. 173, 1887.

Lhywd (Luidius), E. (1760) (1699) Lithophylacii Britannici ichnographia. *Oxford*, 1760. (First edition published in 1699.)

Limpricht, K. G. (90) Die Laubmoose. *Rabenhorst's Kryptogamen Flora*, vol. IV. *Leipzig*, 1890.

Lindenberg, J. B. W. (39) Species Hepaticarum. *Bonn*, 1839.

Lindley, J. and **Hutton, W.** (31) The fossil Flora of Great Britain. *London*, 1831—37.

Vol. I. Pls. 1—79 ; 1831—33.
Vol. II. Pls. 80--156 ; 1833—35.
Vol. III. Pls. 157—230 ; 1837.

Linnarsson, J. G. O. (69) On some fossils found in the Eophyton Sandstones at Lugnås in Sweden. *Geol. Mag.* vol. VI. p. 393, 1869.

Lister, A. (94) A monograph of the Mycetozoa, being a descriptive Catalogue of the species in the Herbarium of the British Museum. *London,* 1894.

Ludwig, R. (57) Fossile Pflanzen aus der jüngsten Wetterauer Braunkohle. *Palaeontograph,* vol. v. p. 81, 1857.

—— (59) Fossile Pflanzen aus der ältesten Abtheilung der Rheinisch-Wetterauer Tertiär-Formation. *Palaeontograph,* vol. VIII.

Luerssen, C. (89) Die Farnpflanzen oder Gefässbündelkryptogamen. *Rabenhorst's Krypt. Flora,* vol. III. *Leipzig,* 1889.

Lydekker, R. (89) *Vide* Nicholson.

Lyell, C. (29) On a recent formation of freshwater limestone in Forfarshire. *Trans. Geol. Soc.* vol. II. [2] p. 73, 1829.

—— (45) Travels in North America. *London,* 1845.

—— (55) A manual of Elementary Geology. *London,* 1855, (Edit. 5).

—— (67) Principles of Geology. *London,* 1867, (Edit. 10).

—— (78) Elements of Geology. *London,* 1878, (Edit. 3).

M'Coy, F. (47) On the Fossil Botany and Zoology of the rocks associated with the coal of Australia. *Annals Mag. Nat. Hist.* vol. xx. p. 145, 1847.

Mahr, — (68) Über *Sphenophyllum Thoni,* eine neue Art aus dem Steinkohlengebirge von Ilmenau. *Zeit. deutsch. geol. Ges.* vol. xx. p. 433, 1868.

Mantell, G. (44) Medals of Creation. *London,* 1844.

Marsh, O. C. (71) Notice of a fossil forest in the Tertiary of California. *Amer. Journ. Sci.* vol. I. [3] p. 266, 1871.

Martin, W. (09) Petrificata Derbiensia. *Wigan,* 1809.

Maskell, W. M. (87) The Scale-insects. *Wellington,* 1887.

Massalongo, E. G. (51) Sopra le piante fossili dei terreni terziarj del Vicentino, Osservazioni, &c. *Padova,* 1851.

—— (59) Studii sulla flora fossile e geologia stratigrafica del Senigalliese. *Imola,* 1859.

Matthew, G. F. (89) The Cambrian organisms in Acadia. *Trans. R. Soc. Canada* (Sect. IV.) 1889, p. 135.

Meek, F. B. (73) Report of the Geological Survey of Ohio. Vol. I.

Meschinelli, A. (92) Fungi fossiles. In P. A. Saccardo's Sylloge Fungorum. Vol. x. p. 741, *Patavii,* 1892.

Michelin, H. (40) Iconographie zoophytologique. *Paris,* 1840—47.

Migula, W. (90) Die Characeen. *Rabenhorst's Kryptogamen Flora,* vol. v. *Leipzig,* 1890.

Milde, J. (67) Monographia Equisetorum. *Nova Act. Acad. Caes. Leop. Car.* vol. XXXII. p. 1, 1867.

Morris, J. (54) A catalogue of British fossils. *London,* 1854, (Edit. 2).

Morton, G. H. (91) The geology of the country around Liverpool. *London,* 1891, (Edit. 2).

Moseley, H. N. (75) Notes on freshwater algae obtained at the boiling springs at Furnas, St. Michael's, Azores, and the neighbourhood. With notes by W. T. Thiselton-Dyer. *Journ. Linn. Soc.* vol. XIV. (Botany) p. 321, 1875.

Mougeot, A. (44) *Vide* Schimper.

—— (52) Essai d'une flore du nouveau Grès Rouge des Vosges. *Epinal*, 1852.

Munier-Chalmas, E. (77) Observations sur les algues calcaires appartenant au groupe des Siphonées verticillées (Dasycladées Harv.) et confondues avec les Foraminifères. *Compt. Rend.* vol. LXXXV. p. 814, 1877.

—— (79) Observations sur les algues calcaires confondues avec les Foraminifères et appartenant au groupe des Siphonées dichotomes. *Bull. Soc. Géol.* [3] vol. VII. p. 661, 1879.

Murchison, R. I. (29) On the Coal-field of Brora in Sutherlandshire, and some other stratified deposits in the North of Scotland. *Trans. Geol. Soc.* [2] vol. II. p. 293.

—— (72) Siluria. (Edit. 5.) *London*, 1872.

Murray, G. (92) On a fossil alga belonging to the genus *Caulerpa*, from the Oolite. *Phycological Memoirs*, pt. I. p. 11, 1892.

—— (95) An introduction to the study of seaweeds. *London*, 1895.

—— (95²) Calcareous pebbles formed by Algae. *Phycol. Mem.* pt. III. p. 74, 1895.

—— (95³) A new part of *Pachytheca*. *Ibid.* p. 71, 1895.

—— (97) On the reproduction of some marine Diatoms. *Proc. R. Soc. Edinburgh*, vol. XXI. p. 207, 1897.

Murray, G. and Blackman, V. H. (97) Coccospheres and Rhabdospheres. *Nature*, vol. LV. p. 510, 1897.

Murray, J. and Renard, A. (91) *Vide* "Challenger."

Nathorst, A. G. (80) Reseberättelse. *Öfversigt Kongl. Vetenskaps-Akad. Förh.* 1880, no. 5, p. 23.

—— (81) Om spår af några evertebrerade djur m. m. och deras paleontologiska betydelse. *K. Svensk. Vet.-Akad. Hand.* vol. XVIII. no. 7, 1881.

—— (81²) Om Aftryck af Medusar i Sveriges Kambriska layer. *K. Svensk. Vet.-Akad. Hand.* vol. XIX. no. 1, 1881.

—— (83¹) Quelques remarques concernant la question des algues fossiles. *Bull. Soc. Géol.* [3] vol. XI. p. 452, 1883.

—— (83²) Fossil Algae. (Letter.) *Nature*, vol. XXVIII. p. 52, 1883.

—— (86) Nouvelles observations sur des traces d'animaux et autres phénomènes d'origine purement mécanique décrits comme "algues fossiles." *Kongl. Svensk. Vetenskaps-Akad. Hand.* vol. XXI. no. 14. *Stockholm*, 1886.

—— (90) Beiträge zur mesozoischen Flora Japans. *Denkschr. k. Akad. Wiss. Wien*, vol. LVIII. 1890.

Nathorst, A. G. (94) Zur paläozoischen Flora der arktischen Zone. *Kongl. Svensk. Vetenskaps-Akad. Hand.* vol. XXVI. no. 4, 1894.

—— (94²) Sveriges Geologi. *Stockholm*, 1894.

Naunyn, B. (96) A treatise on Cholelithiasis. (Translation by A. E. Garrod.) *New Sydenham Society, London*, 1896.

Neumayr, M. (83) Über klimatische Zonen während der Jura- und Kreidezeit. *Denkschr. k. Ak. Wiss. Wien*, vol. XLVII. 1883.

Newberry, J. S. (88) Fossil fishes and fossil plants of the Triassic rocks of New Jersey and the Connecticut Valley. *U. S. Geol. Surv. (Monographs)*, vol. XIV. 1888.

—— (91) The genus *Sphenophyllum. Journ. Cincinnati Soc. Nat. Hist.* 1891, p. 212.

Nicholson, H. A. (69) On the occurrence of plants in the Skiddaw Slates. *Geol. Mag.* vol. VI. p. 494, 1869.

Nicholson, H. A. and Etheridge, R. junr. (80) A monograph of the Silurian fossils of the Girvan district in Ayrshire. *Edinburgh*, 1880.

Nicholson, H. A. and Lydekker, R. (89) A manual of Palaeontology. *London*, 1889.

Nicol, W. (34) Observations on the structure of recent and fossil Coniferae. *Edinb. New Phil. Journ.* vol. XVI. p. 137, 1834.

Noll, F. (95) Lehrbuch der Botanik für Hochschulen. Strasburger, Noll, Schimper und Schenck. *Jena*, 1895, p. 248 (Edit. 2).

Parkinson, J. (11) Organic Remains of a former world. *London*, 1811. (Vol. I. 1811, vol. II. 1808, vol. III. 1811.)

Parsons, J. (1757) An account of some fossil fruits and other bodies found in the island of Sheppey. *Phil. Trans.* vol. L. pt. I. p. 396, 1757.

Paterson (41) Description of *Pothocites Grantoni*, a new fossil vegetable from the Coal Formation. *Trans. Bot. Soc. Edinburgh*, vol. I. p. 45, 1841.

Penhallow, D. P. (89) On *Nematophyton* and allied forms from the Devonian (Erian) of Gaspé and Bay des Chaleurs. *Trans. R. Soc. Canada*, vol. VI. sect. 4 (1888).

—— (93) Notes on *Nematophyton crassum. Proc. U. S. Nat. Museum*, vol. XVI. p. 115, 1893.

—— (96) *Nematophyton Ortoni*, n. sp. *Annals Bot.* vol. X. p. 41, 1896.

Petzholdt, A. (41) Über Calamiten und Steinkohlenbildung. *Dresden* and *Leipzig*, 1841.

Pfitzer, E. H. H. (67) Über die Schutzscheide. *Pringsh. Jahrb. Wiss. Bot.* vol. VI. p. 292, 1867—68.

—— (71) Untersuchungen über Bau und Entwickelung der Bacillariaceen. *Hanstein's Bot. Abhand.* 1871.

Philippi, R. A. (37) Beweis das die Nulliporen Pflanzen sind. *Wiegmann, Archiv.* vol. I. Jahrg. 3, p. 387, 1837.

Phillips, J. (29) Illustrations of the Geology of Yorkshire. *York,* 1829.
—— (75) *Ibid.* (Edit. 3) edited by R. Etheridge. *London,* 1875.
Phillips, W. (93) The breaking of the Shropshire Meres. *Midland Naturalist,* vol. XVI. p. 56, 1893.
Plot, R. (1705) The Natural History of Oxfordshire. *Oxford,* 1705.
Potonié, R. (87) Die fossile Pflanzen-Gattung *Tylodendron. Jahrb. k. Preuss. geol. Landesanst.* 1887, p. 313.
—— (90) Der im Lichthof der Königl. Geol. Landesanstalt und Bergakademie aufgestellte Baumstumpf mit Wurzeln aus dem Carbon des Piesberges. *Ibid.* 1890, p. 246.
—— (92) Der äussere Bau der Blätter von *Annularia stellata* (Schloth.) mit ausblicken auf *Equisetites zeaeformis* (Schloth.) Andrae, und auf die Blätter von *Calamites varians* Sternb. *Ber. deutsch. bot. Ges.* vol. X. Heft 8, p. 561, 1892.
—— (93) Die Flora des Rothliegenden von Thüringen. (Th. I.) *Jahrb. k. Preuss. geol. Landesanst.* vol. IX. 1893.
—— (94) Über die Stellung der Sphenophyllum im System. *Ber. deutsch. bot. Ges.* Bd. XII. Heft 4, p. 97, 1894.
—— (96) Die floristische Gliederung des deutschen Carbon und Perm. *Abhand. k. Preuss. geol. Landesanst.* (N. F.) Heft. XXI. p. 1, 1896.
—— (96²) Palaeophytologische Notizen. *Naturwiss. Wochenschrift,* vol. XI. no. 10, p. 115, 1896.
Quekett, J. (54) Lectures on Histology. Vols. I. and II. *London,* 1852—54.
Raciborski, M. (94) Flora Kopalna ogniotrwalych Glinek Krakowskich. *Pamięt. Akad. Umiejętnosci,* 1894.
Reinsch, P. F. (81) Neue Untersuchungen über die Mikrostruktur der Steinkohle. *Leipzig,* 1881.
Renault, B. (73) Recherches sur l'organisation des Sphenophyllum et des Annularia. *Ann. Sci. Nat.* [5] vol. XVIII. p. 5, 1873.
—— (76) Recherches sur la fructification de quelques végétaux. *Ann. Sci. Nat.* [6] vol. III. p. 5, 1876.
—— (76²) Nouvelles recherches sur la structure des Sphenophyllum. *Ann. Sci. Nat.* [6] vol. IV. p. 277, 1876.
—— (82) Cours de botanique fossile. Vol. II. *Paris,* 1882.
—— (85) Recherches sur les végétaux fossiles du genre Astromyelon *Ann. Sci. Géol.* vol. XVII. 1885.
—— (93) Bassin Houiller et Permien d'Autun et d'Epinac. (*Atlas.*) *Etudes des gîtes minéraux de la France,* Fasc. IV. *Paris,* 1893.
—— (95) Note sur les cuticles de Tovarkovo. *Soc. Hist. Nat. d'Autun,* 1895.
—— (95²) Sur quelques Bactéries des temps primaires. *Autun,* 1895.
—— (96) Bassin Houiller et Permien d'Autun et d'Epinac. (*Text.*) *Etudes des gîtes minéraux de la France,* Fasc. IV. *Paris,* 1896.
—— (96²) Notice sur les travaux scientifiques. *Autun,* 1896.

Renault, B. (96³) Recherches sur les Bactériacées fossiles. *Ann. Sci. Nat.* [8] vol. II. p. 275, 1896.

Renault, R. and **Bertrand** (94) Sur une bactérie coprophile de l'époque Permienne. *Compt. Rend.* vol. CXIX. p. 377, 1894.

Renault, R. and **Zeiller, R.** (88) Études sur le terrain houiller de Commentry. *St Étienne*, 1888—90. Pt. I. pp. 1—366, 1888. Pt. II. pp. 367—746, 1890.

"Report" (62) Report of the Jury trial in the action of declaration, &c. at the instance of Mr and Mrs Gillespie, of Torbanehill, against Messrs James Russell and Son, Coal-masters, Falkirk. *Edinburgh*, 1862.

Reuss, A. E. (61) Über die fossile Gattung *Acicularia* d'Archiac. *Sitzb. d. k. Akad. Wiss.* Bd. XLIII. Abth. I. p. 7, 1861.

Rodway, J. (95) In the Guiana Forest. *London*, 1895.

Römer, A. (54) Beiträge zur geologischen Kenntniss des nordwestlichen Harzgebirges von Friedrich. *Palaeontograph.* vol. III. p. 1, 1854.

Römer, F. (62) Beiträge zur geologischen Kenntniss des nordwestlichen Harzgebirges. *Palaeontograph.* vol. IX. 1862.

—— (70) Geologie von Oberschlesien. *Breslau*, 1870.

Rosanoff, S. (66) Recherches anatomiques sur les Mélobésiées. *Mém. Soc. Sci. nat. Cherbourg*, vol. XI. 1865.

Rose, C. B. (55) On the discovery of parasitic borings in fossil fish-scales. *Trans. mic. soc.* (*N. S.*) vol. III. p. 7, 1855.

Rosenvinge, L. K. (93) Grönlands Havalger. Conspectus florae Groenlandicae. Pt. III. p. 765. Grönlands Havalger. (*Meddelelser om Grönland*. III. *Supplt.* 3), 1887—94.

Rothpletz, A. (80) Die Flora und Fauna der Carbon Culmformation bei Hainischen in Sachsen. *Bot. Cent.* vol. II. (*Gratis Beilage* 3), 1880.

—— (90) Über *Sphaerocodium Bornemanni*, eine neue fossile Kalkalge aus den Raibler Schichten der Ostalpen. *Bot. Cent.* vol. XLI. p. 9, 1890.

—— (91) Fossile Kalkalgen aus den Familien der Codiaceen und der Corallineen. *Zeit. deutsch. geol. Ges.* vol. XLIII. p. 295, 1891.

—— (92) Über die Bildung der Oolithe. *Bot. Cent.* vol. LI. p. 265, 1892.

—— (92²) Über die Diadematiden-Stacheln und *Haploporella fasciculata* aus dem Oligocaen von Astrupp. *Bot. Cent.* vol. LII. p. 235, 1892.

—— (94) Ein geologischer Querschnitt durch die Ost-Alpen nebst Anhang über die sogenannte Glarner Doppelfalte. *Stuttgart*, 1894.

—— (96) Über die Flysch-Fucoiden und einige andere fossile Algen sowie über liassische Diatomeen führende Hornschwämme. *Zeit. deutsch. geol. Ges.* vol. XLVIII. p. 854, 1896.

Royle, J. F. (39) Illustrations of the Botany and other branches of the Natural History of the Himalayan Mountains and of the flora of Cashmere. *London*, 1839.

Rultey, F. (92) Notes on Crystallites. *Min. Mag.* vol. IX. p. 261, 1892.

Salter, J. W. (63) On some fossil crustacea from the Coal-Measures and Devonian rocks of British North America. *Quart. Journ. Geol. Soc.* vol. XIX. p. 75, 1863.

—— (73) A catalogue of Cambrian and Silurian fossils. *Cambridge*, 1873.

Saporta, de G. (68) Prodrome d'une flore fossile dès travertins anciens de Sézanne. *Mém. Soc. Géol. France*, vol. VIII. [2] 1865—68.

—— (72) Plantes Jurassiques. *Paléont. Franç. Végétaux*, vols. I—IV. 1872—1891.

—— (73) *Ibid.* vol. I. Algues, Équisétacées, Characées, Fougères, 1873.

—— (75) *Ibid.* vol. II. Cycadées, 1875.

—— (77) Sur la découverte d'une plante terrestre dans la partie moyenne du terrain Silurien. *Compt. Rend.* vol. LXXXV. p. 500, 1877.

—— (79) Le monde des plantes avant l'apparition de l'homme. *Paris*, 1879.

—— (81) *Vide* Saporta and Marion.

—— (82) A propos des algues fossiles. *Paris*, 1882.

—— (84) Les organismes problématiques des anciennes mers. *Paris*, 1884.

—— (86) Nouveaux documents relatifs à des fossiles végétaux et à des traces d'invertébrés associés dans les anciens terrains. *Bull. Soc. bot. France*, vol. XIV. [3] p. 407, 1886.

—— (91) Plantes Jurassiques. Vol. IV. *Paris*, 1891.

Saporta, de G. and **Marion, A. F.** (81) L'Évolution du règne végétal. Les Cryptogames. *Paris*, 1881.

Schenk, A. (65) *Vide* Schoenlein.

—— (67) Die fossile Flora der Grenzschichten des Keupers und Lias Frankens. *Wiesbaden*, 1867.

—— (83) Pflanzliche Versteinerungen. *Richthofen's "China,"* vol. IV. *Berlin*, 1883.

—— (88) Die fossile Pflanzenreste. (*Handbuch der Botanik.*) *Breslau*, 1888.

—— (90) *Vide* Schimper.

Scheuchzer, J. T. (1723) Herbarium diluvianum. *Lugduni Batavorum*, 1723.

Schiffner, V. and **Müller, C.** (95) Engler und Prantl ; Die natürlichen Pflanzenfamilien. Teil I. Abt. 3. *Leipzig*, 1895.

Schimper, W. P. (65) Euptychium Muscorum Neocaledonicorum genus novum et genus *Spiridens. Nova Acta Acad. Caes. Leop. Car.* vol. XXXII. 1865.

—— (69) Traité de Paléontologie végétale. *Paris*, vol. I. 1869, vol. II. 1870—72, vol. III. 1874.

—— (74) *Ibid.* Atlas.

Schimper, W. P. and **Koechlin-Schlumberger, J.** (62) Mémoire sur le Terrain de transition des Vosges. *Mém. Soc. Sci. Nat. Strasbourg*, vol. v. 1862.

Schimper, W. P. and **Mougeot** (44) Monographie des plantes fossiles du grès bigarré de la Chaîne des Vosges. *Leipzig*, 1844.

Schimper, W. P. and **Schenk, A.** (90) Palaeophytologie. *Zittel's Handbuch der Palaeontologie*, Abt. II. *Munich* and *Leipzig*, 1890.

Schlotheim, E. F. von. (04) Beschreibung merkwürdiger Kräuter-Abdrücke und Pflanzen-Versteinerungen. *Gotha*, 1804.

—— (20) Die Petrefactenkunde auf ihrem jetzigen Standpunkte durch die Beschreibung seiner Sammlung versteinerter und fossiler Über-reste des Thier- und Pflanzenreichs der Vorwelt erläutert. *Gotha*, 1820.

—— (22) Nachträge zur Petrefactenkunde. *Gotha*, 1822—23.

Schlüter, C. (79) *Coelotrochium Decheni*, eine Foraminifere aus dem Mitteldevon. *Zeit. deutsch. geol. Ges.* vol. XXXI. p. 668, 1879.

Schmalhausen, J. (79) Beiträge zur Juraflora Russlands. *Mém. Acad. imp. St Pétersbourg*, [7] vol. XXVII. no. 4, 1879.

Schmitz, F. (97) Rhodophyceae. Engler und Prantl; Die natürlichen Pflanzenfamilien. Teil I. Abt. 2. *Leipzig*, 1897.

Schoenlein, J. L. and **Schenk, A.** (65) Abbildungen von fossilen Pflanzen aus dem Keuper Frankens. *Wiesbaden*, 1865.

Schröter, J. (89) Engler und Prantl; Die natürlichen Pflanzenfamilien. Teil I. Abth. 1. 1889.

Schulze, C. F. (1755) Kurtze Betrachtung derer Kräuterabdrücke im Steinreiche. *Dresden* and *Leipzig*, 1755.

Schulze, F. (55) Über das Vorkommen wohlerhaltener Cellulose in Braunkohle und Steinkohle. *Bericht. k. Preuss. Akad. Wiss. Berlin*, 1855, p. 676.

Schütt, F. (93) Das Pflanzenleben der Hochsee. *Kiel* and *Leipzig*, 1893. (From Hensen's Plankton-Expedition.)

—— (96) Bacillariales. Engler und Prantl; Die natürlichen Pflan-zenfamilien. Teil I. Abt. 1. b. 1896.

Schweinfurth, G. (82) Zur Beleuchtung der Frage über den verstein-erten Wald. *Zeit. deutsch. geol. Ges.* vol. XXXIV. p. 139, 1882.

Schwendener, S. (74) Das mechanische Princip im anatomischen Bau der Monocotylen. 1874.

Scott, D. H. (94) (95) (96) *Vide* Williamson, W. C.

—— (96) An introduction to structural Botany. Pt. II. Flowerless plants. *London*, 1896.

—— (96²) Presidential address. (Liverpool Meeting.) *British Assoc. Report*, p. 992, 1896.

—— (97) On the structure and affinities of fossil plants from the Palaeozoic rocks. On *Cheirostrobus*, a new type of fossil cone from the Lower Carboniferous strata (Calciferous Sandstone Series). *Phil. Trans. R. Soc.* vol. CLXXXIX. B. p. 1, 1897.

Seeman, B. (65) A gigantic *Equisetum*. *Journ. Bot.* vol. III. p. 123, 1865.

Seward, A. C. (88) On *Calamites undulatus*. *Geol. Mag.* vol. v. p. 289, 1888.

—— (89) *Sphenophyllum* as a branch of *Asterophyllites*. *Mem. and Proc. Lit. Phil. Soc. Manchester*, vol. III. [4] p. 1, 1889—90.

—— (94) Algae as rock-building organisms. *Science Progress*, vol. II. no. 7, p. 10, 1894.

—— (94^2) Catalogue of the Mesozoic plants in the Department of Geology, British Museum. The Wealden Flora, Pt. I. *London*, 1894.

—— (95) *Ibid.* Pt. II. *London*, 1895.

—— (95^2) Coal: its structure and formation. *Science Progress*, vol. II. p. 355, 1895.

—— (95^3) Notes on *Pachytheca*. *Proc. Phil. Soc. Cambridge*, vol. VIII. pt. 5, p. 278, 1895.

—— (96) Notes on the geological history of Monocotyledons. *Annals Bot.* vol. x. p. 205, 1896.

—— (96^2) *Vide* Feilden, H. W.

—— (97) On *Cycadeoidea gigantea*, a new Cycadean stem from the Purbeck beds of Portland. *Quart. Journ. Geol. Soc.* vol. LIII. p. 22, 1897.

—— (97^2) On the association of *Sigillaria* and *Glossopteris* in South Africa. *Ibid.* p. 315, 1897.

Sharpe, S. (68) On a remarkable incrustation in Northamptonshire. *Geol. Mag.* vol. v. p. 563, 1868.

Sherborn, C. D. (93) An index to the genera and species of the Foraminifera. *Smithsonian Miscellaneous Collections*, 856. *Washington*, 1893.

Shrubsole, W. H. and **Kitton, F.** (81) The Diatoms of the London Clay. *Journ R. Mic. Soc.* [2] vol. I. p. 381, 1881.

Skertchley, J. B. J. (77) The geology of the Fenland. *Mem. Geol. Surv.* 1877.

Smith, W. G. (77) A fossil *Peronospora*. *Gard. Chron.* Oct. 20, 1877, p. 499.

Sollas, W. J. (86) On a specimen of slate from Bray-Head traversed by the structure known as *Oldhamia radiata*. *Proc. R. Soc. Dublin* (N.S.) vol. v. pp. 355, 358, 1886—87.

Solms-Laubach, Graf zu. (81) Die corallinen Algen des Golfes von Neapel und der angrenzenden Meeres-Abschnitte. *Fauna und Flora des Golfes von Neapel*, vol. IV. *Leipzig*, 1881.

—— (91) Fossil Botany. *Oxford*, 1891.

—— (93) Über die Algen-genera *Cymopolia*, *Neomeris* und *Bornetella*. *Ann. Jard. Buitenzorg*, vol. XI. p. 61, 1893.

—— (95) William Crawford Williamson. (Obit. notice.) *Nature*, vol. LII. p. 441, 1895.

Solms-Laubach, Graf zu. (95²) Über Devonische Pflanzenreste aus den Lenneschiefern der Gegend von Gräfrath am Niederrhein. *Jahrb. k. preuss. geol. Landesanst. Berlin,* 1895, p. 67.

—— (95³) Monograph of the Acetabularieae. *Trans. Linn. Soc.* [2] vol. v. pt. 1. p. 1, 1895.

—— (95⁴) *Bowmanites Römeri,* eine neue Sphenophylleen-Fructification. *Jahrb. k. k. geol. Reichsanst. Wien,* vol. XLV. Heft 2, p. 225, 1895.

—— (96) Über die seinerzeit von Unger beschriebenen Struktur bietenden Pflanzenreste des Unterculm von Saalfeld in Thüringen. *Abhand. k. Preus. Geol. Landesanst.* (N.F.) Heft XXIII. *Berlin,* 1896.

—— (97) Über die in den Kalksteinen des Culm von Glätzisch Falkenberg in Schlesien enthaltenen Structur bietenden Pflanzenreste. *Bot. Zeit.* Heft XII. p. 219, 1897.

Sorby, H. C. (79) Presidential Address. *Proc. Geol. Soc.* 1878—79, p. 56.

Spencer, J. (81) *Astromyelon* and its affinities. *Geol. Polytech. Soc.* W. R. Yorks. 1881, p. 439.

Sprengel, A. (28) Commentatio de Psarolithis. *Halle,* 1828.

Stefani, K. de. (82) Vorläufige Mittheilung über die rhätischen Fossilen der apuanischen Alpen. *Verhand. k. k. Geol. Reichsanst. Wien,* 1882, p. 96.

Steinhauer, H. (18) On fossil Reliquia of unknown vegetables in the Coal Strata. *Trans. Phil. Soc. America,* N. S. vol. I. p. 265, 1818.

Steinmann, G. (80) Zur Kenntniss fossiler Kalkalgen (Siphoneae). *Neues Jahrb.* vol. II. p. 130, 1880.

Stenzel, G. (86) *Rhizodendron oppoliense,* Göpp. *Jahresbericht. Schles. Ges. Vat.-Cult.* Ergänzheft LXIII. 1886.

Sternberg, C. von. (20) Versuch einer geognostisch-botanischen Darstellung der Flora der Vorwelt. Fasc. I. *Leipzig,* 1820.

—— (21) *Ibid.* Fasc. II.

—— (25) *Ibid.* Fasc. IV.

—— (38) *Ibid.* Fasc. VIII.

Sterzel, J. T. (82) Über die Fruchtähren von *Annularia Sphenophylloides* (Zenk.). *Zeit. deutsch. geol. Ges.* vol. XXXIV. p. 685, 1882.

—— (86) Die Flora des Rothliegenden im nordwestlichen Sachsen. *Pal. Abh. Dames und Kayser,* vol. III. Heft 4, 1886.

—— (93) Die Flora des Rothliegenden im Plauenscher Grunde bei Dresden. *Abhand. k. Sächs. Ges. Wiss.* vol. XIX. *Leipzig,* 1893.

Stokes, C. (40) Notice respecting a piece of recent wood partly petrified by carbonate of lime, with some remarks on fossil wood. *Trans. Geol. Soc.* [2] vol. v. p. 207, 1840.

436 LIST OF WORKS

Stolley, E. (93) Über Silurische Siphoneen. *Neues Jahrb.* vol. II.
p. 135, 1893.

Strasburger, E. (91) Über den Bau und die Verrichtungen der Lei-
tungsbahnen in den Pflanzen. *Histologische Beiträge*, Heft III.
Jena, 1891.

Strickland, H. E. and Hooker, J. D. (53) On the distribution of
organic contents of the Ludlow bone-bed in the districts of
Woolhope and May Hill. *Quart. Journ. Geol. Soc.* vol. IX. p. 8,
1853.

Stur, D. (75) Die Culm-Flora. Heft 1. Die Culm-Flora des mährisch-
schlesischen Dachschiefers. Heft 2. Die Culm-Flora des Ostrauer
und Waldenburger Schichten. *Abhand. k. k. geol. Reichsanst. Wien*,
vol. VIII. 1875—77.

—— (85) Über die in Flötzen-reiner Steinkohle enthaltenen Stein-
Rundmassen und Torf-Sphärosiderite. *Jahrb. k. k. geol. Reichsanst.
Wien*, vol. XXXV. Heft 3, p. 613, 1885.

—— (87) Die Carbon-Flora der Schatzlarer Schichten. Abt. II. Die
Calamarien. *Abhand. k. k. geol. Reichsanst. Wien*, vol. XI. Abt. 2,
1887.

Suckow, G. A. (1784) Beschreibung einiger merkwürdigen Abdrücke
von der Art der sogenannten Calamiten. *Hist. Comm. Acad. elect.
Sci. litt. Theod.-Palat.* vol. V. p. 355. *Mannheim*, 1784.

Tenison-Woods, J. E. (83) On the Fossil Flora of the Coal deposits of
Australia. *Proc. Linn. Soc. New South Wales*, vol. VIII. p. 37.
Sydney, 1884.

Thiselton-Dyer, W. T. (72) On some fossil wood from the Lower
Eocene. *Geol. Mag.* vol. IX. p. 241, 1872.

—— (91) Note on Mr Barber's paper on *Pachytheca. Annals Bot.*
vol. V. p. 223, 1890—91.

—— (95) Presidential address. *Brit. Assoc. Report, Ipswich*, p. 836,1895.

Thomas, K. (48) On the Amber of East Prussia. *Annals Mag.* vol. II.
[2] p. 369, 1848.

van Tieghem, P. (77) Sur le *Bacillus amylobacter* et son rôle dans le
putréfaction des tissues végétaux. *Bull. Soc. bot.* vol. XXIV. p. 128,
1877.

—— (79) Sur le ferment butyrique (*Bacillus amylobacter*) à l'époque
de la houille. *Compt. Rend.* vol. LXXXIX. p. 1102, 1879.

—— (91) Traité de Botanique. *Paris*, 1891.

Tilden, J. E. (97) On some algal characteristics of the Yellowstone
Park. *Bot. Gazette*, vol. XXIV. p. 194, 1897.

Treub, M. (88) Notice sur la nouvelle Flore de Krakatoa. *Ann. jard.
bot. Buitenzorg*, vol. VII. p. 213, 1888.

Turner, D. (11) Historia Fucorum. *London*, 1811.

Unger, F. (40) Über die Struktur der Calamiten und ihre Rangordnung
im Gewächsreiche. *Flora*, Jahrg. XIII. p. 654, 1840.

Unger, F. (44) Ein Wort über Calamiten und Schachtelhalmähnliche Pflanzen der Vorwelt. *Bot. Zeit.* Jahrg. 2, p. 177, 1844.

—— (50) Genera et species Plantarum fossilium. *Vienna*, 1850.

—— (58) Beiträge zur näheren Kenntniss des Leithakalkes. *Denksch. k. Akad. Wiss. München*, vol. XIV. p. 13, 1858.

Vaillant, —. (1719) Caractères de quatorze genres de plantes. *Hist. Acad. Roy. Sciences,* Ann. 1719, p. 9. *Paris,* 1721.

Vinci, Leonardo da. (83) Literary works. Compiled and edited from the original MSS. by J. P. Richter. *London,* 1883.

Vines, S. H. (95) A student's text-book of Botany. *London,* 1895.

Vogelsang, H. (74) Die Krystalliten. *Bonn,* 1874.

Volkmann, G. A. (1720) Silesia subterranea. *Leipzig,* 1720.

Wallace, A. R. (86) The Malay Archipelago. *London,* 1886.

Walther, J. (85) Die Gesteinbildenden Kalkalgen des Golfes von Neapel und die Entstehung structurloser Kalke. *Zeit. deutsch. geol. Ges.* vol. XXXVII. p. 329, 1885.

—— (88) Die Korallenriffe der Sinaihalbinsel. *Abh. math. phys. Cl. K. Sächs. Ges.* vol. XIV. 1888.

Ward, L. F. (84) Sketch of Palaeobotany. *U. S. Geol. Surv. Fifth Ann. Report. Washington,* 1883—84.

—— (92) Principles and methods of geologic correlation by means of fossil plants. 1892.

—— (96) Fossil plants of the Wealden. *Science,* June 12, p. 869, 1896.

Warming, E. (96) Lehrbuch der ökologischen Pflanzengeographie. (German edit.) *Berlin,* 1896.

Watelet, Ad. (66) Description des plantes fossiles du Bassin de Paris. *Paris,* 1866.

Wedl, C. (59) Über die Bedeutung der in den Schalen von manchen Acephalen und Gasteropoden vorkommenden Canäle. *Sitzbericht. Akad. Wiss. Wien,* vol. XXXIII. p. 451, 1859.

Weed, W. H. (87) The formation of Travertine and siliceous Sinter. *U. S. Geol. Surv. Annual Rep.* IX. 1887—88.

Weiss, C. E. (76) Steinkohlen-Calamarien, mit besonderer Berücksichtigung ihrer Fructificationen. *Abhand. geol. Specialkarte von Preussen und den Thüringischen Staaten,* Bd. II. Heft 1, 1876.

—— (84) Steinkohlen-Calamarien. *Ibid.* Bd. v. Heft 2, 1884.

Westermaier, M. and **Ambronn, H.** (81) Beziehungen zwischen Lebensweise und Structur der Schling- und Kletterpflanzen. *Flora,* 1881, p. 417.

Wethered, E. (93) On the microscopic structure of the Wenlock limestone. *Quart. Journ. Geol. Soc.* vol. XLIX. p. 236, 1893.

Whitney, J. D. and **Wadsworth, M. E.** (84) The Azoic system. *Bull. Mus. Comp. Zool. Harvard Coll.* VII. 1884.

Wille, N. (97) Die Chlorophyceae. Engler und Prantl; Die natürlichen Pflanzenfamilien. Teil I. Abt. 2. *Leipzig,* 1897.

Williamson, W. C. (71) On the organization of the Fossil Plants of the Coal-Measures. Memoir I. *Calamites. Phil. Trans. R. Soc.* vol. CLXI. p. 477, 1871.

—— (71²) On the structure of the woody zone of an undescribed form of Calamite. *Proc. Lit. Phil. Soc. Manchester*, vol. IV. [3] p. 155, 1871.

—— (71³) On a new form of Calamitean Strobilus from the Lancashire Coal-Measures. *Proc. Lit. Phil. Soc. Manchester*, vol. IV. [3] p. 248, 1871.

—— (73) On the organization, &c. Mem. IV. *Dictyoxylon, Lyginodendron* and *Heterangium. Phil. Trans.* vol. CLXIII. p. 377, 1873.

—— (74) *Ibid.* Mem. V. *Asterophyllites. Phil. Trans.* vol. CLXIV. p. 41, 1874.

—— (76) On the organization of *Volkmannia Dawsoni*, an undescribed verticillate strobilus from the Lower Coal-Measures of Lancashire. *Proc. Lit. Phil. Soc. Manchester*, vol. V. [3] p. 28, 1876.

—— (78) On the organization, &c. Mem. IX. *Phil. Trans.* vol. CLXIX. p. 319, 1878.

—— (80) *Ibid.* Mem. X. *Phil. Trans.* vol. CLXXI. p. 493, 1880.

—— (81) *Ibid.* Mem. XI. *Phil. Trans.* vol. CLXXII. p. 283, 1881.

—— (82) *Helophyton Williamsonis. Nature*, vol. XXV. p. 124, 1882.

—— (83) On some anomalous Oolitic and Palaeozoic forms of vegetation. *R. Instit. Great Britain.* (*Weekly Evening Meeting*), Feb. 16, 1883.

—— (83²) On the organization, &c. Mem. XII. *Phil. Trans.* vol. CLXXIV. p. 459, 1883.

—— (85) On some undescribed tracks of invertebrate animals from the Yoredale rocks. *Proc. Lit. Phil. Soc. Manchester*, vol. X. [3] p. 19, 1885.

—— (87) A monograph of the morphology and histology of *Stigmaria ficoides. Palaeont. Soc. London*, 1887.

—— (87²) On the relations of *Calamodendron* to *Calamites. Proc. Lit. Phil. Soc. Manchester*, vol. X. [3] p. 255, 1887.

—— (88) On some anomalous cells developed within the interior of the vascular and cellular tissues of the fossil plants of the Coal-Measures. *Annals Bot.* vol. I. 315, 1888.

—— (88²) On the organisation &c. Mem. XIV. The true fructification of *Calamites. Phil. Trans.* vol. CLXXIX. p. 47, 1888.

—— (89) *Ibid.* Mem. XV. *Phil. Trans.* vol. CLXXX. p. 155, 1889.

—— (91) General, morphological and histological Index to the Author's Collective Memoirs on fossil plants of the Coal-Measures. Pt. I. *Proc. Lit. Phil. Soc. Manchester*, vol. IV. [3] 1891.

—— (91²) On the organisation &c. Mem. XVIII. *Phil. Trans.* vol. CLXXXII. B. p. 255, 1891.

—— (92) The genus *Sphenophyllum. Nature*, vol. XLVII. p. 11, 1892.

Williamson, W. C. (96) Reminiscences of a Yorkshire Naturalist (edited by Mrs Crawford Williamson). *London,* 1896.

Williamson, W. C. and **Scott, D. H.** (94) Further observations on the organisation of the fossil plants of the Coal-Measures. Pt. I. *Calamites, Calamostachys* and *Sphenophyllum. Phil. Trans.* vol. CLXXXV. p. 863, 1894.

—— (95) *Ibid.* Pt. II. The roots of *Calamites. Phil. Trans.* vol. CLXXXVI. p. 683, 1895.

—— (96) *Ibid.* Pt. III. *Lyginodendron* and *Heterangium. Phil. Trans.* vol. CLXXXVI. p. 703, 1896.

Wilson, J. S. G. (87) The Diatomite of Cuithir. *Min. Mag.* vol. VII. p. 35, 1887.

Witham, H. (31) A description of a fossil tree discovered in the quarry at Craigleith near Edinburgh. *Nat. Hist. Soc. Northumberland, Durham and Newcastle-on-Tyne,* 1831.

—— (33) The internal structure of fossil vegetables found in the Carboniferous and Oolitic deposits of Great Britain. *Edinburgh,* 1833.

Woodward, H. (72) A monograph of the British fossil Crustacea belonging to the order Merostomata. *Palaeont. Soc.* 1866—78.

Woodward, H. B. (87) The Geology of England and Wales. *London,* 1887.

—— (95) The Jurassic rocks of Britain. *Mem. Geol. Surv.* vol. V. 1895.

Woodward, J. (1695) An essay toward a Natural History of the Earth. *London,* 1695.

—— (1728) A catalogue of the additional English fossils in the collection of J. Woodward, M.A. *London,* vol. II. 1728.

—— (1729) An attempt towards a Natural History of the fossils of England ; in a catalogue of the English fossils in the collection of J. Woodward, M.A. *London,* vol. I. 1729.

Young, G. and **Bird, J.** (22) A Geological Survey of the Yorkshire Coast. *Whitby,* 1822.

Young, J., Glen, D. C. and **Kidston, R.** (88) Notes on a section of Carboniferous strata containing erect stems of fossil trees &c. in Victoria Park, Whiteinch. *Trans. Geol. Soc. Glasgow,* vol. VIII. p. 227, 1888.

Zeiller, R. (80) Végétaux du terrain houiller de la France. *L'explication de la carte géologique de la France,* vol. IV. *Paris,* 1880.

—— (82) Observations sur quelques cuticles fossiles. *Ann. Sci. Nat.* vol. XIII. [6] p. 217, 1882.

—— (84) Sur des traces d'insectes simulant des empreintes végétales. *Bull. Soc. Géol. France,* [3] vol. XII. p. 676, 1884.

—— (88) Bassin houiller de Valenciennes. Description de la flore fossile. *Études Gîtes Min. France. Paris,* 1888.

—— (88) *Vide* Renault, B.

Zeiller, R. (91) Sur la valeur du genre *Trizygia*. *Bull. Soc. Géol. France*, vol. XIX. [3] p. 673, 1891.

—— (92) Sur les empreintes du sondage de Douvres. *Compt. Rend.* Oct. 24, 1892.

—— (92²) Bassin houiller et permien de Brive. *Études Gîtes Min. France. Paris*, 1892.

—— (93) Étude sur la constitution de l'appareil. fructificateur des Sphenophyllum. *Mém. Soc. Géol. France*, 1893, p. 3.

—— (94) Note sur la flore des Couches permiennes de Trienbach (Alsace). *Bull. Soc. Géol. France*, vol. XXII. [3] p. 193, 1894.

—— (95) Notes sur la flore des Gisements houillers de la Rhune et d'Ibantelly (Basses-Pyrénées). *Bull. Soc. Géol. France*, vol. XXIII. [3] p. 482, 1895.

—— (95²) Sur la flore des dépôts houillers d'Asie Mineure et sur la présence, dans cette flore, du genre *Phyllotheca*. *Compt. Rend.* vol. CXX. p. 1228, 1895.

—— (96) Remarques sur la flore fossile de l'Altaï. *Bull. Soc. Géol. France*, vol. XXIV. [3] p. 466, 1896.

Zenker, F. C. (33) Beschreibung von *Galium sphenophylloides*. *Neues Jahrb. Min.* 1833, p. 398.

Zigno, A. de (56) Flora fossilis formationis Oolithicae. Vol. I. *Padova*, 1856.

INDEX.

CAMBRIDGE : PRINTED BY J. AND C. F. CLAY, AT THE UNIVERSITY PRESS.